Advances in Italian Robotics

Advances in Italian Robotics

Special Issue Editors

Giulio Rosati
Giovanni Boschetti
Giuseppe Carbone

MDPI • Basel • Beijing • Wuhan • Barcelona • Belgrade • Manchester • Tokyo • Cluj • Tianjin

Special Issue Editors
Giulio Rosati
University of Padova
Italy

Giovanni Boschetti
University of Padova
Italy

Giuseppe Carbone
University of Calabria
Italy

Editorial Office
MDPI
St. Alban-Anlage 66
4052 Basel, Switzerland

This is a reprint of articles from the Special Issue published online in the open access journal *Robotics* (ISSN 2218-6581) (available at: https://www.mdpi.com/journal/robotics/special_issues/Italianrobotics).

For citation purposes, cite each article independently as indicated on the article page online and as indicated below:

LastName, A.A.; LastName, B.B.; LastName, C.C. Article Title. *Journal Name* **Year**, *Article Number*, Page Range.

ISBN 978-3-03928-931-8 (Pbk)
ISBN 978-3-03928-932-5 (PDF)

Contents

About the Special Issue Editors

Giulio Rosati is a Full Professor of Machine Mechanics and Robotics in the Department of Industrial Engineering (DII) of the University of Padova, Italy. He is the Prospective Coordinator of the Doctoral School in Industrial Engineering of Padova University, Associate of IFToMM, Italy, and of the Italian Institute of Nuclear Physics (INFN). He is responsible for the Robotics and Automation Labs of DII.

Giovanni Boschetti is currently an Associate Professor of Mechanisms and Machine Science with the Department of Management and Engineering of the University of Padova, where he teaches Industrial Robotics and Mechanics of Machines. He is President of the degree course in Mechatronics Engineering. His main research interests are in kinematics, dynamics, performance evaluation, and industrial applications of serial and parallel industrial manipulators, cable direct driven robots, and collaborative robotics. He is an active member of the IFToMM Italy Association; he was the chief organizer of its first international conference. He is currently associate editor of the /International Journal of Mechanics and Control/.

Giuseppe Carbone has a Master cum laude from the University of Cassino (Italy) where he also completed his Ph.D. studies being a key member of LARM (Laboratory of Robotics and Mechatronics) for about 20 years. From 2015 to 2017, he was a Senior Lecturer at Sheffield Hallam University (U.K.) and a member of the executive board of Sheffield Robotics. Since 2018, he has been Associate Professor at DIMEG, University of Calabria, Italy. Prof. Carbone has received several awards, including three IFToMM "Young Delegate" Awards and two JSPS Awards in Japan. His research interests cover aspects of mechanics of manipulation and grasp, mechanics of robots, and mechanics of machinery, with more than 300 published papers and over 10 patents. He has edited/co-edited four books with Springer International Publisher and one book with Elsevier. He has participated in or coordinated more than 20 research projects at the national and international levels, including 7th European Framework and Horizon 2020. Currently, he is Chair of IFToMM Technical Committee on Robotics and Mechatronics, Member of the Executive Board of Directors of the International Society of Bionic Engineering, and Treasurer of the IFToMM Italy Society.

robotics

MDPI

Editorial

Advances in Mechanical Systems Dynamics

Giulio Rosati [1,*], Giovanni Boschetti [2] and Giuseppe Carbone [3]

1 Department of Industrial Engineering, University of Padova, 35131 Padova, Italy
2 Department of Management and Engineering, University of Padova, 36100 Vicenza, Italy; giovanni.boschetti@unipd.it
3 DIMEG, University of Calabria, 87036 Cosenza, Italy; giuseppe.carbone@unical.it
* Correspondence: giulio.rosati@unipd.it; Tel.: +39-049-827-6809

Received: 4 March 2020; Accepted: 5 March 2020; Published: 9 March 2020

1. Introduction

Nowadays, robotics is developing at a much faster pace than ever in the past, both inside and outside industrial environments. Service robotics [1], surgical and rehabilitation robotics [2–4], assistive robotics, and other novel application fields are becoming more and more significant, not only from technological and economical viewpoints, but also in terms of their daily life and social implications. Even the implementation and role of robots in production lines and other traditional frames is being widely revised, towards novel flexible [5,6] and agile [7] manufacturing systems. Moreover, novel architectures such as cable robots [8], devices for handling of horticulture products [9] and other service robotics tasks are also being widely investigated. In this context, research on machine and robot mechanics, modelling, design, and control is going to play an increasingly central role, as outlined for example in [10].

This Special Issue aims at disseminating the latest research achievements, findings, and ideas in the robotics field, with particular attention to the Italian scenario. This Issue includes revised and substantially extended versions of selected papers that have been presented at IFIT2018, the 2nd International Conference of the italian branch of the International Federation for the Promotion of Mechanism and Machine Science (IFToMM ITALY). However, we have also strongly encouraged the submission of additional contributions from researchers working in this field who did not participate to the IFIT 2018 Conference, in order to further widen the field coverage.

2. Advances in Italian Robotics

This journal special issue includes papers belonging to a broad range of disciplines, such as robotic manipulation, variable stiffness actuation, mobile system, social robotics, optimization of robotic tasks, compliance property of robot, biomedical device, collaborative robotics, trajectory planning and wearable robotics.

In the first paper, the authors outline the influence of electric power quality on the performance of a robotic device. Namely, voltage dip effects are addressed from an experimental viewpoint by focusing on robotic grasping applications [11]. A specific case study is reported, by means of a three-fingered robotic hand. The main goal of paper [12] is to introduce an original two-DoFs planar variable-stiffness mechanism, characterized by an orthogonal arrangement of the actuation units to favor the isotropy. Such a device combines the concepts of a one-DoF agonist-antagonist variable-stiffness mechanism and the rigid planar parallel and orthogonal kinematics leading to an innovative solution. The authors of paper [13] present the modeling and the validation of a novel family of climbing robots that are capable of adhering to vertical surfaces by means of permanent magnetic elements. The robotic system is composed of two modules, the master and the follower carts, which are arranged in a sandwich configuration. Accordingly, the surface to be climbed is interposed between the master and follower modules. Palli and Pirozzi in [14] present a robotized cabling of switchgears with main focus at a gripper with tactile sensors for the wire manipulation. In particular, the developed gripper is experimentally tested to assess its success rate during wire manipulation.

A key challenge in the Human-Robot Interaction (HRI) field is to provide robots with cognitive and affective capabilities, by developing architectures that let them establish empathetic relationships with users. Nocentini et al. in [15] propose a survey of multiple models that have been proposed in the literature as referring to three key aspects. Namely, the development of adaptive behavioral models, the design of cognitive architectures, and the ability to establish empathy with the user. Another emerging technology for assistive robotics is reported in [16]. In particular, this paper addresses a low-cost mechanical design solution exploiting compliant actuation at the shoulder joint to increase safety in human-robot cooperation of an Industrial Exoskeleton for Advanced Human Empowering in Heavy Parts Manipulation Tasks. Authors of paper [17] address a specific industrial application on assembly kitting lines studying the subsystems that compose a hybrid flexible assembly workcell. In particular, the authors investigated the possibility and performance of replacing a conventional weighting device with a vision on inspection system. The paper [18] focuses at robot compliance modeling for achieving a compensation of small position and orientation errors of the end-effector as well as reducing chatter vibrations. In this paper, joint compliances of a serial six-joint industrial robot are identified with a novel modal method making use of specific modes of vibration dominated by the compliance of only one joint. Then, in order to represent the effect of the identified compliances on robot performance in an intuitive and geometric way, a novel kinematic method based on the concept of "Mozzi axis" of the end-effector is presented and discussed.

Menga and Chirardi propose a control of the sit-to-stand transfer of a biped robotic device [19]. The control has been synthesized analyzing the basic laws of dynamics by considering a two-phase dynamic setting, with an external force disturbance affecting the center of pressure under the feet. The paper objectives are threefold: identifying the major dynamical determinants of the exercise; sythesizing an automatic control for an autonomous device; proposing an innovative approach for the rehabilitation process with an exoskeleton. A similar approach is later developed as referring to a device for postural rehabilitation [20]. In paper [21], Martelli et al. propose an analysis of ankle mechanical properties for the design of an exoskeleton to be suitable for both adults and children. Experimental tests have carried out on 16 young adults and 10 children for the evaluation of ankle mechanical impedance and kinematic performance. Ankle impedance was measured by imposing stochastic torque perturbations in dorsi-plantarflexion and inversion-eversion directions. Kinematic performance was assessed by asking participants to perform a goal-directed task. Magnitude and anisotropy of impedance were computed using a multiple-input multiple-output system. These findings are considered for a proper development of robotic devices.

Over the last decade, the market has seen the introduction of a new category of robots—collaborative robots (or "cobots")—designed to physically interact with humans in a shared environment, without the typical barriers or protective cages used in traditional robotics systems. The paper [22] provides an overview of collaborative robotics towards manufacturing applications. In paper [23] Bottin and Rosati address the challenging problem of trajectory planning and optimization by considering a redundant serial robot and a set of Cartesian via points. The proposed method is based on a search of suboptimal paths as based on graph theory and the Dijkstra algorithm, allowing performing a reasonably wide search of the suboptimal path within a reasonable computation time. Malvezzi et al., in [24] address the topic of augmenting the human hand with robotic extra fingers for a compensatory and rehabilitation purposes on patients with upper limb impairments. The paper [25] outlines several solutions with one or two extra fingers. Underactuation and compliance are considered as design choices that can reduce the device complexity and weight, maintaining the adaptability to different grasped objects.

This Special Issue follows other two special issues that were published in the International Journal of Mechanics and Control, whose content is available at www.jomac.it. In particular, Marco Ceccarelli et al. have contributed a paper on a novel parallel mechanism for a biped robot leg application [25].

Paolo Gallina et al. introduced the concept of Anti-Hedonistic Machine (AHM), which is designed to "prevent people from doing something" [26]. Giulio Reina et al. reported a novel architecture of robotic hand designed for prosthesis purposes with under-actuation features [27]. A kinematic and quasi-static analysis of a class of Quick-Release hooks was presented by Luca Bruzzone et al. in [28]. Paolo Boscariol et al. introduce a novel design of an electromechanical clamp for portable ultrafiltration device [29]. Giovanni Boschetti et al. reported a novel failure recovery strategy for direct driven cable robots [30]. Francesco Biral et al. propose an analytical model for tractors having suspended front axle with a combination of a four-bar linkage mechanism and a hydraulic system [31]. Ilaria Palomba et al. report a technique for the reduction of nonlinear models of flexible-link multibody systems through an equivalent rigid-link system method [32]. A novel design of a human powered press for straw bale construction is proposed by Giuseppe Quaglia et al. as optimized for underserved communities [33]. Francesco Timpone et al. proposed a method for on-line estimation of tyre/road friction forces by considering various road conditions [34]. Sforza et al. presented a literature overview on electric vehicles with independent drivetrains [35]. Giannoccaro et al. presented control aspects for an active suspension system [36] while paper [37] deals with the modelling and simulation of vehicle lateral dynamic behavior. Aspragkathos et al. presented a control strategy for robots with flexible beams [38], while paper [39] provides insight on the energy efficiency of a parallel manipulator when considering compliant elements. Finally, paper [40] addresses the application of robots for surgical craniotomy operations.

3. Final Remarks

This Special Issue contains valuable research works focused at advances in robotics, covering a wide area of application areas. This collection shows the high research interest in these topics with high impact and potential for future developments.

Author Contributions: Conceptualization, G.R., G.B., G.C.; writing—original draft preparation, G.R., G.B., G.C.; writing—review and editing, G.R., G.B., G.C.; supervision, G.R. All authors have read and agreed to the published version of the manuscript.

Funding: This research received no external funding.

Conflicts of Interest: The authors declare no conflict of interest.

References

1. Rubio, F.; Valero, F.; Llopis-Albert, C. A review of mobile robots: Concepts, methods, theoretical framework, and applications. *Int. J. Adv. Rob. Syst.* **2019**, *16*, 2019. [CrossRef]
2. Rosati, G. The place of robotics in post-stroke rehabilitation. *Expert Rev. Med. Devices* **2010**, *7*, 753–758. [CrossRef] [PubMed]
3. Masiero, S.; Poli, P.; Rosati, G.; Zanotto, D.; Iosa, M.; Paolucci, S.; Morone, G. The value of robotic systems in stroke rehabilitation. *Expert Rev. Med. Devices* **2014**, *11*, 187–198. [CrossRef] [PubMed]
4. Copilusi, C.; Ceccarelli, M.; Carbone, G. Design and numerical characterization of a new leg exoskeleton for motion assistance. *Robotica* **2015**, *33*, 1147–1162. [CrossRef]
5. Rosati, G.; Faccio, M.; Carli, A.; Rossi, A. Fully flexible assembly systems (F-FAS): A new concept in flexible automation. *Assembly Autom.* **2013**, *33*, 8–21. [CrossRef]
6. Boschetti, G. A Picking Strategy for Circular Conveyor Tracking, 2016, 81, 241–255. *J. Intell. Robot. Syst. Theory Appl.* **2016**, *81*, 241–255. [CrossRef]
7. Barbazza, L.; Faccio, M.; Oscari, F.; Rosati, G. Agility in assembly systems: A comparison model. *Assembly Autom.* **2017**, *37*, 411–421. [CrossRef]
8. Boschetti, G.; Trevisani, A. Cable robot performance evaluation by Wrench exertion capability. *Robotics* **2018**, *7*, 15. [CrossRef]
9. Boschetti, G.; Carbone, G. Advances in Italian Mechanism Science. *Int. J. Mech. Control* **2017**, *18*, 1.
10. Russo, M.; Ceccarelli, M.; Corves, B.; Hüsing, M.; Lorenz, M.; Cafolla, D.; Carbone, G. Design and test of a gripper prototype for horticulture products. *Rob. Comput. Integr. Manuf.* **2017**, *44*, 266–275. [CrossRef]

11. Carbone, G.; Ceccarelli, M.; Fabrizi, C.; Varilone, P.; Verde, P. Effects of Voltage Dips on Robotic Grasping. *Robotics* **2019**, *8*, 28. [CrossRef]
12. Malosio, M.; Corbetta, F.; Ramìrez Reyes, F.; Giberti, H.; Legnani, G.; Molinari Tosatti, L. On a Two-DoF Parallel and Orthogonal Variable-Stiffness Actuator: An Innovative Kinematic Architecture. *Robotics* **2019**, *8*, 39. [CrossRef]
13. Seriani, S.; Scalera, L.; Caruso, M.; Gasparetto, A.; Gallina, P. Upside-Down Robots: Modeling and Experimental Validation of Magnetic-Adhesion Mobile Systems. *Robotics* **2019**, *8*, 41. [CrossRef]
14. Palli, G.; Pirozzi, S. A Tactile-Based Wire Manipulation System for Manufacturing Applications. *Robotics* **2019**, *8*, 46. [CrossRef]
15. Nocentini, O.; Fiorini, L.; Acerbi, G.; Sorrentino, A.; Mancioppi, G.; Cavallo, F. A Survey of Behavioral Models for Social Robots. *Robotics* **2019**, *8*, 54. [CrossRef]
16. Mauri, A.; Lettori, J.; Fusi, G.; Fausti, D.; Mor, M.; Braghin, F.; Legnani, G.; Roveda, L. Mechanical and Control Design of an Industrial Exoskeleton for Advanced Human Empowering in Heavy Parts Manipulation Tasks. *Robotics* **2019**, *8*, 65. [CrossRef]
17. Comand, N.; Minto, R.; Boschetti, G.; Faccio, M.; Rosati, G. Optimization of a Kitting Line: A Case Study. *Robotics* **2019**, *8*, 70. [CrossRef]
18. Doria, A.; Cocuzza, S.; Comand, N.; Bottin, M.; Rossi, A. Analysis of the Compliance Properties of an Industrial Robot with the Mozzi Axis Approach. *Robotics* **2019**, *8*, 80. [CrossRef]
19. Menga, G.; Ghirardi, M. Estimation and Closed-Loop Control of COG/ZMP in Biped Devices Blending CoP Measures and Kinematic Information. *Robotics* **2019**, *8*, 89. [CrossRef]
20. Menga, G.; Ghirardi, M. Control of the Sit-To-Stand Transfer of a Biped Robotic Device for Postural Rehabilitation. *Robotics* **2019**, *8*, 91. [CrossRef]
21. Martelli, F.; Taborri, J.; Del Prete, Z.; Palermo, E.; Rossi, S. Quantifying Age-Related Differences of Ankle Mechanical Properties Using a Robotic Device. *Robotics* **2019**, *8*, 96. [CrossRef]
22. Matheson, E.; Minto, R.; Zampieri, E.G.G.; Faccio, M.; Rosati, G. Human-Robot Collaboration in Manufacturing Applications: A Review. *Robotics* **2019**, *8*, 100. [CrossRef]
23. Bottin, M.; Rosati, G. Trajectory Optimization of a Redundant Serial Robot Using Cartesian via Points and Kinematic Decoupling. *Robotics* **2019**, *8*, 101. [CrossRef]
24. Malvezzi, M.; Iqbal, Z.; Valigi, M.C.; Pozzi, M.; Prattichizzo, D.; Salvietti, G. Design of Multiple Wearable Robotic Extra Fingers for Human Hand Augmentation. *Robotics* **2019**, *8*, 102. [CrossRef]
25. Russo, M.; Ceccarelli, M. Kinematic Design of a Novel Robotic Leg Mechanism with Parallel Architecture. *Int. J. Mech. Control* **2017**, *18*, 3–8.
26. Scalera, L.; Gallina, P.; Gasparetto, A.; Seriani, S. Anti-Hedonistic Machines. *Int. J. Mech. Control* **2017**, *18*, 9–16.
27. Zappatore, G.A.; Reina, G.; Messina, A. Analysis of a Highly Underactuated Robotic Hand. *Int. J. Mech. Control* **2017**, *18*, 17–23.
28. Bruzzone, L.; Berselli, G.; Bilancia, P.; Fanghella, P. Quasi-Static Models of a Four-Bar Quick-Release Hook, I. *Int. J. Mech. Control* **2017**, *18*, 25–32.
29. Boscariol, P.; Boschetti, G.; Caracciolo, R.; Neri, M.; Richiedei, D.; Ronco, C.; Trevisani, A. Design Optimization of a Safety Clamp for Portable Medical Devices. *Int. J. Mech. Control* **2017**, *18*, 33–39.
30. Boschetti, G.; Passarini, C.; Trevisani, A. A Recovery Strategy for Cable Driven Robots in Case of Cable Failure. *Int. J. Mech. Control* **2017**, *18*, 41–48.
31. Biral, F.; Riccardo Pelanda, R.; Cis, A. Longitudinal Dynamic Model of an Agricultural Tractor with Front Suspension: Anti-Dive Behaviour Analysis. *Int. J. Mech. Control* **2017**, *18*, 49–58.
32. Palomba, I.; Richiedei, D.; Trevisani, D. Reduction Strategy at System Level for Flexiblelink Multibody Systems. *Int. J. Mech. Control* **2017**, *18*, 59–68.
33. Franco, W.; Quaglia, G.; Ferraresi, C. Appropriate Design of Human Powered Press for Straw Bale Construction in Poor Contexts. *Int. J. Mech. Control* **2017**, *18*, 69–76.
34. Sharifzadeh, M.; Timpone, F.; Senatore, A.; Farnam, A.; Akbari, A.; Russo, M. Real Time Tyre Forces Estimation for Advanced Vehicle Control. *Int. J. Mech. Control* **2017**, *18*, 77–83.
35. Sforza, A.; Lenzo, B.; Timpone, F. A State-Of-The-Art Review On Torque Distribution Strategies Aimed At Enhancing Energy Efficiency For Fully Electric Vehicles With Independently Actuated Drivetrains. *Int. J. Mech. Control* **2019**, *20*, 3–15.

36. Giannoccaro, N.I.; Reina, G.; Rizzo, L. Fuzzy logic controller for active suspension systems of intelligent vehicle. *Int. J. Mech. Control* **2019**, *20*, 17–30.
37. Perrelli, M.; Cosco, F.; Carbone, G.; Mundo, D. "Evaluation Of Vehicle Lateral Dynamic Behaviour According To Iso-4138 Tests By Implementing A 15-Dof Vehicle Model And An Autonomous Virtual Driver. *Int. J. Mech. Control* **2019**, *20*, 31–38.
38. Aspragkathos, S.N.; Sakellariou, J.S.; Koustoumpardis, P.N.; Aspragathos, N.A. Vibration control of flexible beams manipulated by industrial robots via a stochastic AR-based control system. *Int. J. Mech. Control* **2019**, *20*, 39–47.
39. Scalera, L.; Carabin, G.; Palomba, I.; Vidoni, R.; Wongratanaphisan, T. Energy efficiency in a 4-DOF parallel robot featuring compliant elements. *Int. J. Mech. Control* **2019**, *20*, 49–57.
40. Essomba, T.; Sandoval, J.; Laribi, M.A.; Wu, C.-T.; Breque, C.; Zeghloul, S.; Richer, J.-P. Burr hole craniotomy on cadavers for the design of teleoperated robot: motion specifications and interaction forces. *Int. J. Mech. Control* **2019**, *20*, 59–64.

robotics

MDPI

Article

Effects of Voltage Dips on Robotic Grasping

Giuseppe Carbone [1,*], Marco Ceccarelli [2], Christopher Fabrizi [3], Pietro Varilone [3] and Paola Verde [3]

1 DIMEG, University of Calabria, 87036 Cosenza, Italy
2 LARM2, University of Rome Tor Vergata, 00133 Rome, Italy; marco.ceccarelli@uniroma2.it
3 LASE: Laboratory of Electric Systems, DIEI, University of Cassino and South Latium, 03043 Cassino, Italy; fabrizichristopher94@gmail.com (C.F.); pietro.varilone@unicas.it (P.V.); verde@unicas.it (P.V.)
* Correspondence: giuseppe.carbone@unical.it

Received: 18 February 2019; Accepted: 8 April 2019; Published: 11 April 2019

Abstract: This paper addresses the effects of electric power quality on robotic operations. A general overview is reported to highlight the main characteristics of electric power quality and it's effects on a powered system by considering an end-user's viewpoint. Then, the authors outline the influence of voltage dip effects by focusing on robotic grasping applications. A specific case study is reported, namely that of LARM Hand IV, a three-fingered robotic hand which has been designed and built at LARM in Cassino, Italy. A dedicated test rig has been developed and set up to generate predefined voltage dips. Experimental tests are carried out to evaluate the effects of different types of voltage dip on the grasping of objects.

Keywords: robotic hands; grasping; electric power quality; voltage dips effects

1. Introduction

Nowadays, electronic equipment and computing devices are used in most types of industrial machines and robotic devices. They are key systems for the successful implementation of most industrial processes. However, the wide use of electronics makes this equipment more vulnerable to disturbances in terms of power quality (PQ). PQ is related to several disturbances that include, among others, momentary interruptions, voltage dips or sags, swells, transients, harmonic distortion, electrical noise, and flickering lights [1]. In general, the electrical power grid is designed to deliver power reliably with the aim of maximizing the amount of power available to customers. However, PQ disturbances are not always taken into consideration despite the fact that they can significantly affect industrial production, as well as permanently damage expensive equipment, costing industrial plants millions of dollars [2]. In order to minimize these costs, it is critical for industrial customers to understand how PQ can affect the operation of their systems and how it is possible to mitigate the effects of PQ disturbances [1–3].

The international framework of the actual standards on PQ is based on the norms of the International Electro-technical Commission (IEC), which is accepted as a worldwide reference. Moreover, national or supranational committees give further indications on the maximum limits to be imposed on PQ disturbances. For example, the European Committee for Electrotechnical Standardization (CENELEC) is the European reference, while Comitato Elettrotecnico Italiano (CEI) is the Italian national reference for adopting IEC and CENELEC standards. The above-mentioned bodies have released the norm EN 50160 that defines the European and Italian standards for the PQ in terms of voltage dips and other voltage disturbances [4]. Similarly, the norms EN 61000-4-11 and EN 61000-4-34 are adopted worldwide [4–7].

PQ is gaining significance also in robotics, and specifically in applications of service robotics. Voltage dips, also defined by the equivalent term voltage sags, are recognized as one of the most severe disturbances that can affect the operation of industrial devices. The detrimental effects of voltage dips can result both in the tripping of the protective devices with the equipment shut down and in the malfunction of a device. The latter constitutes a sort of failure that determines a far from normal or satisfactory functionality. Both these typologies of effects have significant economic impacts on a system's operation and productivity. These costs depend on many factors that are linked to the type of manufacturing activity and to the extent of the affected area [3].

Among other industrial devices, robots certainly suffer for the presence of voltage dips in the supply voltage. This is particularly critical in the case of collaborative robots which are penetrating several new applications also thanks to the publication of a collaborative robotics reference standard [8]. In fact, the recent ISO (International Organization for Standardization) norm establishes a novel regulatory framework allowing a wide spread of collaborative robots in industrial and civil environments. The close interaction among robots and humans makes safety one of the most significant aspects of robot design and operation. Clearly, the effects of the voltage dips in the supply voltage can significantly influence robot performance as well as generate potentially critical safety issues, such as missing operations or unpredictable robot behaviors.

The case of robot grasping is quite significant, since the performance of an end-effector is considered to be the most important contribution to achieving the successful manipulation of an object. Several researchers have addressed the design of grasping devices with solutions ranging from simple end-effectors (suction cups, electromagnetic devices) to finger grippers for handling specific objects, and even complex multi-purpose robotic hands [9–13]. It appears very significant to investigate the effects of power quality on a robot grasping, since a grasping failure implies a failure of the whole robotic manipulation procedure. Moreover, this can have strong safety implications, especially in collaborative robotics tasks, as mentioned in [8].

This paper addresses the effects of the voltage dips on the performance of robotic grasping. A specific case of study is reported as referring to LARM Hand IV, a three-fingered robotic hand which has been designed and built at LARM at the University of Cassino [14–17]. A dedicated test rig has been designed and set up to generate predefined voltage dips to experimentally investigate their effect on the grasping of objects with different sizes. Experimental tests are carefully analysed and discussed to demonstrate the influence of the voltage dips on the grasping performance, as well as to propose some mitigation actions to avoid safety implications during the grasping.

2. Main Characteristics of the Voltage Dips

The term PQ embraces a wide set of disturbances that can affect the voltage and/or current [3]. The disturbances are categorized in two groups: the variations and events [4]. Each group represents a different type of phenomena and different ways of treating the disturbances [5,6]. The variations and events are due to the interaction between the power supply and the devices installed at the customers' premises.

Variations are minor changes from the ideal value of voltage or current that show a relatively slow reduction in value. The level of variations can be measured continuously and at predefined instants of time. Examples of variations are the voltage amplitude variations and the waveform distortion. Events can have large deviations from the ideal value and they can occur suddenly. Events cannot be measured continuously because they may occur occasionally. A trigger condition is needed to measure these events. In the group of events affecting the supply voltage, voltage dips are one of the most severe disturbances that can affect especially industrial end-users. Several devices are significantly vulnerable to voltage dips. The main detrimental effects of the voltage dips are the tripping of protected devices and the degradation of the performance of a device.

A voltage dip is defined as a "sudden reduction of the supply voltage, below 90% and above 1% of the declared voltage, followed by a voltage recovery after a short period of time" [4]. Figure 1 plots the time of a voltage affected by a dip. In Figure 1, the main characteristic quantities of a voltage dip are expressed as the amplitude with the symbol Vr, and the duration with the symbol Δt as shown in Figure 1. The amplitude of a voltage dip is the minimum value of the RMS voltage during an event; it is known also as the residual voltage. The duration of a voltage dip is the time elapsed when the voltage falls below the threshold value, which is assumed to be 90% of the rated value. Further quantities can characterize a voltage dip, like the number of involved phases, the phase angle jump, or the symmetry of the voltage dips on the phases.

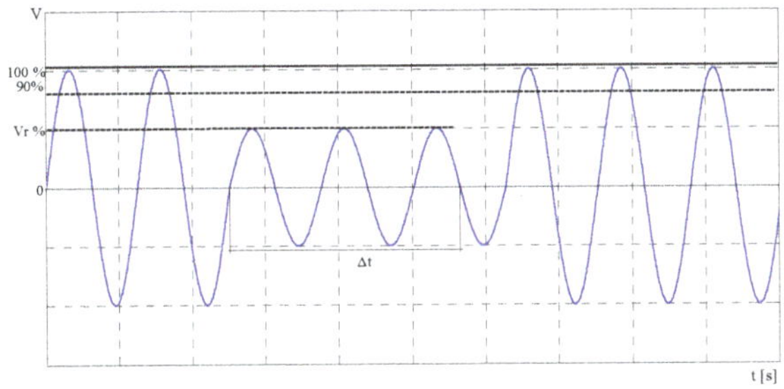

Figure 1. Example of a real voltage dip in industrial frames [2].

In transmission and distribution systems, most voltage dips originate with the short circuits and further causes include the start of a large motor and the insertion of a large transformer, or of a high power load as can frequently happen in industrial systems. In the transmission and distribution systems, the dips more frequently originate with short circuits in some nodes of the electrical network. In the presence of a symmetrical solid short circuit in a specific node, two main phenomena happen. In the node where the short circuit occurs, the voltage is equal to zero and in the other nodes electrically close to it, the voltage is affected by the sudden reduction that represents a voltage dip. This phenomenon lasts until the protection device clears the short circuit.

The framework of the actual standards on the limits of voltage dips is mainly referred to the IEC and the CENELEC norms. In particular, IEC 6100-4-11 [7] states the immunity test for the devices to define its operation class with reference to the EMC (Electro Magnetic Compatibility) and two main classes are defined which are the Class II and the Class III. The main standard of the CENELEC is the EN50160 that indicates the voltage characteristics of the electricity supply by public distribution network. In particular, for the voltage dips, this standard proposes the table shown in Table 1 to classify them according to residual voltage and duration.

Table 1 refers to all the voltage dips that can be recorded in a node. It allows for immediately ascertaining the performance of a node in a considered period, typically at least one year. Actually, the trends of future standardization activities on the voltage dips are towards a limitation of the number of voltage dips that can be tolerated at any node of a system in a defined time period as the year. The limits could be expressed using a table similar to that in Table 1. In such a case, any number of cells would express the boundary of the performance of the power supply that any customer should expect. Summarizing, the most important characteristics of a voltage dip are the amplitude and the duration.

Table 1. Classification of the voltage dips according to residual voltage and duration [EN50160].

	Duration [ms]				
Residual Voltage u [%]	**$10 \leq t \leq 200$**	**$200 < t \leq 500$**	**$500 < t \leq 1000$**	**$1000 < t \leq 5000$**	**$5000 < t \leq 60,000$**
$90 > u \geq 80$	A1	A2	A3	A4	A5
$80 > u \geq 70$	B1	B2	B3	B4	B5
$70 > u \geq 40$	C1	C2	C3	C4	C5
$40 > u \geq 5$	D1	D2	D3	D4	D5
$5 > u$	X1	X2	X3	X4	X5

3. Main Features of LARM Hand IV

Several designs have been developed for LARM Hand at LARM in Cassino, as detailed for example in [14–17]. The LARM Hand prototypes have three one-DOF (Degree of Freedom) human-like fingers. Their main features are low-cost design and easy operation. Only one motor is needed to drive each finger. Its torque is applied to the first link of its driving mechanism as indicated in Figure 2 with C_m. One of the most complex design issues for LARM Hand has been the design of a suitable driving mechanism that can be embedded in the finger body and remains within the finger body also during its movement, as shown in the scheme of Figure 2. This patented linkage-based driving mechanism of LARM Hand is described in full detail in references [14–17].

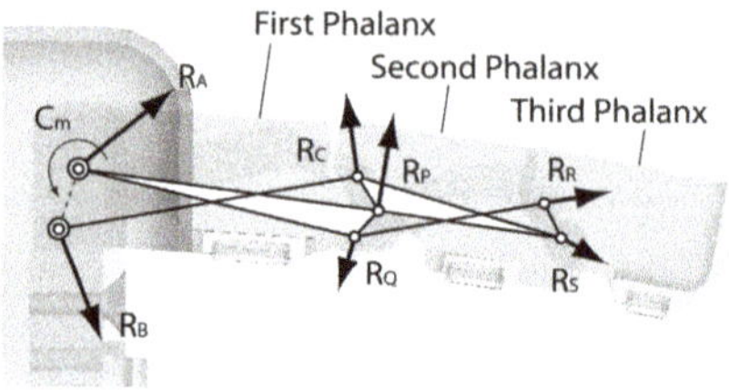

Figure 2. A CAD model of LARM hand with its transmission mechanism and reactions in joints.

The LARM Hand IV, shown in Figure 3, is equipped with three force sensors on each finger for measuring the grasping force on each phalanx while its operation is achieved by means of a low-cost PLC, which directly drives the three DC motors. A simple control logic is achieved by using a reference force threshold and by limiting the motor input current as it is directly linked with the motor output torque. It is worth noting that a firm grasp is achieved when all forces are in equilibrium. Therefore, the input torque has to be regulated to ensure a firm grasp as function of several parameters including the external force acting on the object, and the position, size, and shape of the grasped object. In this paper, we investigate how voltage dips influence the grasping of an object while using LARM Hand IV.

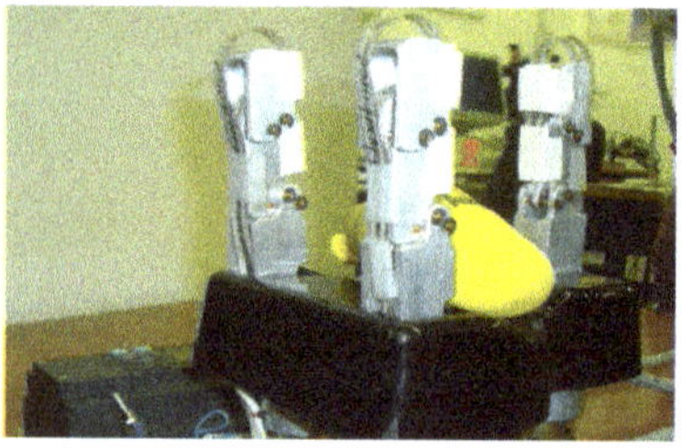

Figure 3. A prototype of LARM Hand IV.

4. Test Rig Set-Up

Experimental activities have been carried out to validate the effects of voltage dips by developing a dedicated test rig. The proposed test rig set up is outlined in Figure 4.

The main components of the proposed test rig are:

(1) An AC Power Source model AMX 360 by Pacific Power Source Inc. (Irvine, CA, USA)
(2) A dual stabilized Power Supply model 32DP32 by Elind (Milano, Italy);
(3) A current sensor INA 219 by Texas Instruments (Dallas, TX, USA);
(4) A LARM hand IV prototype, Figure 3;
(5) An Arduino Mega board (Tilburg, the Netherlands);
(6) A data acquisition board DAQ NI-6009 USB by National Instruments (Austin, TX, USA);

(7) A laptop with USB ports.

The voltage generation system is used to emulate the role of the electric network behavior. In fact, it can generate any predefined waveform. The voltage generation system is connected to a dedicated PC via GPIB. Its operation is managed by using a dedicated software called UPC Manager. This device is used to generate pre-defined voltage dips. The stabilized power supply that has been used is a high-performance power supply with fast recovery time and low current ripple. A Texas Instruments current sensor INA 219 has been selected for measuring the power supply to the LARM Hand. This has been selected due to its low cost and easy operation in combination with a cheap Arduino Mega board for data collection. Furthermore, the sensor resolution is suitable for the current absorbed by LARM Hand. Instead, a National Instruments DAQ NI-6009 USB has been used for collecting voltage outputs. This has been chosen for its convenient features in terms of cost, user-friendliness, and performance for the data acquisition of the analog data outputs generated by the four FSR piezo resistive force sensors that are on the LARM Hand.

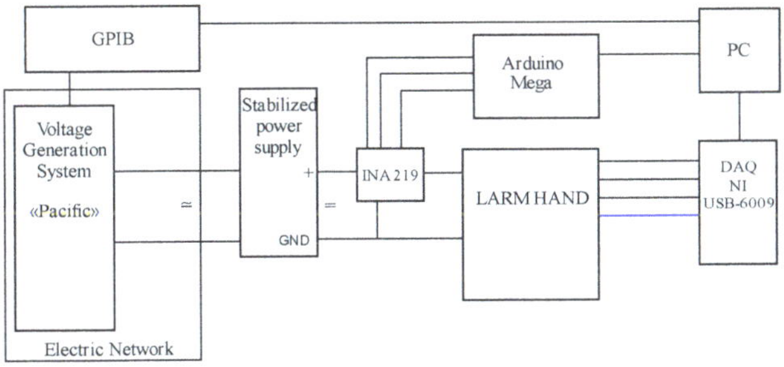

Figure 4. A scheme of the proposed test rig.

5. The Proposed Testing Procedure

The built test rig is used to perform a set of experimental tests that have been selected for describing the most significant cases that can occur in the event of a voltage dip. For each test the LARM hand is set to start at an open position of the three fingers, and after about 5 s the LARM hand starts a phase with a closing operation of all fingers. This phase ends with the grasping of an object. As soon as contact is established between the object and fingers, the force sensors convert the grasping force into a voltage signal. The last phase consists of the opening of all fingers. Figure 5 shows a photo sequence of the testing phases where Figure 5a is the starting phase with fingers fully open; Figure 5b is showing the closing phase; Figure 5c shows the phase in which fingers are in contact with the object. The letters reported in Figure 6 and Table 2 summarize the effects on the most significant cases that can occur. Plots of the experimental tests are then reported for each case to demonstrate what is the corresponding effect on the grasping.

(a) (b) (c)

Figure 5. A sequence of the operation phases of LARM Hand during testing: (**a**) starting phase with fingers fully open; (**b**) closing phase; (**c**) fingers in contact with the object.

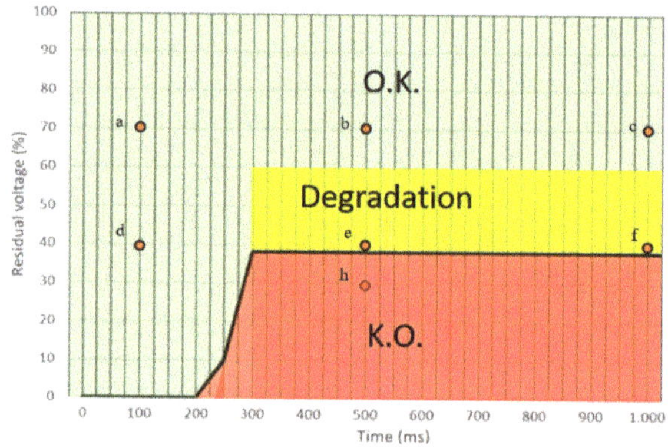

Figure 6. Experimental results in terms of a vulnerability curve for the whole test rig. (points a, b, c, d, e, f, h refer to the cases that have been tested and reported within this paper).

Table 2. List of value of the voltages dip used in the test cases that are reported in Figure 6.

Residual Voltage %	Voltage Dip Duration (ms)	Test Case
70%	100	a
70%	500	b
70%	1000	c
40%	100	d
40%	500	e
40%	1000	f
30%	500	h

A first set of experimental tests can be defined as vulnerability tests. They are aimed at identifying the vulnerability curve as reported in Figure 6. In particular, the vulnerability curve reports the level of voltage supply conditions that are critical for the operation of LARM Hand. Namely, the vulnerability curve in Figure 6 identifies a set of voltage supply conditions in which the LARM hand is not able to run properly due to a voltage dip. It is worth noting the vulnerability depends on the combination of the duration of the voltage dip and the residual voltage percentage. For example, a voltage dip of 250 ms will prevent the successful operation of LARM hand if the residual voltage is less than 10%. But, a voltage dip of 250 ms will not produce any effect on the operation of LARM Hand if the residual voltage is higher than 10%. Similarly, a voltage dip of 300 ms will prevent the successful operation of LARM hand if the residual voltage is less than 38%. However, a voltage dip of 250 ms will not produce any effect on the operation of LARM Hand if the residual voltage is higher than 38%. In other words: any voltage dip, whose characteristics (residual voltage percentage and duration in ms) are below the vulnerability curve is a K.O. condition for the system; otherwise, any voltage dip whose characteristics are below the vulnerability curve is an O.K. condition for the system. It is important to note that there is a region close to the vulnerability curve where the LARM hand will be running, but a degradation of performance can be expected. The precision in identifying the O.K. and K.O. areas can be assumed to be comparable with the accuracy of sensors (1% Full Scale). After the above-mentioned vulnerability tests, specific tests have been carried out with the LARM hand by considering operation conditions being close to the vulnerability curve. The following aspects have been considered as performance parameters for the behavior of LARM Hand during voltage dips:

(1) Absorbed current by LARM hand;
(2) Tension output from the force sensor.

The tests that have been carried out can be divided into two main cases:

A. Normal condition with LARM hand being is supplied by a nominal sinusoidal voltage without disturbances;

B. Voltage Dip condition with LARM hand being supplied by a sinusoidal voltage with voltage dips.

The type A case has been investigated to obtain the nominal performance of the LARM hand in order to have reference nominal output data. The type B cases aims to evaluate the effect of voltage dips on the grasping performance of LARM Hand. In particular, type B cases have been investigated by considering the voltage dip cases, which are reported in Table 2. Results of type A and type B cases are reported in the following section.

6. Testing Results

A first set of experimental results refers to the nominal performance of LARM hand as defined in the type A testing case in the previous section. In particular, it has been possible to collect the current absorbed by LARM Hand in nominal conditions as reported in the plot of Figure 7. Moreover, the output grasping force has been obtained as measured by the force sensors on each finger of LARM hand in nominal conditions as shown in Figure 8. The experimental data in Figures 7 and 8 are collected by using two different data acquisition boards, as shown in the scheme of Figure 4. The data sampling is synchronized using a common trigger while different sampling rates have been selected according to the different characteristics of the collected data.

Referring to Figure 7 it is possible to observe that in the first line segment the current value absorbed from the system has an average value of 750 mA. The current negative peak occurs in relation with the closing operation. This operation ends with a stabilized current value (660 mA) allowing a firm grasp. It is of note that a firm grasp is defined in terms a static equilibrium of the grasped object with no relative motion with respect to the fingers in contact with the object. The time period is identified from the experimental data by identifying the grasping force changes and getting the corresponding time interval. From the collected data the closing operation lasts for 531 ms corresponding to the time needed to reach about 6 N from 0 N.

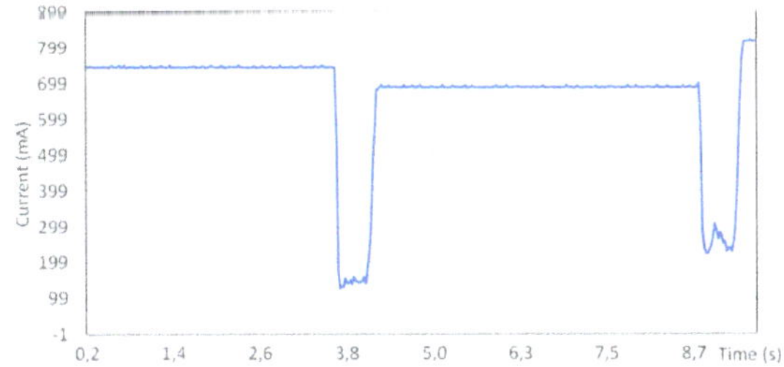

Figure 7. Experimentally measured absorbed current by LARM hand in nominal conditions.

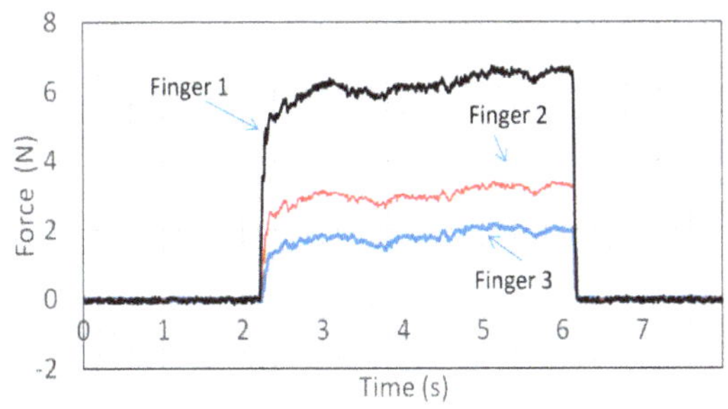

Figure 8. Experimentally measured grasping forces in nominal conditions.

Moreover, Figure 8 shows that the measured grasping forces are all zero until the finger touch the object during the grasping. Then, the grasping force quickly grows until a firm grasp is achieved. At this time oscillations of the grasping force are measured due to small motions of the object as well as due to small changes in the contact point between the object and the sensors as well as small joint clearances on LARM Hand. Additionally, Figure 8 shows that the measured grasping forces on finger 2 and finger 3 are similar to each other while the grasping force on finger 1 is nearly twice as much as on finger 2 or finger 3. This is due to the design of LARM hand where finger 1 is placed opposite to both finger 2 and finger 3 so that it needs to apply twice as much force to balance the combined forces due to finger 2 and finger 3.

The following set of experiments refer to cases a, b, and c in Table 2 where voltages dips with residual tension of 70% have been considered. For these cases, any duration of the voltage dip did not generate an appreciable variation on the grasping performance of LARM Hand. Given the similar performance of cases a, b, and c in Table 2, only case c is reported here. This case refers to a voltage dip with residual voltage of 70% and duration of 1000 ms. The measured plots for this case are reported in Figures 9 and 10. In particular, Figure 9 shows a comparison of the current absorbed by LARM Hand in nominal conditions and during a voltage dip of 70% and duration of 1000 ms. Figure 10 shows a comparison of one of the measured grasping forces by LARM Hand in nominal conditions and during a voltage dip of 70% with duration of 1000 ms. No significant differences can be identified in this case.

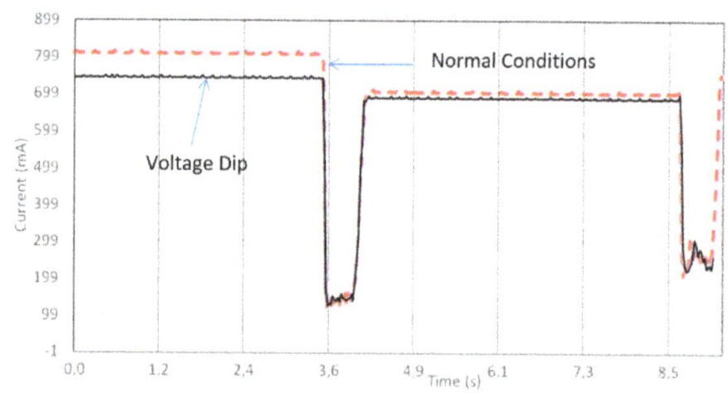

Figure 9. Experimentally measured absorbed current with a voltage dip 70% with duration 1000 ms (case c in Table 2).

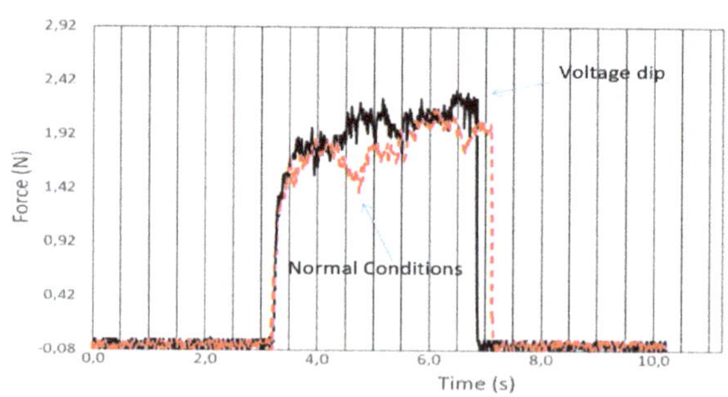

Figure 10. Comparison of the grasping force in normal conditions and voltage dip conditions 70% with duration 1000 ms (case c in Table 2).

The tests with voltage dip with residual tension of 40% and duration of 100 ms (case d in Table 2) have not shown any significant difference as compared with cases a, b, and c. Therefore, the related plots have not been reported in this paper.

The next considered case refers to a voltage dip with residual tension of 40% and duration of 500 ms (case e in Table 2). In Figure 11 it is possible to note the effect that this type of voltage dip has on the current absorbed by the LARM Hand. This result is even clearer in the zoomed view that is shown in Figure 12, where the finger closing operation phase is shown. In particular, it is possible to identify in this plot a heavy ripple current. This ripple causes a significant degradation of the grasping performance, introducing a relevant delay in achieving the grasp. The produced grasping delay is longer than the duration of the voltage dip as the system takes some time to recover from the voltage dip and to go back to nominal operation conditions. The plot of the grasping force that is reported in Figure 13 shows clearly the effect of the measured current. In fact, the comparison between the measured forces with voltage dip and in nominal conditions shows a very significant change of the grasping force in terms of a step in the grasping force that significantly affects the achievement of a firm grasp. The first contact between the finger and the grasped object shows a similar value of grasping force, but after the voltage dip ends the system starts to get more current and this causes a grasping force increase on the object with the need of achieving a new grasping equilibrium condition and a higher risk of losing the firm grasp of the object, as indicated by the increase of the grasping force after the voltage dip.

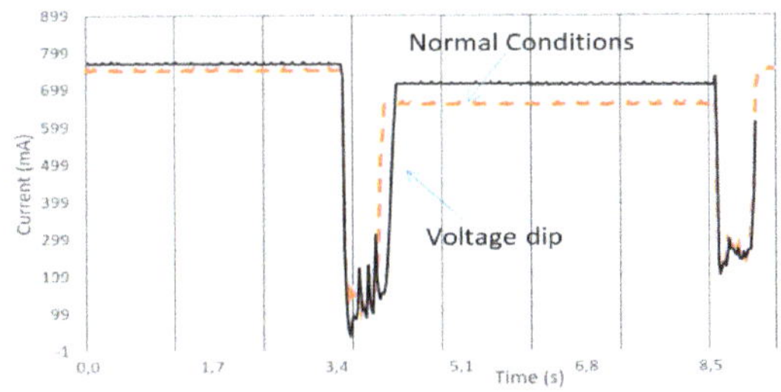

Figure 11. Experimentally measured absorbed current with a voltage dip 40% with duration 500 ms (case e in Table 2).

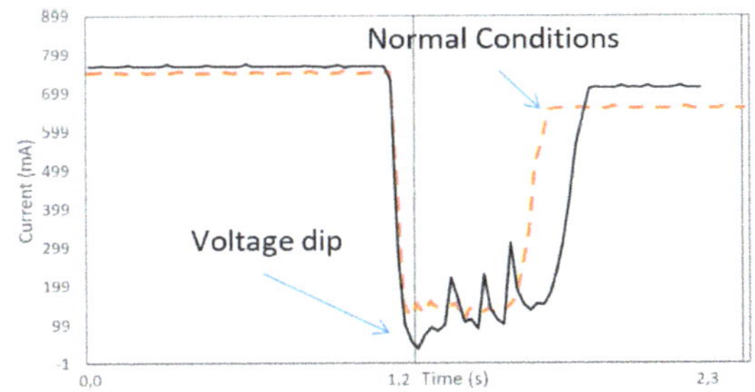

Figure 12. A zoomed view of the absorbed current in Figure 11 with a voltage dip 40% with duration 500 ms (case e in Table 2) during the finger closing operation.

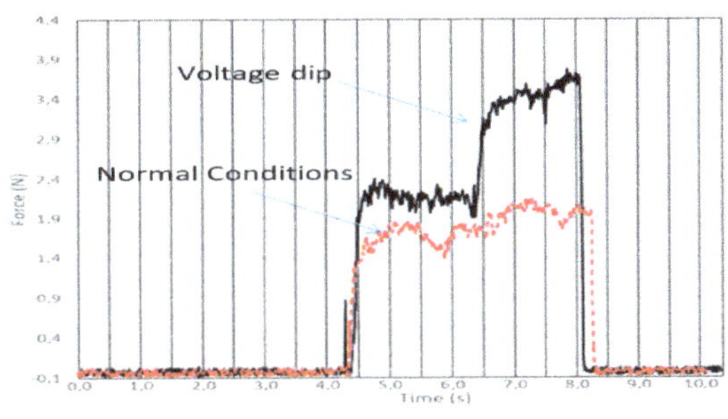

Figure 13. Comparison of the grasping force in normal conditions and voltage dip conditions 70% with duration 1000 ms (case e in Table 2).

The next considered case refers to a voltage dip with residual tension of 40% and duration of 1000 ms (case f in Table 2). In Figure 14 it is possible to note the effect that this type of voltage dip has on the current absorbed by the LARM Hand, which is similar to case e in Table 2. Figure 15 shows a zoomed view of Figure 14 by referring to the finger closing operation phase. In this plot there is a significant current ripple causing a significant degradation of the grasping performance and introducing a relevant delay in achieving the grasping. The produced grasping delay is significantly longer than the duration of the voltage dip, as the system takes some time to recover from the voltage dip and to go back to nominal operation conditions. The plot of the grasping force that is reported in Figure 16 shows clearly the effect of the measured current. In fact, the comparison between the measured forces with voltage dip and in nominal conditions shows a very significant change of the grasping force in terms of a step and a delay of about 140% in the grasping force. This effect significantly affects the achievement of a firm grasp although this is smoother than the previously analyzed case e in Table 2.

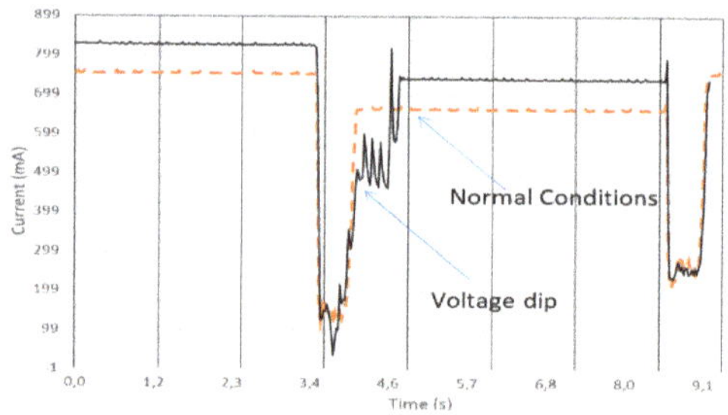

Figure 14. Experimentally measured absorbed current with a voltage dip 40% with duration 1000 ms (case f in Table 2).

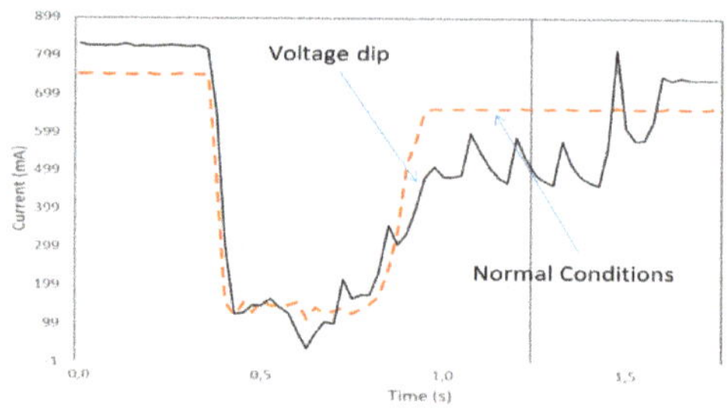

Figure 15. A zoomed view of the absorbed current with a voltage dip 40% with duration 1000 ms (case f in Table 2) during the finger closing operation.

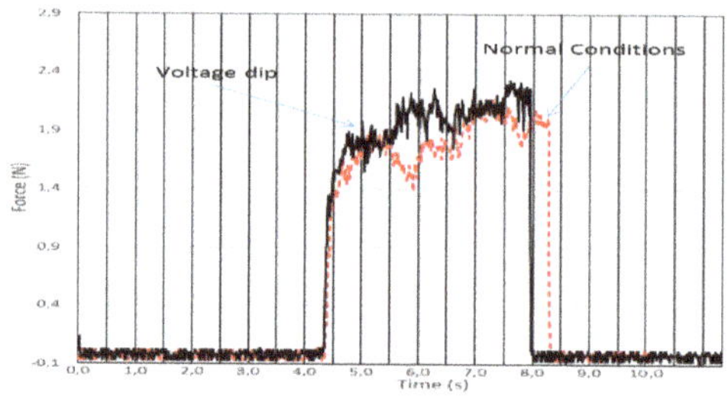

Figure 16. Comparison of the grasping force in normal conditions and voltage dip conditions 70% with duration 1000 ms (case f in Table 2).

In the last case the voltage dip has such characteristics to fall fully below the vulnerability curve shown in Figure 6. This is the heaviest voltage dip in terms of residual voltage, which has been experimentally tested with residual voltage of 30% and duration 500 ms (case h in Table 2). The measured absorbed currents for this case are shown in Figure 17. This plot shows the current absorbed by the LARM Hand is drastically modified as compared with the nominal case. The reason is that this voltage dip causes the re-initialization of the system and the release of the grasped object with a complete manipulation failure.

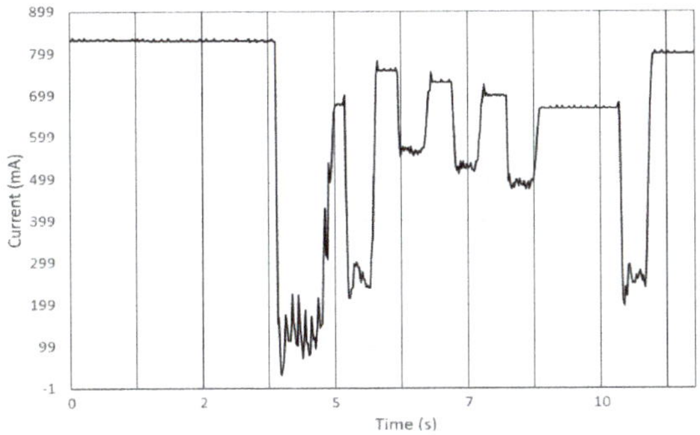

Figure 17. Experimentally measured absorbed current with a voltage dip conditions 30% with duration 500 ms (case h in Table 2); in this case, the LARM Hand fully turns off and re-initializes after the voltage dip.

Results of the reported experimental tests prove the significant influence of voltage dips on the grasping performance. In particular, tests have identified four main cases:

- Voltage dips above the vulnerability curve, Figure 6, (residual tension between 90% and 70% of the nominal value and duration under 500 ms): They do not generate any significant degradation of the performance.
- Voltage dips below the vulnerability curve, Figure 6, (residual tension below 40% of the nominal value and duration above 400 ms): They generate a quite significant performance degradation with a temporary complete system shut down. This case is the most undesired one as the grasping/manipulation process is stopped with potential economic implications in an industrial production as well as potential safety implications if the grasping is performed during human-robot collaborative tasks.
- Voltage dips close to the vulnerability curve shown in Figure 6, (cases between the previous ones): In these cases, there is a progressive degradation of performances. The most evident effect is current ripple and a delay in current supply that is amplified in terms of evident delays and steps in the grasping forces that significantly affect the firm grasping as well as that can produce damages on grasped objects due to unexpected overloadings.
- With voltage dip with 40% of residual voltage and duration of 500 ms cause a delay of the closing operation, which goes from the initial 531 ms in normal conditions to 700 ms, the closing time increasing by 40%. Even the absorbed current is affected by a big variation the maximum difference from the one in normal condition and the one in presence of voltage dip is 500 mA. Further, the grasping force suffers from sudden step variations in magnitude that can cause significant problems, especially while grasping very delicate objects.

The investigated operation of LARM hand can be considered as a case study offering some general rules. In particular, this gives insight on the significance and potential usage of the vulnerability curve for predicting and addressing the behavior of any robotic grasping device with respect to power quality.

7. Conclusions

This paper is focused on offering an evaluation of the effects of voltage dips on robot operation by looking at a case study a prototype LARM hand. The main characteristics of electric supply failures are addressed to outline both the occurrences and effects on a robotic system. The specific analysis and testing on grasping by the LARM hand shows that electric voltage dips can significantly affect the robotic operation. A specific vulnerability curve has been identified to predict the effects on the robot operation that are produced by the electric power supply input in terms of duration and magnitude of a voltage dip.

Author Contributions: Conceptualization, M.C., G.C., P.V. (Pietro Varilone) and P.V. (Paola Verde); methodology, M.C., G.C., P.V. (Pietro Varilone) and P.V. (Paola Verde); validation, C.F., and G.C.; writing—original draft preparation, G.C. and C.F.; writing—review and editing, G.C., M.C., P.V. (Pietro Varilone) and P.V. (Paola Verde)

Funding: This project received no specific funding.

Conflicts of Interest: Authors declare no conflict of interest.

References

1. Bollen, M.H.J. *Understanding Power Quality—Voltage Sags and Interruptions*; IEEE Press: Piscataway, NJ, USA, 2000.
2. Di Fazio, A.R.; Duraccio, V.; Varilone, P.; Verde, P. Voltage Sags in the Automotive Industry: Analysis and Solutions. *Electr. Power Syst. Res.* **2014**, *110*, 25–30. [CrossRef]
3. Bollen, M.H.J.; Verde, P. A Framework for Regulation of RMS Voltage and Short-Duration under and Overvoltages. *IEEE Trans. Power Deliv.* **2008**, *23*, 2105–2112. [CrossRef]
4. European Norms. *EN 50160, Voltage Characteristics of Electricity Supplied by Public Distribution Systems*. Available online: https://www.orgalim.eu/position-papers/en-50160-voltage-characteristics-electricity-supplied-public-distribution-system (accessed on 10 April 2019).
5. IEEE Standards. *IEEE Standard Definitions for the Measurement of Electric Power Quantities Under Sinusoidal, Non-sinusoidal, Balanced, or Unbalanced Conditions*. IEEE Std 1459–2010 (Revision of IEEE Std 1459–2000). Available online: https://standards.ieee.org/standard/1459-2010.html (accessed on 10 April 2019).
6. IEEE Standards. *IEEE Recommended Practice for Monitoring Electric Power Quality*. IEEE Std 1159–2009 (Revision of IEEE Std 1159–1995). Available online: https://standards.ieee.org/standard/1159-2009.html (accessed on 10 April 2019).
7. EMC Standards. *Electromagnetic Compatibility (EMC)—Part 4–11: Testing and Measurement—Voltage Dips, Short Interruptions and Voltage Variations Immunity Tests*. IEC 6100-4-11. Available online: https://webstore.iec.ch/publication/4162 (accessed on 10 April 2019).
8. International Organization for Standardization ISO Webpage. *ISO 15066–2:2016 Robots and Robotic Devices—Collaborative Robots*. Available online: https://www.iso.org/standards.html (accessed on 10 April 2019).
9. Carbone, G. *Grasping in Robotics*; Springer: Dordrecht, The Netherlands, 2013.
10. Russo, M.; Ceccarelli, M.; Corves, B.; Hüsing, M.; Lorenz, M.; Cafolla, D.; Carbone, G. Design and Test of a Gripper Prototype for Horticulture Products. *Robot. Comput. Integr. Manuf.* **2017**, *44*, 266–275. [CrossRef]
11. Yao, S.; Ceccarelli, M.; Carbone, G.; Dong, Z. Grasp configuration planning for a low-cost and easy-operation underactuated three-fingered robot hand. *Mech. Mach. Theory* **2018**, *129*, 51–69. [CrossRef]
12. Gal, I.-A.; Bucur, D.; Vladareanu, L. DSmT Decision-Making Algorithms for Finding Grasping Configurations of Robot Dexterous Hands. *Symmetry* **2018**, *10*, 198. [CrossRef]
13. Vlădăreanu, V.; Dumitrache, I.; Vlădăreanu, L.; Sacală, I.S.; Tonţ, G.; Moisescu, M.A. Versatile Intelligent Portable Robot Control Platform Based on Cyber Physical Systems Principles. *Stud. Inform. Control* **2015**, *24*, 409–418. [CrossRef]
14. Carbone, G.; Iannone, S.; Ceccarelli, M. Regulation and control of LARM Hand III. *Robot. Comput. Integr. Manuf.* **2010**, *26*, 202–211. [CrossRef]
15. Carbone, G.; Ceccarelli, M. Experimental tests on feasible operation of a finger mechanism in the LARM hand. *Mech. Based Des. Struct. Mach.* **2008**, *36*, 1–13. [CrossRef]

16. Ceccarelli, M.; Rodriguez, N.E.N.; Carbone, G. Design and Tests of a Three-Finger Hand with 1-dof Articulated Fingers. *Robot. Int. J.* **2006**, *24*, 183–196. [CrossRef]
17. Carbone, G.; González, A. Numerical Simulation of the Grasp Operation by LARM Hand IV, A Three Finger Robotic Hand. *Robot. Comput. Integr. Manuf.* **2011**, *27*, 450–459. [CrossRef]

robotics

MDPI

Article

On a Two-DoF Parallel and Orthogonal Variable-Stiffness Actuator: An Innovative Kinematic Architecture

Matteo Malosio [1,*,†,‡], Francesco Corbetta [2,‡], Francisco Ramìrez Reyes [2,‡], Hermes Giberti [3], Giovanni Legnani [1,4] and Lorenzo Molinari Tosatti [1]

1 Istituto di Sistemi e Tecnologie Industriali Intelligenti per il Manifatturiero Avanzato, Consiglio Nazionale delle Ricerche, via Alfonso Corti 12, 20133 Milano, Italy; giovanni.legnani@stiima.cnr.it or giovanni.legnani@unibs.it (G.L.); lorenzo.molinaritosatti@cnr.it (L.M.T.)

2 Dipartimento di Meccanica, Politecnico di Milano, Via la Masa 1, 20156 Milano, Italy; francesco1.corbetta@mail.polimi.it (F.C.); franciscoalberto.ramirez@mail.polimi.it (F.R.R.)

3 Dipartimento di Ingegneria Industriale e dell'Informazione, Università di Pavia, Via ferrata 5, 27100 Pavia, Italy; hermes.giberti@unipv.it

4 Dipartimento di Ingegneria Meccanica e Industriale, Università di Brescia, Piazza del Mercato 15, 25121 Brescia, Italy

* Correspondence: matteo.malosio@cnr.it; Tel.: +39-0341-235-0204

† Current address: Istituto di Sistemi e Tecnologie Industriali Intelligenti per il Manifatturiero Avanzato, Consiglio Nazionale delle Ricerche, c/o Polo Universitario di Lecco, via G. Previati 1/E, 23900 Lecco, Italy.

‡ These authors contributed equally to this work.

Received: 19 April 2019; Accepted: 23 May 2019; Published: 27 May 2019

Abstract: Variable-Stiffness Actuators are continuously increasing in importance due to their characteristics that can be beneficial in various applications. It is undisputed that several one-degree-of-freedom (DoF) solutions have been developed thus far. The aim of this work is to introduce an original two-DoF planar variable-stiffness mechanism, characterized by an orthogonal arrangement of the actuation units to favor the isotropy. This device combines the concepts forming the basis of a one DoF agonist antagonist variable stiffness mechanism and the rigid planar parallel and orthogonal kinematic one. In this paper, the kinematics and the operation principles are set out in detail, together with the analysis of the mechanism stiffness.

Keywords: parallel kinematic architecture; agonist-antagonist variable-stiffness actuator; tendon-driven mechanism; stiffness analysis; planar movements

1. Introduction

A Variable-Stiffness Actuator (VSA) permits the adjustment of both the position and stiffness of the load [1]. The fundamental aspects of a VSA are: (i) real-time adjustable stiffness requiring neither force sensors nor transmission backdrivability; (ii) robustness to external perturbations and unpredictable model errors; (iii) adaptability and force accuracy in the interaction with the operator; (iv) suitable for direct interaction with humans in the presence of safety requirements, limiting the force of collisions in the event of a malfunction or unexpected movements.

The above aspects, as well as the onset of new field of applications, are making them increasingly viable solutions. In fact, a first attempt to organize the numerous available technologies and solutions presented in the scientific literature was considered necessary and fundamental to establish a common language for designers and potential users [2]. Different types of VSA actuation schemes have been developed thus far [1,3]. In particular, it is worth mentioning, the so-called principle of operation of the agonist-antagonist [4], commonly used in such devices and adapted to the kinematic architecture set out in the present work.

Usually these are made up of two Series Elastic Actuators (SEA) arranged in parallel to a mobile mass as shown in Figure 1. SEAs are constituted by a rigid actuator and a non-linear spring assembled serially [5]. The nonlinearity of the elastic element is necessary to allow the adjustment of the VSA stiffness [6]. Non-linear springs can be realized in different ways. An example of these, are non-homogeneous coil springs, obtained by varying the pitch or the diameter along the axis. With the same objective an alternative solution would be to employ cams with variable-radius, e.g., [7,8]. In fact, the vast majority of the VSAs developed thus far are characterized by a one-DoF actuation scheme. To the authors' knowledge only one two-DoF mechanism enabling planar movements has been developed up to now [9]. This is a cable-driven device arranged on a triangular framework, with three non-linear SEAs. Due to the presence of three tendons, the possibility to adjust the stiffness of the load along different directions is limited. Moreover, each tendon is controlled independently by two actuators, thus requiring six of these.

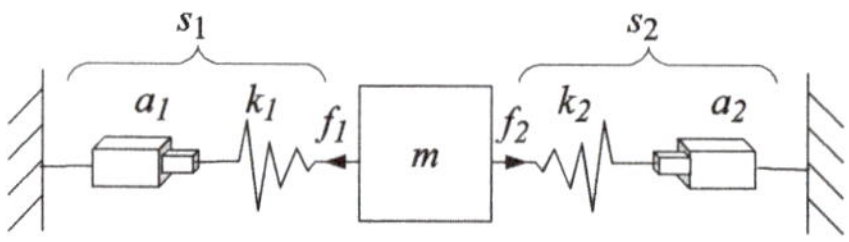

Figure 1. Schematic representation of a one-DoF agonist-antagonistic VSA: two antagonist non-linear SEAs s_i, made up of a rigid actuator a_i and a compliant element with stiffness k_i exerting opposite forces f_i, control the position and the stiffness of a mobile mass m.

To take the existing architecture one step further and enhance the isotropy in terms of stiffness and decoupling of degrees of freedom, an original kinematics two-DoF planar solution based on two orthogonal tendon-driven VSA actuation schemes was developed and is presented in this work. Its original kinematics allows one to tune, at the same time, both the equilibrium position and the stiffness of a mobile platform along two orthogonal directions. The orthogonal configuration is an arrangement that helps to reach an isotropic behavior in the entire workspace [10–12].

Moreover, the architecture has been conceived with the aim of minimizing the number of actuators required. In fact, the architecture presented within this work uses only four actuators, being the minimum number required to control both the equilibrium position and the mechanical stiffness of a mobile mass along two orthogonal directions. Combining the typical modularity of VSA antagonist actuators, the architecture presented herein is essentially made up of two orthogonal VSAs, as shown in the scheme set out in Figure 2, within which each one of these are characterized by two antagonist identical submodules.

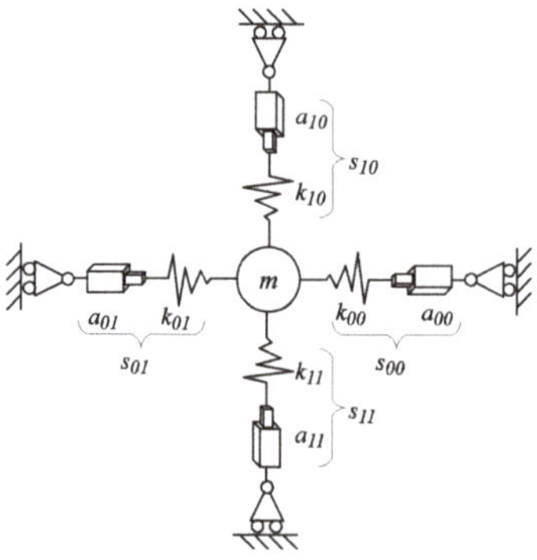

Figure 2. Schematic representation of a two-DoF agonist-antagonist VSA: two one-DoF VSAs are arranged orthogonally with respect to a mobile mass.

The paper is organized as follows: the nomenclature is listed in Table 1; the kinematic architecture is presented in Section 2; the force and stiffness analysis is described in Section 3; conclusions are drawn in Section 4.

Table 1. Table of symbols.

Symbol	Description
$P_{...}$	Geometric point
$\{f\}$	Reference frame. $\{w\}$ denotes the global reference frame.
$\mathbf{P}_{...(f)} =, [P_{...,x(f)}, P_{...,y(f)}]^T$	Coordinates x and y of point $P_{...}$ WRT $\{f\}$. If no $\{f\}$ is indicated, $\{w\}$ is understood.
$\dot{\mathbf{P}}$	First time derivative of $\mathbf{P}$
$\mathbf{O}_f$	Origin of $\{f\}$
$\{\mathbf{e}_{f,x}, \mathbf{e}_{f,y}\}$	Axes of $\{f\}$
$d(P, Q)$	Distance between point P and Q
$\mathbf{v} = [v_{x(f)}, v_{y(f)}]^T, \mathbf{v} = v\mathbf{u}_v$	Vector $\mathbf{v}$, with components $v_{x(f)}$ and $v_{y(f)}$ WRT $\{f\}$, of magnitude v and direction $\mathbf{u}_v$
$\mathbf{v}(Q, P) = \mathbf{Q} - \mathbf{P}$	Vector from P to Q applied in P.
$\mathbf{u}(Q, P) = \frac{\mathbf{v}(Q,P)}{\lvert\mathbf{v}(Q,P)\rvert}$	Unit vector from P toward Q.
$\mathbf{u}(\mathbf{v}) = \frac{\mathbf{v}}{\lvert\mathbf{v}\rvert}$	Unit vector of $\mathbf{v}$.

2. Kinematics

Let us refer to Figure 3 and to Figure 4 for a graphical representation of the kinematic architecture under consideration. The former is a simplified scheme of two main embedded submodules. The latter is a completed scheme of all the main components and sets out all the required symbols for a complete geometrical and analytic description.

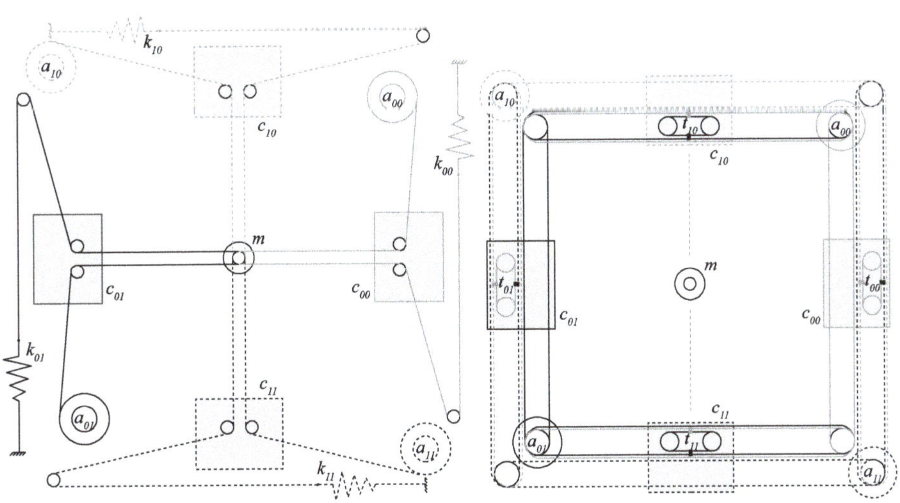

(a) Tendons subsystem **(b)** Carriages subsystem

Figure 3. Simplified representations of the two main submechanisms embedded in the presented architecture. Continuous lines refer to the two horizontal antagonist actuators, dotted lines refer to the two vertical antagonist actuators. (**a**) The tendons subsystem includes two orthogonal pairs of antagonistic non-linear SEA. The direction of the tendons is defined by four carriages c_{ij} which embeds deflecting pulleys. (**b**) The carriages subsystem depicts the principle of operation of the tendon-based system which guarantees that the tendons are constantly orthogonal if no external force is applied to m.

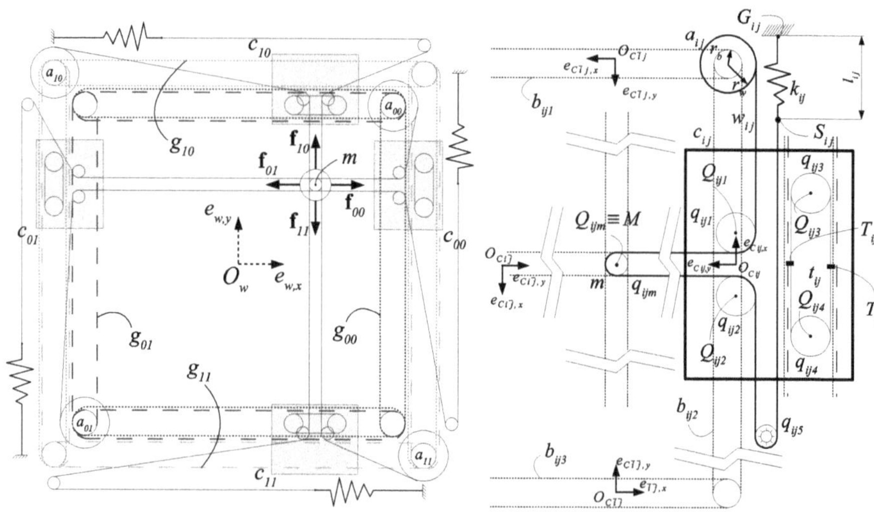

(a) Global view **(b)** Carriage c_{ij}

Figure 4. Top view of the mechanism. Modules g_{00} (dotted black lines) and g_{01} (dashed black lines) apply pulling forces along the horizontal direction. Similarly, g_{10} (dotted grey lines) and g_{11} (dashed grey lines) apply pulling forces along the vertical direction.

The two-DoF planar VSA presented in this work is made up of four almost identical actuation modules, orthogonal and antagonistic in pairs, identified by g_{ij}, where $i \in \{0,1\}$ and $j \in \{0,1\}$. The subscripts (ij) indicate hereafter both the direction and the orientation with respect to $\{w\}$, being the global planar reference frame. This notation has been chosen for the sake of generality. Nevertheless, referring to Figure 4, the subscript i is the actuation direction and, specifically, $i = 0$ and $i = 1$ correspond to $i = \mathbf{e}_{w,x}$ and $i = \mathbf{e}_{w,y}$, respectively, whereas, the subscript j refers to its orientation and, specifically, $j = 0$ and $j = 1$ correspond to the same or the opposite orientation of the specified i unit vector, respectively. Moreover, let us define $\bar{i}$ and $\bar{j}$ the complement to 1 of i and j, respectively. Accordingly, given that g_{ij} refers to, generically, one of the four actuation modules, $g_{\bar{i}j}$ and $g_{\bar{i}\bar{j}}$ refer to the two actuation modules perpendicular to g_{ij} and antagonists between them.

A mobile body m is connected to the ground by means of four wires w_{ij} tensioned by non-linear springs k_{ij}, each of these belonging to g_{ij}. Wires w_{ij} are wrapped around four pulleys concentric to the center of m, indicated by M.

By hypothesizing that the inertial and external forces acting on m are disregarded, this system configuration shows the characteristic of applying to m a set of forces $\mathbf{f}_{ij}$, parallel to i with orientation j, antagonist and orthogonal in pairs.

The generic module g_{ij} is made up of:

- a rotational actuator a_{ij} characterized by the rotational coordinate θ_{ij} and the torque τ_{ij};
- a non-linear spring k_{ij}, assumed to be identical $\forall i, j$, with length $l_{ij} = d(S_{ij}, G_{ij}) = \Delta l_{ij} + l_{ij,eq}$, indicating G_{ij} as ground point, S_{ij} the free endpoint of w_{ij}, $l_{ij,eq}$ the preload length of the spring, and Δl_{ij} the length variation from the preload length;
- a wire w_{ij} actuated by a_{ij} at one of its endpoints, exerting on M the pulling force $\mathbf{f}_{ij}$ due to the elongation of k_{ij};
- a carriage c_{ij}, ensuring the correct direction of $\mathbf{f}_{ij}$;
- a set of synchronous belts b_{ijl}, with $l \in \{1,2,3\}$, with the task of synchronizing the motion of the carriages $c_{\bar{i}j}$ and $c_{\bar{i}\bar{j}}$ with respect to M thus assuring the orthogonality of the antagonist force pairs ($\mathbf{f}_{\bar{i}j}$ and $\mathbf{f}_{\bar{i}\bar{j}}$) acting on the end effector.

As can be seen from Figure 4a, the only difference between g_{0j} and g_{1j} (for each j) is the length of belts b_{ijl} for assembly purposes. We can moreover denote:

- $\mathbf{f}_m$ the external force applied to m;
- v_i each couple (g_{i0}, g_{i1}) constituting an antagonistic VSA.

Referring to Figure 5 and ignoring inertial effects, the position of m (i.e., M) can be obtained by solving the equilibrium equation:

$$\sum_{i=1}^{2}\sum_{j=1}^{2} \mathbf{f}_{ij} + \mathbf{f}_m = \mathbf{f}_s + \mathbf{f}_m = \mathbf{0}. \tag{1}$$

where $\mathbf{f}_s$ denotes the total force exerted by the springs on m.

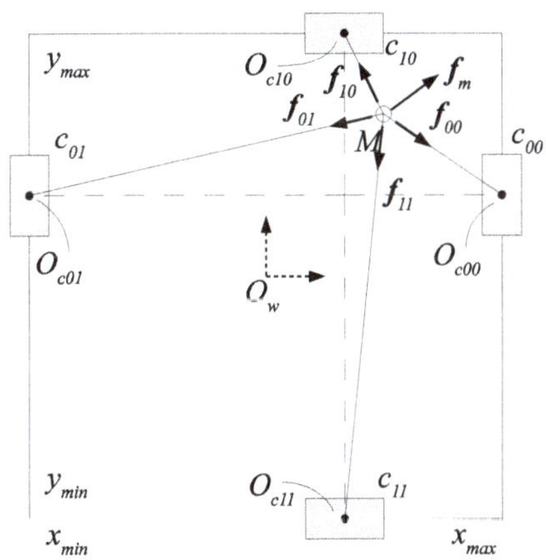

Figure 5. Representation of forces applied to m in a generic configuration of the mechanism.

The mechanical transmission of this architecture, constituted by c_{ij}, w_{ij} and b_{ijl}, ensures that the forces applied to m are orthogonal and antagonistic in pairs, as set out below (Figure 4):

$$\mathbf{f}_{01} = -\mathbf{f}_{00}, \qquad \mathbf{f}_{11} = -\mathbf{f}_{10}, \qquad \mathbf{f}_{00} \perp \mathbf{f}_{10}, \tag{2}$$

and assuming that all k_{ij} are identical and no external forces are applied, i.e., $\mathbf{f}_m = \mathbf{0}$. This is an aspect which can be anticipated to control both the equilibrium position and the stiffness of m along two orthogonal directions in performing planar movements.

We will now focus on the generic carriage c_{ij} to illustrate the transmission mechanism represented in Figure 4b where $\{c_{ij}\}$ denotes its local reference frame. Each c_{ij} includes four pulleys q_{ijk}, with $k \in \{1 \dots 4\}$, each center being indicated by Q_{ijk}. Pulleys q_{ij1} and q_{ij2} are configured so that $\mathbf{v}(Q_{ij1}, O_{c_{ij}}) = -\mathbf{v}(Q_{ij2}, O_{c_{ij}})$, and $\mathbf{e}_{c_{ij},x} = \mathbf{u}(Q_{ij1}, O_{c_{ij}})$. Pulley q_{ijm} is fixed to m and centered in M, pulley q_{ij5} is affixed to the ground. It must be pointed out that $q_{ijm} \forall i, j$ are four pulleys (one per g_{ij}) centered in M, coaxial, and placed side by side along their axis. By hypothesizing $\mathbf{f}_{ij}$ as coplanar, notwithstanding the slight displacement which might occurs among pulleys for construction reasons (Figure 6b).

The wire w_{ij}, connected to the ground point G_{ij} through k_{ij} at its endpoint S_{ij}, partially enveloped in series on q_{ij1} q_{ijm} q_{ij2} q_{ij5}, applies the pulling force $\mathbf{f}_{ij}$ on m.

The transmission mechanism of two antagonistic modules (Figure 6) has the objective of synchronizing c_{ij} among them so that

$$M \in \mathbf{e}_{c_{ij},y} \quad \forall i, j \tag{3}$$

in this way (2) is respected, since $\mathbf{f}_{ij} \parallel \mathbf{u}(O_{c_{ij}}, M)$.

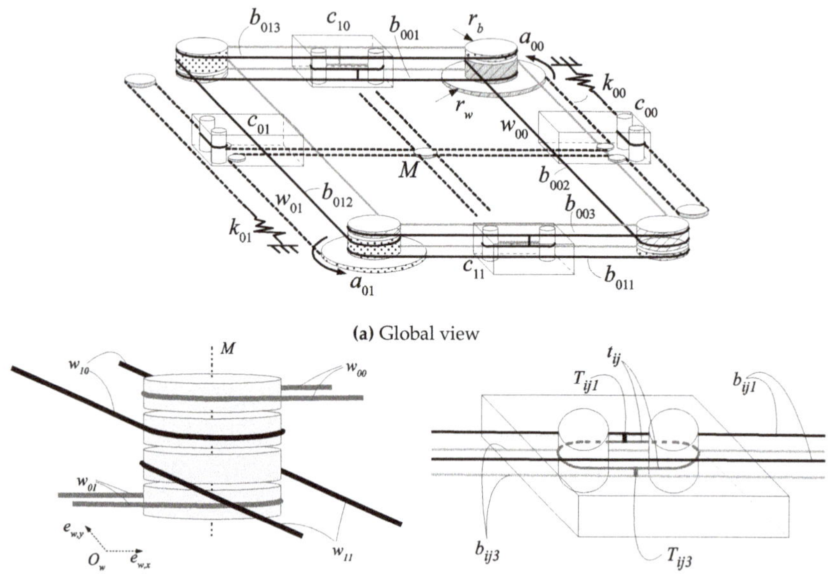

(a) Global view

(b) End effector **(c)** Carriage

Figure 6. Three-dimensional representation of the VSA v_0. Greyed pulleys are drive pulleys actuated by a_{00}, white pulleys are driven pulleys. Dashed pulleys are connected to a_{00}, dotted pulleys are connected to a_{01}. Conceptually it represents even v_1, neglecting a few mechanical details due to the different lengths of the belts.

Referring to Figure 4b and Figure 7, it is

$$\dot{M}_i = \frac{\dot{\theta}_{i0} r_w - \dot{l}_{i0}}{2} = -\frac{\dot{\theta}_{i1} r_w - \dot{l}_{i1}}{2} \quad \forall i, \tag{4}$$

where $\dot{M}_i$ denotes the velocity of M along the i direction.

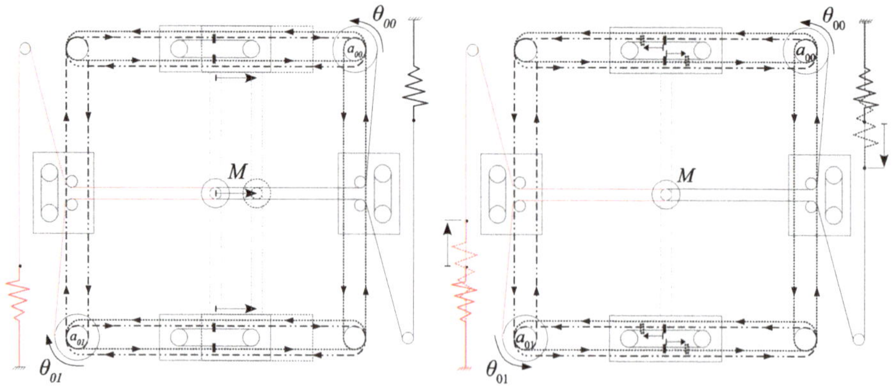

(a) $\dot{\theta}_{00} = -\dot{\theta}_{01}$: M moves, stiffness is constant **(b)** $\dot{\theta}_{00} = \dot{\theta}_{01}$: M is steady, stiffness changes

Figure 7. Effects of opposite or same rotation velocities of a_{00} and a_{01}. The red and black tendons are antagonist and control the position and the stiffness along the horizontal direction.

If no external force is applied to m, the result is $|\mathbf{f}_{ij}| = |\mathbf{f}_{i\bar{j}}|$ by reason of (2) above and as the $\mathbf{f}_{ij}$ are exerted by k_{ij}, the result is $l_{i0} = l_{i1} = l_i$. Differentiating with respect to the time, one obtains $\dot{l}_{i0} = \dot{l}_{i1} = \dot{l}_i$. Therefore, (4) can be written in the following form

$$\dot{l}_i = r_w(\dot{\theta}_{i0} + \dot{\theta}_{i1})/2 \tag{5}$$

which, if substituted in (4), will result in

$$\dot{M}_i = \frac{(\dot{\theta}_{i0} - \dot{\theta}_{i1})r_w}{4} \tag{6}$$

Referring to Figure 6, r_w,r_b indicate the radius of the pulleys enveloped by w_{ij},b_{ijl} and actuated through a_{ij}, respectively. If B_{ijl} is a generic point of the belts b_{ijl}, the following relationship results:

$$\left|\dot{\mathbf{B}}_{ijl}\right| = r_b\dot{\theta}_{ij} \quad \forall B_{ijl} \in b_{ijl} \quad \forall i, j. \tag{7}$$

Each c_{ij} includes two pulleys q_{ij3} q_{ij4} enveloped by the belt t_{ij} arranged in a manner that $\mathbf{v}(Q_{ij3}, Q_{ij4}) \parallel \mathbf{e}_{c_{ij},x}$. The velocities of two opposite points $T_{ij1}, T_{ij3} \in t_{ij}$ with respect to the velocity of c_{ij} are set out in (Figure 4b, Figure 7):

$$\dot{\mathbf{O}}_{c_{ij}} = (\dot{\mathbf{T}}_{ij1} + \dot{\mathbf{T}}_{ij3})/2 \quad \forall i, j \tag{8}$$

Therefore, constraining B_{ij1} and $B_{i\bar{j}3}$ (chosen for the sake of convenience), to $T_{\bar{i}j1}$ and $T_{\bar{i}j3}$, respectively, and combining (6), (7) and (8) so that $\dot{M}_i = \dot{O}_{c_{\bar{i}j}} \forall i, j$ in order to satisfy condition (3), the result is

$$r_w = 2r_b, \tag{9}$$

that is used as a design parameter to ensure (2).

It is noteworthy to underline that the architecture presented herein minimizes the total number of actuators required for the functional specification needed. In fact, four is the minimum number of actuators to independently control position and stiffness along two orthogonal directions.

The deformations of tension springs due to the application of an external force to the mobile mass M is set out in Figure 8.

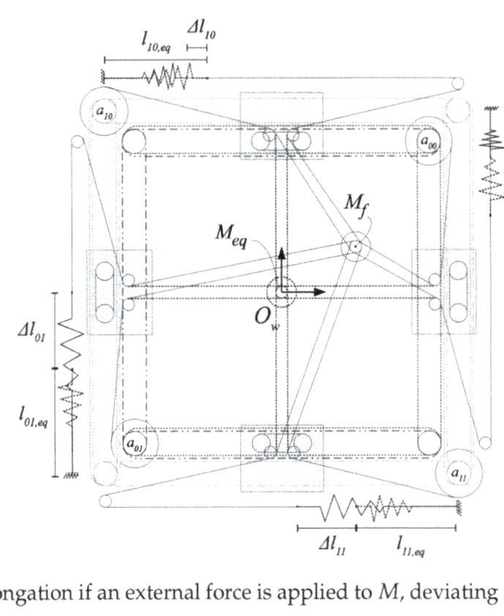

Figure 8. Springs elongation if an external force is applied to M, deviating it to M_f from its equilibrium configuration M_{eq}.

3. Force and Stiffness Analysis

Generally, because of its intrinsic compliance, if $\mathbf{f}_m \neq \mathbf{0}$, m deviates significantly from its equilibrium position, thus causing it not to comply with (2) even if it complies with (9). Consequently, the orthogonality between v_0 and v_1 is no longer respected. Therefore, it is necessary to evaluate the actual mechanical characteristic of m, i.e., the relationship between the displacement and the force $\mathbf{f}_m$, taking into consideration the possibility of large displacements from the equilibrium configuration.

Let us refer to Figure 9, a zoomed view of Figure 5, for symbols and nomenclature used in this force and stiffness analysis. In particular, we have defined $\mathbf{M}_{eq}$ the position of M if $\mathbf{f}_m = \mathbf{0}$, and $\mathbf{M}_f$ the position of M if $\mathbf{f}_m \neq \mathbf{0}$. We can moreover denote by $\mathbf{\Delta M} = \mathbf{M}_f - \mathbf{M}_{eq}$. Given a force $\mathbf{f}_m$, the resulting position of M can be obtained by solving Equation (1). However, to characterize the mechanism and given the intrinsic nonlinearity of the system, it is necessary to evaluate $\mathbf{f}_s$, opposite to $\mathbf{f}_m$, in function of the position of m. It is moreover useful to take into consideration the representation of $\mathbf{f}_s$ both in Cartesian and polar components. For the sake of convenience, we will define:

- $f_{s,x} = \mathbf{f}_s \cdot \mathbf{e}_{w,x}$, x component of $\mathbf{f}_s$ with respect to $\{w\}$;
- $f_{s,y} = \mathbf{f}_s \cdot \mathbf{e}_{w,y}$, y component of $\mathbf{f}_s$ with respect to $\{w\}$;
- $f_{s,r} = \mathbf{f}_s \cdot \mathbf{u}(\mathbf{\Delta M})$, component of $\mathbf{f}_s$ parallel to $\mathbf{\Delta M}$;
- $f_{s,t} = \sqrt{f_s^2 - f_{s,r}^2}$, component of $\mathbf{f}_s$ normal to $\mathbf{\Delta M}$.

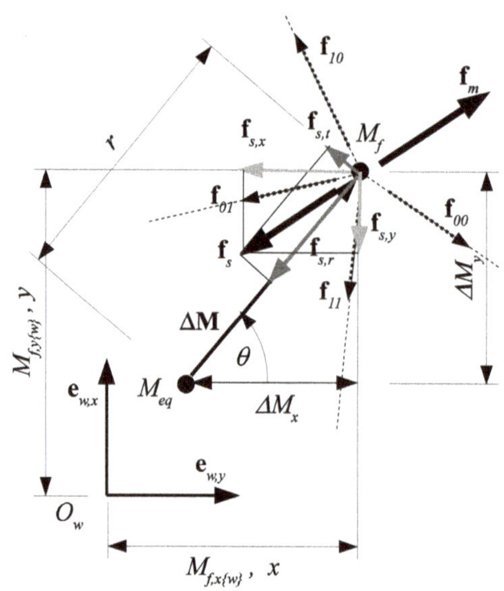

Figure 9. Detailed view of Figure 5 representing forces applied on m.

More particularly, $f_{s,r}$ denotes the amount of force oriented toward the equilibrium point and $f_{s,t}$ denotes the amount of force perpendicular to the displacement vector, due to reasons of non-orthogonality, misalignment, and asymmetry.

The local stiffness of m along different directions can be evaluated by partial derivatives of $\mathbf{f}_s$ with respect to space

$$k_x = \frac{\partial f_{s,x}}{\partial x} \quad k_y = \frac{\partial f_{s,y}}{\partial y} \qquad k_r = \frac{\partial f_{s,r}}{\partial r} \quad k_t = \frac{\partial f_{s,t}}{r \partial \theta}. \tag{10}$$

where $x = M_{f,x(w)}$, $y = M_{f,y(w)}$, $r = |\mathbf{\Delta M}|$ and $\theta = \arctan 2(\Delta M_y, \Delta M_x)$.

Some numerical evaluations are reported and commented hereunder to better understand the mechanical behavior of the mechanism. Without loss of methodological generality, let us refer to a mechanism with unitary quantities by considering the International System of Units. The following assumptions are made:

- referring to Figures 5 and 9 the analyzed workspace is in the interval -1 m$\leq x, y \leq 1$ m;
- k_{ij} is assumed to be a quadratic spring exerting a force $f_{k,ij} = a_k l_{ij}^2 = a_k(\Delta l_{ij} + l_{ij,eq})^2$, with $a_k = 1$ N/m^2;
- $l_{0j,eq} = l_{1j,eq} = 1$ m, i.e., halfway through its available stroke (Figure 4b).

The choice of a quadratic spring is due to the fact that a quadratic force-length function of the actuators in antagonist VSA makes the stiffness of the mobile body independent from the load externally applied to it [6]. This makes it the most common and sought-after spring characteristic in one-DoF *VSA*s. It is important to numerically assess how the asymmetries and non-orthogonalities, which take place by varying the M position, affect this aspect.

Regardless of these assumptions, the reasoning and the methodology behind the simulations presented can be adapted to any type of spring and dimension by using a developed parametric numerical routine.

Referring to Figure 10, let us firstly analyze force components $f_{s,r}, f_{s,t}$ and stiffness k_r in some specific points within the workspace. It is evident that the majority of the isotropic behavior occurs if M_{eq} is placed in the center of the workspace and that the symmetry of this configuration considerably limits any tangential force $f_{s,t}$ which deviates from the line connecting M_f to its equilibrium point M_{eq}.

Approaching the boundaries of the workspace, edge effects and asymmetries are increasingly evident. It is noteworthy that if both M_f and M_{eq} are on one the symmetry axes of the workspace, no tangential force occurs due to existing force symmetries. It is further worth noting that not all the workspace is reachable because of the limits of the k_{ij} stroke. In any event, for the purposes of this work, force and stiffness values for all the points under investigation in the workspace have been plotted. In this way one can obtain a more general overview of force and stiffness variations with the focus being placed on the architecture analysis and overlooking potential limitations due to construction reasons.

Referring to Figure 11, and more specifically to the condition $l_{0j,eq}/l_{1j,eq} = 0.2$ which is indicated by a dot, Figure 12 highlights how, by using two different preload lengths on VSAs v_0 and v_1, one can deform the force field in order to obtain different values of stiffness along two orthogonal directions of the workspace.

Figure 13 shows how the stiffness between two orthogonal directions can be modified by varying the preload length of one of the two VSA and disregarding the other.

By using the force maps, such as the one shown in Figures 10 and 12 one can easily estimate the applied force as a function of the actual displacement of m with respect to its equilibrium position. Moreover, stiffness ellipsis, for specific values of spring preloads, are set out in Figure 14.

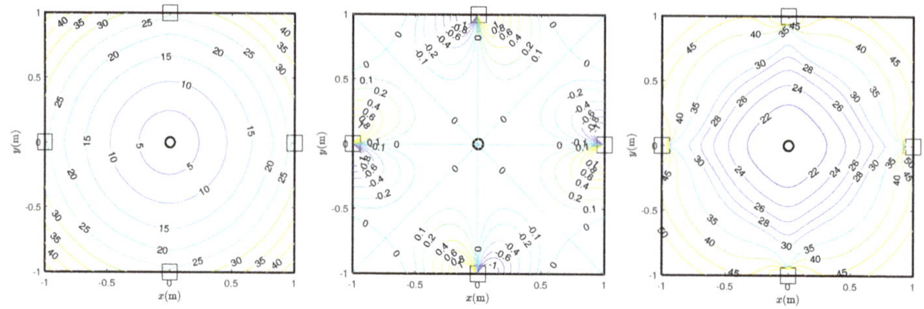

(a) $f_{s,r}$(N); M_{eq} = (0 m, 0 m) **(b)** $f_{s,t}$(N); M_{eq} = (0 m, 0 m) **(c)** k_r(N/m); M_{eq} = (0 m, 0 m)

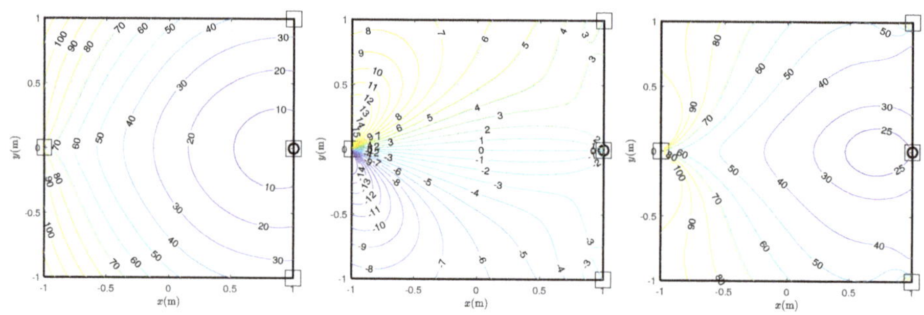

(d) $f_{s,r}$(N); M_{eq} = (1 m, 0 m) **(e)** $f_{s,t}$(N); M_{eq} = (1 m, 0 m) **(f)** k_r(N/m); M_{eq} = (1 m, 0 m)

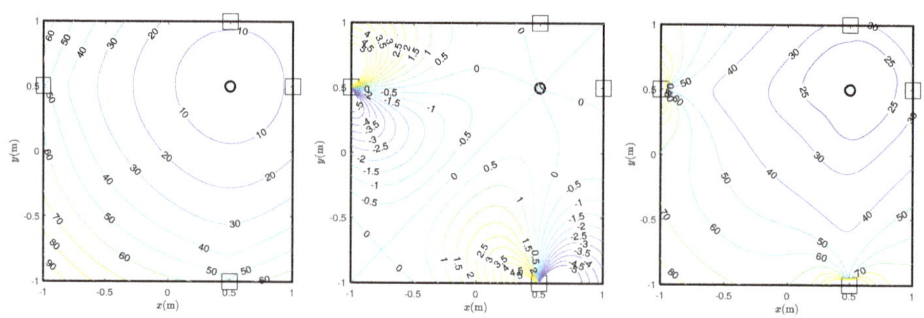

(g) $f_{s,r}$(N); M_{eq} = (0.5 m, 0.5 m) **(h)** $f_{s,t}$(N); M_{eq} = (0.5 m, 0.5 m) **(i)** k_r(N/m); M_{eq} = (0.5 m, 0.5 m)

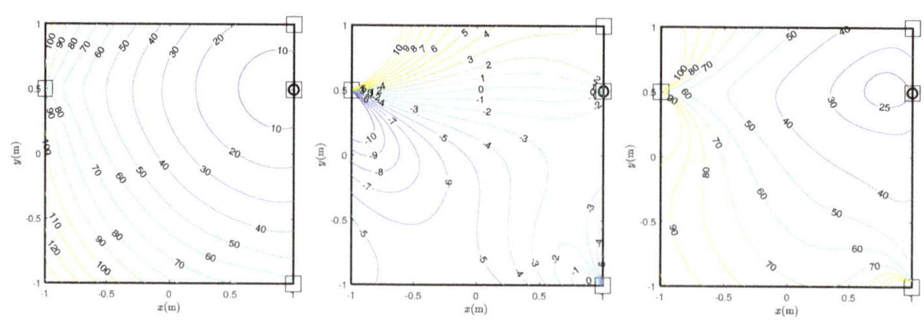

(j) $f_{s,r}$(N); M_{eq} = (1 m, 0.5 m). **(k)** $f_{s,t}$(N); M_{eq} = (1 m, 0.5 m). **(l)** k_r(N/m); M_{eq} = (1 m, 0.5 m)

Figure 10. Radial force $f_{s,r}$, tangential force $f_{s,t}$ and radial stiffness k_r with M_{eq} in some notable points, by assuming $l_{0j,eq} = l_{1j,eq}$ = 1 m (i.e., halfway through its available stroke) and quadratic springs.

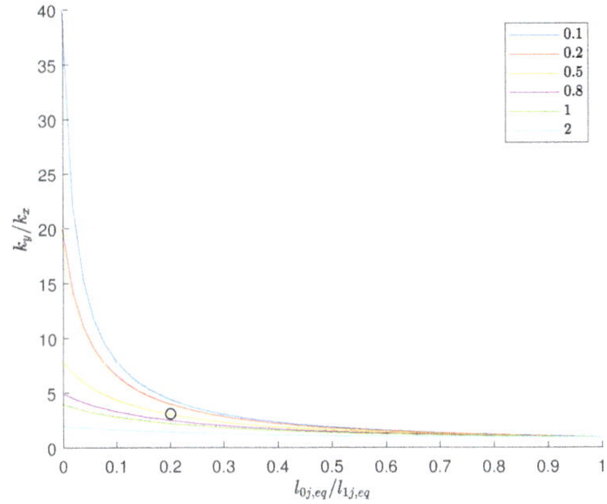

Figure 11. Stiffness ratio along y and x in the center of the workspace, i.e., $M_{eq} = M_f = (0,0)$, as function of the ratio $l_{0j,eq}/l_{1j,eq}$, varying $\Delta l_{1j,eq}, \forall j$ (colored lines). High ratios, i.e., increasing the difference between stiffness along x and y, can be achieved applying short preload lengths. Therefore, high values of forces can be obtained increasing the stiffness of the springs a_k.

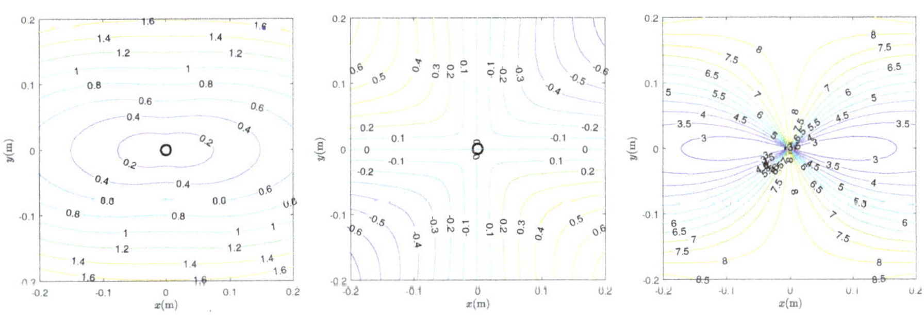

(a) Radial force $f_{s,r}$(N) **(b)** Tangential force $f_{s,t}$(N) **(c)** Radial stiffness k_r(N/m)

Figure 12. Force components and stiffness mapped in the neighborhood of M_{eq} with M_{eq} considered in the center of the workspace, i.e., $M_{eq} = (0,0)$, assuming $l_{0j,eq} = 0.1$ and $l_{1j,eq} = 0.5$.

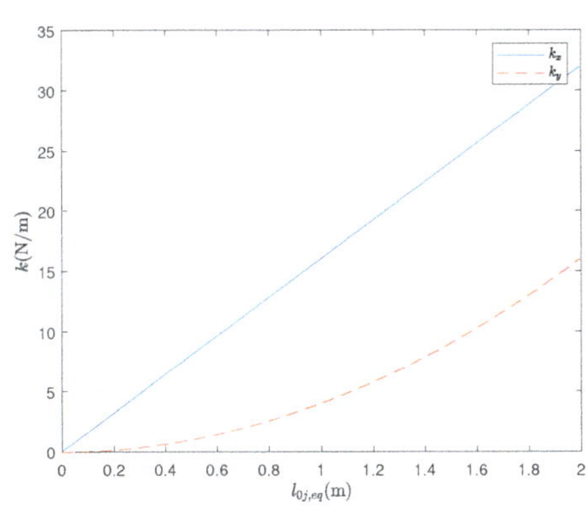

Figure 13. Stiffness along x and y as function of $l_{0j,eq}$ assuming $l_{1j,eq} = 0$ in the center of the workspace, i.e., $M_{eq} = M_f = (0,0)$. It represents how stiffness varies considering only VSA_0 with different preload lengths, neglecting VSA_1.

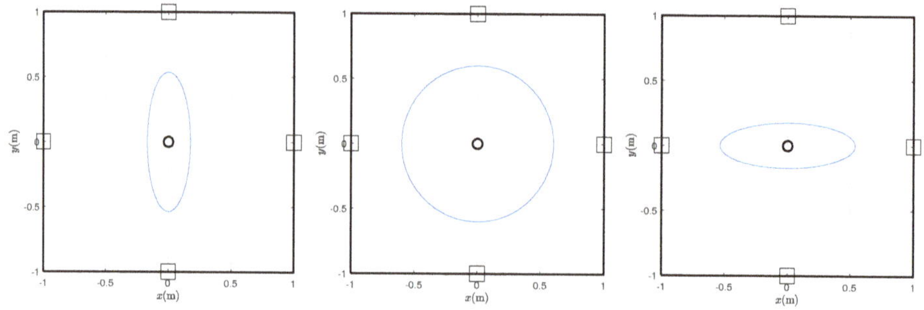

(a) $l_{0j,eq} = 0.1\text{m}, l_{1j,eq} = 0.5\text{m}$ (b) $l_{0j,eq} = 0.5\text{m}, l_{1j,eq} = 0.5\text{m}$ (c) $l_{0j,eq} = 0.5\text{m}, l_{1j,eq} = 0.1\text{m}$

Figure 14. Stiffness ellipsis as function of the preloads $l_{0j,eq}$ and $l_{1j,eq}$.

4. Conclusions

This work presents a novel variable-stiffness mechanical architecture. It features an innovative two-DoF parallel and orthogonal architecture capable of favoring isotropic behavior and tuning stiffness ellipsis along two independent axes. The orthogonality of agonist-antagonist tendons is ensured by a custom-made mechanical transmission. Analytical aspects and numerical simulations have been presented to illustrate and investigate the kinetostatic characteristics of this mechanism.

The mechanism allows configuration of stiffness ellipses with axes approximately parallel to the axes of the structure with very different components. This result is true in the center of the work area and worsens depending on the proximity to the boundaries. This is a general result of the theory and not limited to any particular prototype. This happens despite the use of pairs of quadratic springs, which allow tuning of the stiffness of the mobile body if employed in one-dimensional antagonist VSA, independently from its displacement from the equilibrium position. Asymmetries and non-linearities of force and stiffness characteristics are increasingly evident, the closer the equilibrium point is to the boundaries. This aspect must be considered in developing a control system of a mechatronic device based on the presented architecture, to properly estimate the externally applied force and to control the mechanical stiffness of the mechanism.

To exploit the potentialities of the mechanism it is convenient to add sensors to measure both the positions of the motors and the elongations of the springs. By combining them, it is possible to measure the actual position of the mobile body and to estimate the externally applied force. Additional limit switches could be embedded for resetting purposes if incremental position sensors are used. The actual resolution of the selected sensors will depend on the target actual resolution of a real device. As a general remark, the higher the resolution and the acquisition frequency of the sensors are, the higher is the performance of force-feedback and vibration control algorithms eventually implemented.

Being a cable-driven mechanism, its main limitation is related to the risk of slacking wires, as in other wire-based VSAs. This aspect requires that all the tension springs and wires are enough preloaded, in accordance with the maximum force applied to the load, which should not be greater than the preload.

The current work refers to a general description of the mechanism to analyze it peculiarities, independently from the actual dimensions of a possible prototype. To correctly dimension a real device, it will be required to perform specific calculations based on the presented model.

The mechanism presented in this paper is proposed to be embedded in two-DoF planar haptic devices, for which high mechanical backdrivability is foreseen and where a compliant tunable behavior is required. For this reason, it is embedded within PLANarm, the prototype of a two-DoF end-effector planar device for upper-limb neurorehabilitation (Figure 15).

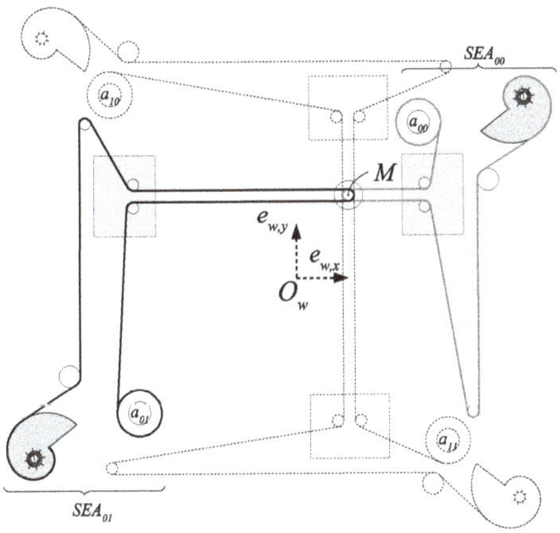

(a) Kinematics

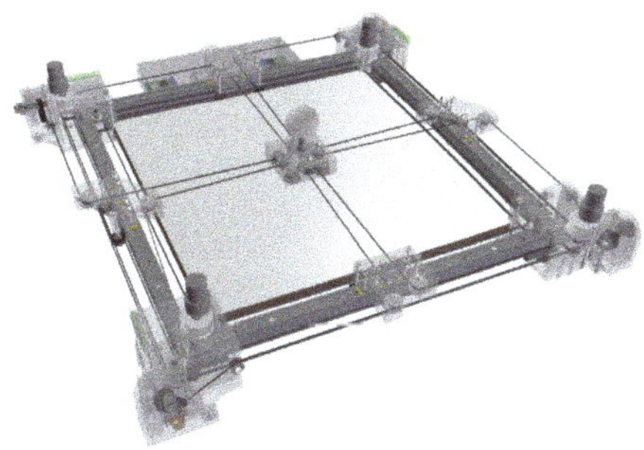

(b) CAD model

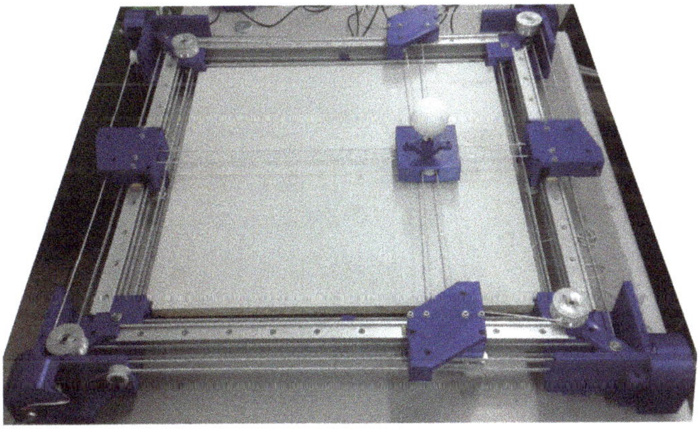

(c) Prototype

Figure 15. PLANarm prototype which exploits the variable-stiffness kinematic architecture presented in this work and embeds cam-based non-linear springs.

Author Contributions: Conceptualization, M.M.; Formal analysis, M.M., F.C. and F.R.R.; Funding acquisition, L.M.T.; Investigation, M.M., F.C. and F.R.R.; Methodology, M.M., F.C., F.R.R. and H.G.; Project administration, L.M.T.; Software, F.C. and F.R.R.; Supervision, H.G., G.L. and L.M.T.; Visualization, M.M., F.C. and F.R.R.; Writing – original draft, M.M., F.C. and F.R.R.; Writing – review & editing, H.G. and G.L.

Funding: This work was partially funded by the Italian Lombardy region within the RIPRENDO@home project, n.18092/RCC del 05/08/2013, and by Fondazione Cariplo within the EMPATIA@Lecco project, Rif. 2016-1428 Decreto Regione Lombardia 6363 del 30/05/2017.

Acknowledgments: The authors would like to thank João Carlos Dalberto and Roberto Bozzi for supporting the mechanical and electrical design of the PLANarm prototype.

Conflicts of Interest: The authors declare no conflict of interest.

References

1. Vanderborght, B.; Albu-Schaeffer, A.; Bicchi, A.; Burdet, E.; Caldwell, D.G.; Carloni, R.; Catalano, M.; Eiberger, O.; Friedl, W.; Ganesh, G.; et al. Variable Impedance Actuators: A Review. *Robot. Auton. Syst.* **2013**, *61*, 1601–1614. [CrossRef]

2. Grioli, G.; Wolf, S.; Garabini, M.; Catalano, M.; Burdet, E.; Caldwell, D.; Carloni, R.; Friedl, W.; Grebenstein, M.; Laffranchi, M.; et al. Variable Stiffness Actuators: The User's Point of View. *Int. J. Rob. Res.* **2015**, *34*, 727–743. [CrossRef]
3. Wolf, S.; Grioli, G.; Friedl, W.; Grebenstein, M.; Hoeppner, H.; Burdet, E.; Caldwell, D.; Bicchi, A.; Stramigioli, S.; Vanderborght, B. Variable Stiffness Actuators: Review on Design and Components. *IEEE/ASME Trans. Mechatron.* **2016**, *21*, 2418–2430. [CrossRef]
4. Laurin-Kovitz, K.; Colgate, J.; Carnes, S. Design of components for programmable passive impedance. In Proceedings of the IEEE International Conference on Robotics and Automation, Sacramento, CA, USA, 9–11 April 1991; pp. 1476–1481. [CrossRef]
5. Pratt, G.; Williamson, M. Series Elastic Actuators. In Proceedings of the 1995 IEEE/RSJ International Conference on Intelligent Robots and Systems. Human Robot Interaction and Cooperative Robots, Pittsburgh, PA, USA, 5–9 August 1995; pp. 399–406. [CrossRef]
6. Jafari, A. Coupling between the Output Force and Stiffness in Different Variable Stiffness Actuators. *Actuators* **2014**, *3*, 270. [CrossRef]
7. Schepelmann, A.; Geberth, K.; Geyer, H. Compact nonlinear springs with user defined torque-deflection profiles for series elastic actuators. In Proceedings of the IEEE International Conference on Robotics and Automation (ICRA), Hong Kong, China, 31 May–7 June 2014; pp. 3411–3416. [CrossRef]
8. Spagnuolo, G.; Malosio, M.; Dinon, T.; Molinari Tosatti, L.; Legnani, G. Analysis and synthesis of LinWWC-VSA, a Variable Stiffness Actuator for linear motion. *Mech. Mach. Theory* **2017**, *110*, 85–99. [CrossRef]
9. Zhou, X.; kook Jun, S.; Krovi, V. Stiffness modulation exploiting configuration redundancy in mobile cable robots. In Proceedings of the IEEE International Conference on Robotics and Automation (ICRA), Hong Kong, China, 31 May–7 June 2014; pp. 5934–5939. [CrossRef]
10. Legnani, G.; Tosi, D.; Fassi, I.; Giberti, H.; Cinquemani, S. The "point of isotropy" and other properties of serial and parallel manipulators. *Mech. Mach. Theory* **2010**, *45*, 1407–1423. [CrossRef]
11. Cinquemani, S.; Giberti, H.; Legnani, G. The generalized jacobian matrix and the manipulators kinetostatic properties. In Proceedings of the 10th Biennial Conference on Engineering Systems Design and Analysis (ASME 2010), Istanbul, Turkey, 12–14 July 2010; pp. 721–729.
12. Molinari Tosatti, L.; Fassi, I.; Legnani, G. Kineto-static optimisation of PKMs. *CIRP Ann. Manuf. Technol.* **2003**, *52*, 337–341. [CrossRef]

robotics

Article

Upside-Down Robots: Modeling and Experimental Validation of Magnetic-Adhesion Mobile Systems

Stefano Seriani [1,*], Lorenzo Scalera [2,3], Matteo Caruso [1], Alessandro Gasparetto [2] and Paolo Gallina [1]

1 Department of Engineering and Architecture, University of Trieste, 34100 Trieste, Italy; matteo.caruso1993@gmail.com (M.C.); pgallina@units.it (P.G.)
2 Polytechnic Department of Engineering and Architecture, University of Udine, 33100 Udine, Italy; lorenzo.scalera@uniud.it (L.S.); alessandro.gasparetto@uniud.it (A.G.)
3 Faculty of Science and Technology, Free University of Bozen-Bolzano, 39100 Bolzano, Italy
* Correspondence: sseriani@units.it

Received: 30 March 2019; Accepted: 29 May 2019; Published: 31 May 2019

Abstract: In this paper, we present the modeling and validation of a new family of climbing robots that are capable of adhering to vertical surfaces through permanent magnetic elements. The robotic system is composed of two modules, the master and the follower carts, which are arranged in a sandwich configuration, with the surface to climb interposed between them. Thanks to this configuration, the mobile robot can climb even nonferromagnetic and curved surfaces; moreover, the master cart is capable of freely moving on the floor by detaching from the follower. In this paper, we propose the mathematical modeling, simulation, and experimental validation of this kind of robots, with particular focus on the transitions between floor and climbing motion.

Keywords: mobile robot; climbing robot; wheeled robot; magnetic adhesion

1. Introduction

In recent years, mobile robots have become increasingly common and have found applications in several different fields. For example, they are employed in the industrial and commercial field for material handling and stocking [1], in agriculture for inspection and monitoring [2,3], in civil contexts for cleaning operations [4], surveillance [5], proximal sensing [6], and service tasks [7], as well in space exploration and hazardous environments [8–10].

In this context, a special family of mobile robots is that of robots capable of climbing on wall surfaces. Climbing robots are often adopted for inspection and safety in process industries, maintenance, and diagnosis in civil constructions, as well as in places where the presence of a human can be very expensive or dangerous due to a hostile environment. Climbing robots have to be lightweight, carry sensors and instrumentation, and have to be provided with a proper locomotion and adhesion system with respect to the surface they have to climb. Several different adhesion methods were investigated and tested in the literature: vacuum or suction cups, gripping to the surface, bioinspired adhesion methods, and magnetic adherence [11].

The use of suction force, generated through the Venturi Principle or by means of vacuum pumps, is the most common approach to provide adhesion to a climbing robot. Vacuum technology is usually light, easy to control, and allows to climb over arbitrary surfaces. The main drawbacks are the required time to generate the vacuum, the problem of carrying suction pumps on board, and the limits given by possible surface irregularities. Examples can be found in References [12–14].

Another approach for climbing robots is gripping to the surface by means of claws, limb extremities, or gripping systems. This method allows to climb more irregular surfaces, such as in References [15,16].

Bioinspired adhesive methods are another solution to guarantee adhesion to wall-climbing robots. These technologies are based on the use of patches of microscopic hairs that ensure dry adhesion thanks to Van der Waals forces, as on the legs of several insects, geckos, or spiders [17]. Examples of bioinspired adhesive-climbing robots can be found in References [18–20].

Magnetic force is another principle for creating adhesion in wall-climbing robots. Permanent magnets have several advantages with respect to other adhesion systems: they are inherently reliable, and low cost and no energy are required for adhesion [21]. Magnetic-adhesion climbing systems work very well when the surfaces to climb are composed of ferromagnetic materials. Robotic systems can either be equipped with permanent magnets to adhere onto the surface combined with wheels to move along it [22,23], or can adopt magnetic wheels that combine locomotion and adhesion [24,25]. In the first case, the presence of an air gap has to be taken into account; thus, the force of adhesion is reduced. On the other hand, magnetic wheels provide higher adhesion forces since they are in direct contact with the ferromagnetic surface. Another approach of wall climbing with the principle of magnetic adhesion is given by crawler-based robots [26–28]. Magnetic adhesion wall-climbing robots were adopted in several applications, such as in nondestructive tests [22,23], or for the inspection of ship hulls [27–29].

If the surface onto which wall-climbing robots have to adhere is nonferromagnetic, a possible solution to using magnetic-adhesion systems is to adopt a robot composed of two modules: a master and a follower cart. They are arranged in a sandwich configuration, with the surface to climb interposed between them. Thanks to this approach, the robotic system can climb even nonferromagnetic and curved surfaces; moreover, the master cart is capable of freely moving on the floor by detaching itself from the follower. The added advantage is that this system does not require power, allowing for more compact, lighter, and cheaper robots; indeed, contrary to some proposed concept that does not rely magnets (e.g., References [12,13,17,30]), this method is passive and does not continuously require power. No examples of wall-climbing robots based on magnetic adhesion and composed of two coupled subsystems can be found in the present literature. There exist some industrial examples of these sandwich robots, e.g., the Windowmate robot produced by RF Co. (http://www.mywindowmate.com), and Windoro produced by Iishim Global Co., Ltd. (http://www.mywindoro.com). These systems require an operator to manually detach the two modules.

The main contributions of this work can be summarized as follows:

- mathematical modeling of a class of magnetic-adhesion mobile robots that are capable of climbing walls thanks to a sandwich configuration. Furthermore, this design allows the master robot to autonomously detach itself from the follower, which is a novel contribution to the state of the art;
- numerical analysis through simulations, taking into account the interaction between the two carts, and between the robot and the support sheet (planar driving, behavior during detachment, and driving on a curved sheet);
- experimental validation of the prototype of an upside-down robot, both in the detachment phase as well as on the curved surface.

Indeed, to the best of the authors' knowledge, the permanent-magnet sandwich-configuration robot presented in this paper is the first that is capable to passively climb on nonferromagnetic surfaces, while at the same time being able to autonomously detach itself from the follower cart when needed.

This paper is an extended version of a preliminary work presented at the Second International Conference of IFToMM Italy, IFIT 2018 [31]; it is organized as follows: in Section 2, the architecture of the robot is briefly described, in Section 3 the mathematical model is presented, and in Section 4 numerical implementation is described along with the main simulation results; experiment validation is presented in Section 5; finally, Section 6 highlights the conclusions and possible future developments of this work.

2. Robot Architecture

The system presented in this paper consists of a robot that is able to crawl and drive on nonmagnetic thin supports, arbitrarily oriented in space. The system, which is illustrated in Figure 1, is composed of two modules, the master and the slave or follower, which are kept magnetically coupled through permanent magnets at either side of the support sheet. This architecture allows for a high degree of mobility through the design of tracks and planes that let the robot climb vertically or even drive upside down. Furthermore, the robot can autonomously discard the slave cart in special slave park areas in order to be free to drive along the horizontal plane as a normal mobile robot would.

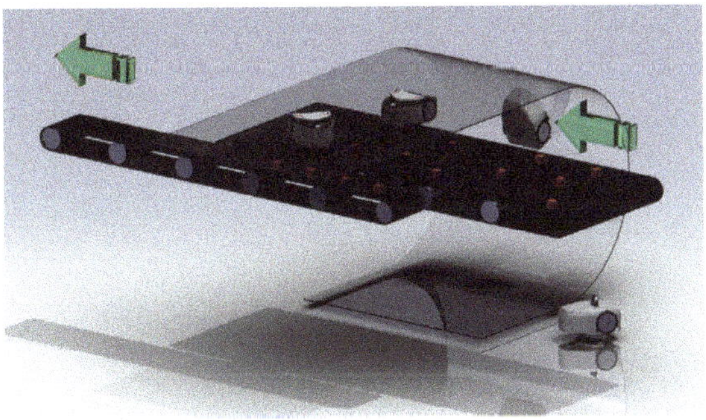

Figure 1. A conceptual illustration of the "upside-down robot" system in a real-world pick-and-place industrial application scenario.

Envisioned applications of this type of architecture include highly compact storage environments such as greenhouses [32] or pharmaceutical warehouses, where complex transfer operations need to be carried out [33]. In these environments, pick-and-place operations and subsequent transfer to remote locations in the industrial floor are usually needed. Normally, these tasks would require at least two manipulators (serial, Delta, or Scara), one at the origin and one at the destination, along with a mobile system, e.g., a mobile robot or rover. On the other hand, the approach here only presented calls for a single mobile robot with climbing capabilities, which can possibly translate into a decrease in cost and complexity of the entire installation.

The proposed system can be described as the interaction between three entities, a robot, a passive cart, and the environment, in the form of the support sheet. These actors exchange magnetic forces, contact forces, and friction, and are subject to gravity. We can break down this complex interaction by analyzing the limited number of possible situations that the robot can encounter:

- upside-down driving phase (A);
- curved-surface driving phase (B);
- slave-decoupling phase (D); and
- slave parking/engaging (E).

Other phases exist that are either a composition or transitions between pairs of the above; furthermore, Phases C and F are considered trivial once the other phases are verified and, as such, need not be discussed in detail. In Figure 2, a diagram of the main phases and main constituents of the system is illustrated.

In the paper, the master robot is generally indicated using the letter "r", while the slave or follower is indicated with "f". The side on which the master drives is called master surface, whereas the opposite is referred to as slave surface. The model geometry is considered as 2D throughout the article, while the model itself can be considered 1D, only considering the misalignment between the master and slave. In order for the system to work correctly, the following conditions have to be satisfied:

- the master robot always has to be pressed against the master surface;
- the master robot has to be able to drive away from the influence of the magnetic field of the slave during parking; and
- the slave robot has to always be pressed against the slave surface except when parked.

In the following section, the model is presented, starting from the magnetic model, and continuing with the pseudostatic model of the robotic system in its main phases, namely, planar and curved-surface driving, and parking.

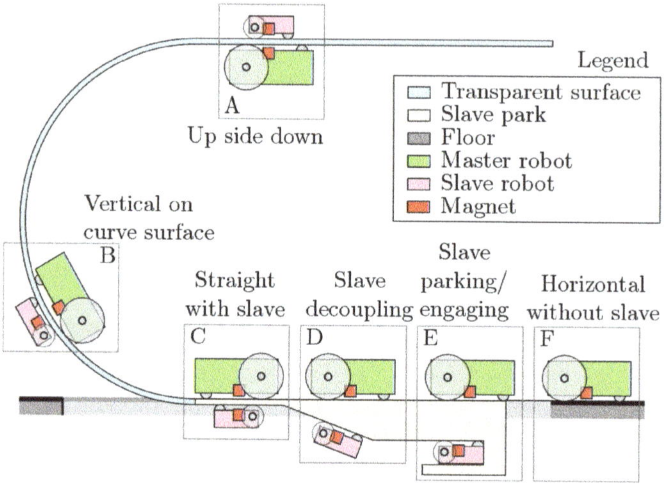

Figure 2. System description and illustration of different operation phases.

3. Mathematical Model

In this section we present the mathematical model of the robot. We describe the model of the magnetic field and of the robot in the case of a planar surface, a cylindrical surface, and during parking and engaging. Finally, we analyze traction and the contact forces. In this preliminary work, a quasistatic model of the robotic system was taken into account.

3.1. Magnetic Model

Two magnetic dipole moments, $\mathbf{m_r}$ and $\mathbf{m_f}$, related to the master robot and follower, respectively, and separated by a vector $\mathbf{r}$, interact by exchanging a magnetic force $\mathbf{F_m}$ and torque $\boldsymbol{\tau}_\mathbf{m}$ [34], as follows:

$$\begin{aligned} \mathbf{F_m} &= \nabla\left(\mathbf{m}_f \cdot \mathbf{B}_r\right) \\ \boldsymbol{\tau}_\mathbf{m} &= \mathbf{m}_f \times \mathbf{B}_r \end{aligned}, \tag{1}$$

where magnetic flux density $\mathbf{B_r}$ of the first dipole moment is defined as follows:

$$\mathbf{B_r} = \frac{\mu}{4\pi}\left(\frac{3\mathbf{r}\left(\mathbf{m}_r \cdot \mathbf{r}\right)}{\|\mathbf{r}\|^5} - \frac{\mathbf{m}_r}{\|\mathbf{r}\|^3}\right), \tag{2}$$

where μ is magnetic permeability. Let the unit vector of $\mathbf{r}$ be $\hat{r} = r/\,\|r\|$; using a truncated Taylor series, a more manageable form for the force was developed by Yung et al. [35]:

$$\begin{aligned} \mathbf{F}_m = \;& \tfrac{3\mu}{4\pi\|\mathbf{r}\|^4}\Big(\left(\left(\hat{\mathbf{r}} \times \mathbf{m}_r\right) \times \mathbf{m}_f + \hat{\mathbf{r}} \times \mathbf{m}_f\right) \times \mathbf{m}_r \\ & -2\hat{\mathbf{r}}(\mathbf{m}_r \cdot \mathbf{m}_f) + 5\hat{\mathbf{r}}((\hat{\mathbf{r}} \times \mathbf{m}_r) \cdot (\hat{\mathbf{r}} \times \mathbf{m}_f))\Big) \end{aligned}. \tag{3}$$

Despite this formulation being approximate, in that it neglects the higher-order Taylor terms, it is sufficiently adherent to the magnetic behavior found in the problem at hand. In Figure 3, the interaction between the two dipoles is shown.

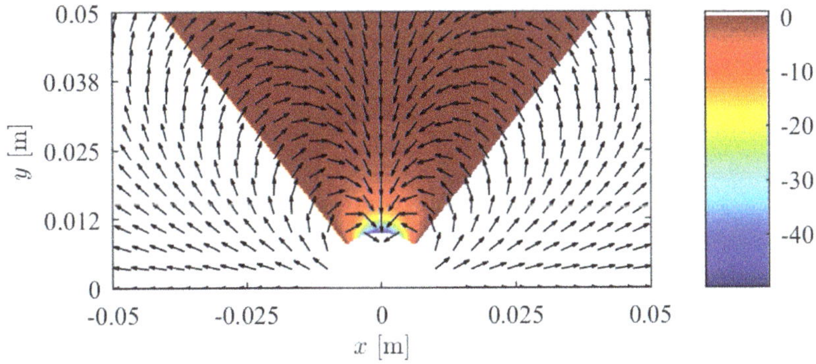

Figure 3. Morphology of $\mathbf{F}_m$ field. In the chart, magnetic moments $\mathbf{m}_r$ and $\mathbf{m_f}$ are vertically aligned. Moment $\mathbf{m}_r$ is placed at coordinates $(0,0)$. Direction of force acting on the second dipole, shown with arrow plot; magnitude of vertical component of the same force, shown in color-coded chart. White area, upward vertical components (i.e., repulsive).

3.2. Planar-Surface Model

In the following, we describe the quasistatic model of the robotic system in the case of planar surface. The most fundamental geometric configuration is that in which the master and slave surfaces are straight and parallel to one another (see Phases A and C); for the sake of generality, we model the surfaces as not necessarily parallel, thus defining the inclination angle γ of the slave surface with respect to the master plane orientation. This can be seen in Figure 4. Based on the same figure, we assume the frame of reference as centered in point $\mathbf{Q}_r$ and integral to the master robot. It is also assumed that the geometry of the master and slave actors is entirely known, and is defined by the following parameters: the contact point of the rear wheel $\mathbf{P}_{r,0}$, of the front one $\mathbf{P}_{r,1}$, the axis of the rear wheel $\mathbf{H}_r$, center of mass $\mathbf{G}_r$, and mass m_r. It is useful to note that, in principle, the points of contact for cylindrical wheels depend on the shape of the surface; in case of a planar surface, their coordinates are trivial to calculate. The geometry and mechanical characteristics of the slave can be defined in the same way with respect to the slave frame of reference (x', y') in $\mathbf{Q}_f$ by the following variables: $\mathbf{P}'_{f,0}$, $\mathbf{P}'_{f,1}$, $\mathbf{H}'_f$, $\mathbf{G}'_f$ and m_f. The pose of the slave with respect to the master can be defined by four variables: the two coordinates of point Ξ^*, angle γ, and position s along the slave surface.

We consider, referring to Figure 4, quantities L, h_r, t and h_f the travel of the robot from the beginning of the ramp, the height of the magnet of the robot, the thickness of the support sheet before the ramp, and the height of the magnet of the follower, respectively; from geometrical considerations,

$$\Xi^* = \mathbf{Q_r} + \begin{bmatrix} -L \\ 0 \end{bmatrix} + \begin{bmatrix} 0 \\ -h_r \end{bmatrix} + \begin{bmatrix} 0 \\ -t \end{bmatrix} + \begin{bmatrix} 0 \\ \frac{h_f}{\cos\gamma} \end{bmatrix}, \tag{4}$$

from which follows

$$\mathbf{Q_f} = \Xi^* + s \begin{bmatrix} \cos\gamma \\ \sin\gamma \end{bmatrix}. \tag{5}$$

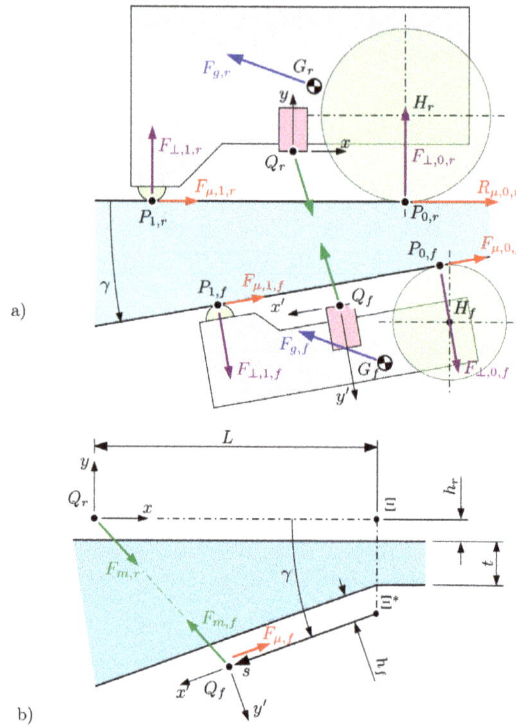

Figure 4. Straight-surface master–slave interaction. (**a**) Physical implementation of the actors; (**b**) geometrical relation between forces and parameter s. Note that friction forces $\mathbf{F}_{\mu,0,f}$ and $\mathbf{F}_{\mu,1,f}$ are represented by a single vector $\mathbf{F}_{\mu,f}$ for simplicity.

In order to determine the pose of the slave in the master frame of reference (x, y), the following transformation matrix can be defined:

$$\mathbf{T} = \begin{bmatrix} \cos(\pi+\gamma) & -\sin(\pi+\gamma) & Q_{f,x} \\ \sin(\pi+\gamma) & \cos(\pi+\gamma) & Q_{f,y} \\ 0 & 0 & 1 \end{bmatrix}, \tag{6}$$

which allows the following:

$$\begin{cases} \mathbf{P}_{f,0} = & \mathbf{T}\mathbf{P}'_{f,0} \\ \mathbf{P}_{f,1} = & \mathbf{T}\mathbf{P}'_{f,1} \\ \mathbf{H}_f = & \mathbf{T}\mathbf{H}'_f \\ \mathbf{G}_f = & \mathbf{T}\mathbf{G}'_f. \end{cases} \tag{7}$$

At this point, from Equations (1)–(3), we can write $F_{m,f} \triangleq f m_r, m_f, r$ and, similarly, $F_{m,r}$, $\tau_{m,f}$ and $\tau_{m,r}$. We introduce force of gravity $F_{g,f} = m_f g$; if we call $F_{\perp,0,f}$ and $F_{\perp,1,f}$ the normal contact forces between the wheels of the slave and the slave surface, we can express the force equilibrium along direction $\hat{y}'$, and the moment equilibrium, as follows:

$$\begin{bmatrix} 1 & 1 \\ M_{F_{\perp 0f}} & M_{F_{\perp 1f}} \end{bmatrix} \begin{Bmatrix} F_{\perp,0,f} \\ F_{\perp,1,f} \end{Bmatrix} = \begin{Bmatrix} -\mathbf{F}_{g,f} \cdot \hat{\mathbf{y}}' - \mathbf{F}_{m,f} \cdot \hat{\mathbf{y}}' \\ -\boldsymbol{\tau}_{m,f} \cdot \hat{\mathbf{z}} - \mathbf{M}_{F_g} \end{Bmatrix}, \tag{8}$$

where, introducing $\hat{\mathbf{z}} = \hat{\mathbf{x}} \times \hat{\mathbf{y}}$, it occurs that $M_{F_{\perp 0f}} = \left(\left(\mathbf{P}_{0,f} - \mathbf{Q}_f\right) \times \hat{\mathbf{y}}'\right) \cdot \hat{\mathbf{z}}$, $M_{F_{\perp 1f}} = \left(\left(\mathbf{P}_{1,f} - \mathbf{Q}_f\right) \times \hat{\mathbf{y}}'\right) \cdot \hat{\mathbf{z}}$ and $M_{F_g} = \left(\left(\mathbf{P}_{1,f} - \mathbf{Q}_f\right) \times \mathbf{F}_{g,f}\right) \cdot \hat{\mathbf{z}}$, which are the moments generated by forces $\mathbf{F}_{\perp,0,f}$, $\mathbf{F}_{\perp,1,f}$, and $\mathbf{F}_{g,f}$. The solution of the linear system in Equation (7) yields contact forces $\mathbf{F}_{\perp,0,f}$ and $\mathbf{F}_{\perp,1,f}$, which can be used to compute the forces caused by friction, according to Coulomb's law, in dynamic conditions:

$$\begin{cases} \mathbf{F}_{\mu,0,f} = \mu_{0,f} \left\| \mathbf{F}_{\perp,0,f} \right\| \\ \mathbf{F}_{\mu,1,f} = \mu_{1,f} \left\| \mathbf{F}_{\perp,1,f} \right\| \hat{\mathbf{x}}' \end{cases} \tag{9}$$

where $\mu_{0,f}$ and $\mu_{1,f}$ are the equivalent dynamic coefficients of friction between the ground and, respectively, the wheel and the front skid of the follower. Variable s determines the position of the slave along the slave surface, as shown in Figure 4b, since the follower is not rigidly connected to the master. In fact, location s_eq, where the slave is stable, is determined by the equilibrium of the external forces (magnetic, gravity, and friction) in direction $\hat{\mathbf{x}}'$ of the support surface, as follows:

$$F_{\varepsilon f}\left(s_{eq}\right) = -\hat{\mathbf{x}}' \cdot \left(\mathbf{F}_{m,f} + \mathbf{F}_{g,f} + \mathbf{F}_{f,\mu,0} + \mathbf{F}_{f,\mu,1}\right) = 0. \tag{10}$$

It is important to note that, while gravity force $F_{g,f}$ is constant, magnetic force $F_{m,f}$ as well as friction $F_{f,\mu,0}$ and $F_{f,\mu,1}$ depend on s, which renders Equation (8) implicit. In order to solve it, thus finding equilibrium point s_{eq}, numerical methods have to be used. In Figure 5, the behavior of $F_\varepsilon(s)$ is plotted.

Once the location of the slave is known, the $\mathbf{F}_{m,r}$ force vector can be calculated for the master robot, as well as $\mathbf{F}_{g,r}$. Hence, the force equilibrium can be stated as $\mathbf{A}_r\mathbf{X}_r = \mathbf{B}_r$, as follows:

$$\mathbf{A_r} = \begin{bmatrix} \hat{\mathbf{F}}_{\perp 0r} \cdot \hat{\mathbf{x}} & \hat{\mathbf{F}}_{\perp 1r} \cdot \hat{\mathbf{x}} & \hat{\mathbf{F}}_{\mu 1r} \cdot \hat{\mathbf{x}} & \hat{\mathbf{R}}_{\mu 0r} \cdot \hat{\mathbf{x}} \\ \hat{\mathbf{F}}_{\perp 0r} \cdot \hat{\mathbf{y}} & \hat{\mathbf{F}}_{\perp 1r} \cdot \hat{\mathbf{y}} & \hat{\mathbf{F}}_{\mu 1r} \cdot \hat{\mathbf{y}} & \hat{\mathbf{R}}_{\mu 0r} \cdot \hat{\mathbf{y}} \\ 0 & \mu_{r1} & -1 & 0 \\ M_{F_{\perp 0r}} & M_{F_{\perp 1r}} & M_{F_{\mu 1r}} & M_{R_{\mu 0}} \end{bmatrix} \tag{11}$$

$$\mathbf{B_r} = \begin{bmatrix} -\left(\mathbf{F}_g + \mathbf{F}_{mr}\right) \cdot \hat{\mathbf{x}} \\ -\left(\mathbf{F}_g + \mathbf{F}_{mr}\right) \cdot \hat{\mathbf{y}} \\ 0 \\ -\left(\left(\mathbf{G}_r - \mathbf{Q}_r\right) \times \mathbf{F}_g\right) \cdot \hat{\mathbf{z}} - \boldsymbol{\tau}_{mr} \cdot \hat{\mathbf{z}} \end{bmatrix} \tag{12}$$

$$\mathbf{X}_r = \begin{bmatrix} F_{\perp 0r} \\ F_{\perp 1r} \\ F_{\mu 1} \\ R_{\mu 0} \end{bmatrix} \tag{13}$$

where $M_{F_{\perp 0r}} = \left(\left(\mathbf{P}_{0r} - \mathbf{Q}_r\right) \times \hat{\mathbf{F}}_{\perp 0r}\right) \cdot \hat{\mathbf{z}}$, $M_{F_{\perp 1r}} = \left(\left(\mathbf{P}_{1r} - \mathbf{Q}_r\right) \times \hat{\mathbf{F}}_{\perp 1r}\right) \cdot \hat{\mathbf{z}}$, $M_{F_{\mu 1r}} = \left(\left(\mathbf{P}_{1r} - \mathbf{Q}_r\right) \times \hat{\mathbf{F}}_{\mu 1r}\right) \cdot \hat{\mathbf{z}}$ and $M_{R_{\mu 0}} = \left(\left(\mathbf{P}_{0r} - \mathbf{Q}_r\right) \times \hat{\mathbf{R}}_{\mu 0}\right) \cdot \hat{\mathbf{z}}$. One should note that unit vectors $\hat{\mathbf{F}}_{\perp 0r}$, $\hat{\mathbf{F}}_{\perp 1r}$, $\hat{\mathbf{F}}_{\mu 1r}$ and $\hat{\mathbf{R}}_{\mu 0}$ are known based on the geometry of the contact surface.

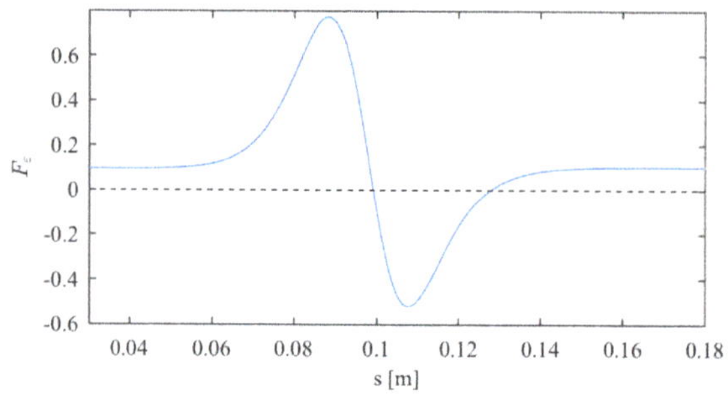

Figure 5. Behavior of $F_{\varepsilon f}$ in relation to master/slave misalignment distance s.

3.3. Cylindrical Surfaces

Similarly to the case of planar surfaces described in Section 3.2, by referring to Figure 6, we assume that frame of reference (x, y) is centered on $\mathbf{Q}_r$ and aligned with the master robot. Contrary to the planar case, when the support surfaces are cylindrical, determining the position of the contact points of the wheels is less trivial. However, based on Figure 6, if we consider the front wheel or skid to be pointlike and described by $\mathbf{P}_{1,r}$, then, through simple geometrical considerations, projection point Ω can be calculated of the axis of the support surfaces; this allows remaining contact point $\mathbf{P}_{0,r}$ to be located. If the frame of reference is (x', y'), then the same considerations apply to the slave for points $\mathbf{P}'_{0,f}$ and $\mathbf{P}'_{1,f}$, and allow the determination of $\mathbf{\Omega}'$.

The pose of the slave with respect to the master frame of reference is given by the geometry of the surfaces (i.e., thickness t and radius ρ) and by angle α that constitutes the misalignment between the master and the slave, as shown in Figure 6c. In fact, in this case, variable α determines the position of the slave along the cylindrical surface in relation to the master. At this point, it is possible to write the transformation matrix from slave frame of reference (x', y') to master (x, y):

$$\mathbf{T} = \begin{bmatrix} 1 & 0 & \Omega_{r,x} \\ 0 & 1 & \Omega_{r,y} \\ 0 & 0 & 1 \end{bmatrix} \begin{bmatrix} \cos\beta & -\sin\beta & 0 \\ \sin\beta & \cos\beta & 0 \\ 0 & 0 & 1 \end{bmatrix} \begin{bmatrix} 1 & 0 & -\Omega_{f,x} \\ 0 & 1 & -\Omega_{f,y} \\ 0 & 0 & 1 \end{bmatrix}, \tag{14}$$

where $\beta = \pi + \alpha_r - \alpha_f$. Therefore, using Equation (7), we can determine $\mathbf{P}_{f,0}$, $\mathbf{P}_{f,1}$, $\mathbf{H}_f$ and $\mathbf{G}_f$ in the (x, y) frame. In order to determine the equilibrium location of the slave robot, we introduce virtual force $\mathbf{F}_{\varepsilon,f}$, which is applied to point $\mathbf{Q}_f$. The slave robot is in equilibrium when $\mathbf{F}_{\varepsilon,f}(\alpha_{eq}) = 0$. The linear system is defined as follows:

$$\mathbf{A}_f \mathbf{X}_f = \mathbf{B}_f \tag{15}$$

$$\mathbf{A}_f = \begin{bmatrix} \hat{\mathbf{F}}_{\perp 0f} \cdot \hat{\mathbf{y}} & \hat{\mathbf{F}}_{\perp 1f} \cdot \hat{\mathbf{y}} & \hat{\mathbf{F}}_{\mu 0f} \cdot \hat{\mathbf{y}} & \hat{\mathbf{F}}_{\mu 1f} \cdot \hat{\mathbf{y}} & \hat{\mathbf{F}}_{\varepsilon f} \cdot \hat{\mathbf{y}} \\ \hat{\mathbf{F}}_{\perp 0f} \cdot \hat{\mathbf{x}} & \hat{\mathbf{F}}_{\perp 1f} \cdot \hat{\mathbf{x}} & \hat{\mathbf{F}}_{\mu 0f} \cdot \hat{\mathbf{x}} & \hat{\mathbf{F}}_{\mu 1f} \cdot \hat{\mathbf{x}} & \hat{\mathbf{F}}_{\varepsilon f} \cdot \hat{\mathbf{x}} \\ \mu_{0f} & 0 & -1 & 0 & 0 \\ 0 & \mu_{1f} & 0 & -1 & 0 \\ M_{F_{\perp 0f}} & M_{\hat{F}_{\perp 1f}} & M_{\hat{F}_{\mu 0f}} & M_{\hat{F}_{\mu 1f}} & 0 \end{bmatrix} \tag{16}$$

$$\mathbf{B_f} = \begin{bmatrix} -\left(\mathbf{F}_{mf} + \mathbf{F}_g\right) \cdot \hat{\mathbf{y}} \\ -\left(\mathbf{F}_{mf} + \mathbf{F}_g\right) \cdot \hat{\mathbf{x}} \\ 0 \\ 0 \\ -\tau_{mf} - M_{F_g} \end{bmatrix} \tag{17}$$

$$\mathbf{X} = \begin{bmatrix} F_{\perp 0f} \\ F_{\perp 1f} \\ F_{\mu 0f} \\ F_{\mu 1f} \\ F_{\varepsilon f} \end{bmatrix} \tag{18}$$

where the moments are $M_{\hat{F}_{\perp 0f}} = \left(\left(\mathbf{P}_{f0} - \mathbf{Q}_f\right) \times \hat{\mathbf{F}}_{\perp 0f}\right) \cdot \hat{\mathbf{z}}$, $M_{\hat{F}_{\perp 1f}} = \left(\left(\mathbf{P}_{f1} - \mathbf{Q}_f\right) \times \hat{\mathbf{F}}_{\perp 1f}\right) \cdot \hat{\mathbf{z}}$, $M_{\hat{F}_{\mu 0f}} = \left(\left(\mathbf{P}_{0f} - \mathbf{Q}_f\right) \times \hat{\mathbf{F}}_{\mu 0f}\right) \cdot \hat{\mathbf{z}}$, $M_{\hat{F}_{\mu 1f}} = \left(\left(\mathbf{P}_{f1} - \mathbf{Q}_f\right) \times \hat{\mathbf{F}}_{\mu 1f}\right) \cdot \hat{\mathbf{z}}$ and $M_{F_g} = \left(\left(\mathbf{G}_f - \mathbf{Q}_f\right) \times \mathbf{F}_g\right) \cdot \hat{\mathbf{z}}$.

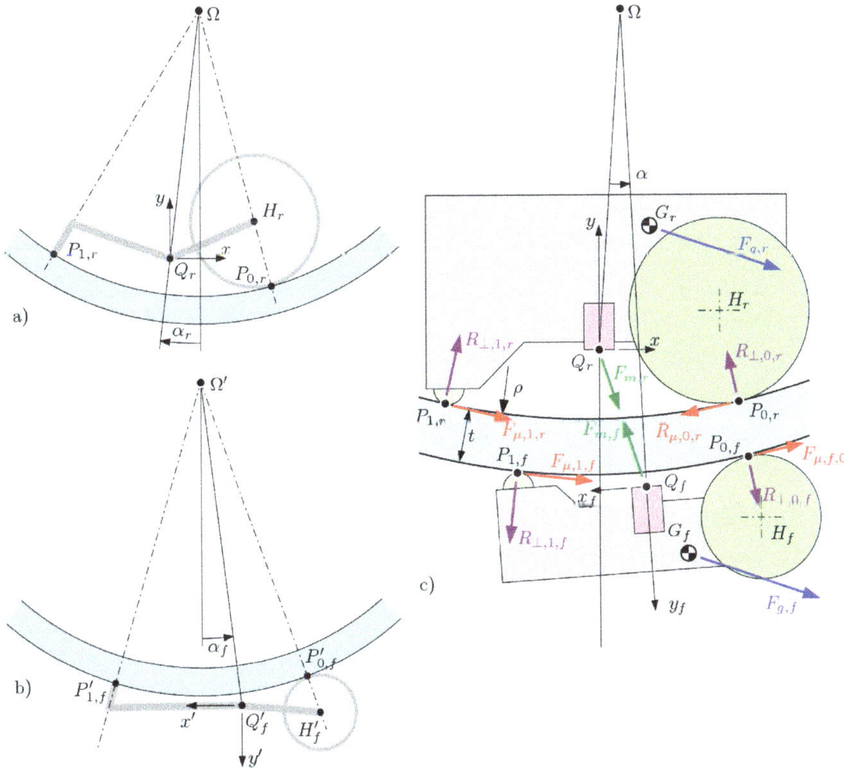

Figure 6. Master robot and slave on a cylindrical support sheet. (**a**) Master with respect to axis $\boldsymbol{\Omega}$; (**b**) slave in its local frame of reference; (**c**) whole system in the frame of reference of the master robot.

Solution $\mathbf{X} = \mathbf{A}^{-1}\mathbf{B}$ of the system, defined in Equation (10), yields information on the forces acting on the slave robot. The behavior of $F_{\varepsilon f}(\alpha)$ can be seen in Figure 7; hence, point α_{eq}, where the $F_{\varepsilon f}(\alpha)$ line intersects with the dashed line, corresponds to equilibrium. The proper solution can be achieved by numerical methods.

Similarly to the case of planar surface, once the slave robot pose is known, it is possible to calculate the forces acting on the master, that is, $\mathbf{F}_{m,r}$. Contact forces can be calculated by balance, as for the planar case. Referring to Equations (11)–(13) related to the planar-surface case, the system for cylindrical surfaces is defined as $\mathbf{A}_r\mathbf{X}_r = \mathbf{B}_r$, with the only difference being the contact and friction force vectors.

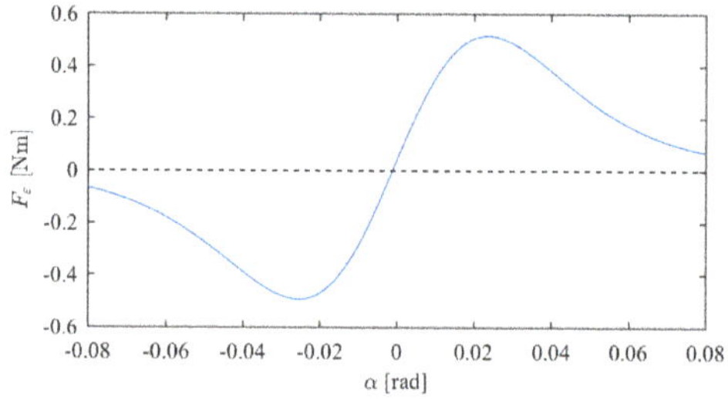

Figure 7. Behavior of $F_{\varepsilon f}$ in relation to master/slave misalignment angle α.

3.4. Slave Parking/Engaging

The last phase that is worth mentioning is the coupling and decoupling of the master from the slave. Referring to Figure 2, the reader can note that, during this procedure, the slave enters a slot on the support structure and is prevented to move horizontally. In fact, in Figure 8, a bumper is highlighted that implements this constraint. In order for the procedure to work, two prerequisites have to be satisfied:

- Magnetic force $F_{m,r}$ acting on the master robot should be sufficiently small to allow it to drive away; indeed, maximum traction force $F_{tr,0,r} = \mu_{wheel} F_{\perp,0,r}$ of the master robot should be larger than $R_{\mu,0,r}$, i.e., the reaction force produced by the external forces on the wheel contact point;
- magnetic force $F_{m,f}$ acting on the slave robot should be sufficiently high (before contact with the bumper) to allow it to both be pulled along by the master robot and remain attached to the ceiling of the support surface; this is to say, and $\exists s : \left(\mathbf{F}_{\varepsilon f}(s) < 0\right) \vee \left(R_{\perp,0,f}(s) > 0\right) \vee \left(R_{\perp,1,f}(s) > 0\right)$.

The geometry of this case is much the same as that described in Section 3.2. We assume that $L = 0$ and $\gamma = 0$; we can also say that $t = t_{park}$. Therefore, we can conclude that, in this case, parameter s describes the horizontal misalignment between master robot and slave. Furthermore, it happens that $R_{hc} = F_{\varepsilon f}$. In Figure 9, a typical behavior of the main values that satisfies the prerequisites is plotted against coordinate s.

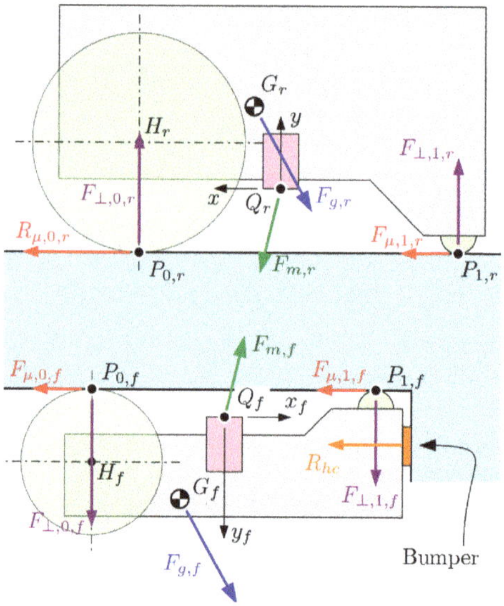

Figure 8. Engaging- or parking-phase configuration.

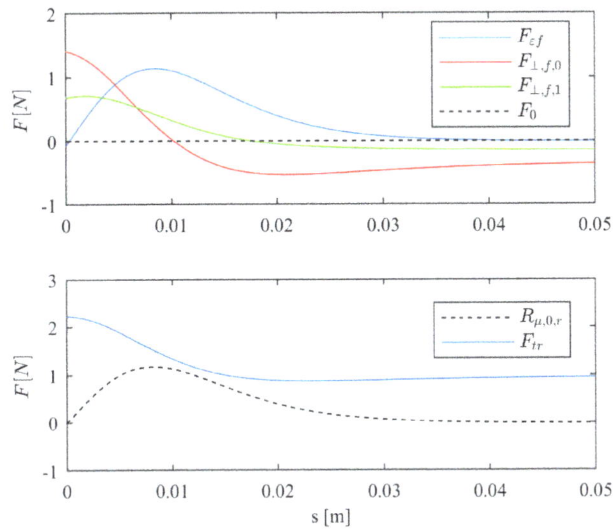

Figure 9. Development of forces as a function of s during the parking or coupling phase. Dashed line: inferior limit for the curves.

3.5. Traction Margin and Contact Forces

As seen in the previous sections, the master robot experiences a set of forces: magnetic, gravitational, and frictional, and their gross effect determines whether the robot is able to correctly perform. In fact, for the three general conditions expressed in Section 3 to be satisfied, the contact forces need to always be positive and, at the same time, the total tangential force on the master robot should be lower than the traction limit of the wheels. While contact forces $R_{\perp,0,r}$ and $R_{\perp,1,r}$ are already determined, it can be useful to define an index for the traction limit; we call this the traction margin and indicate it with T_μ:

$$T_\mu = 1 - \frac{\left\|\mathbf{R}_{\mu,0,r}\right\|}{F_t} = 1 - \frac{\left\|\mathbf{R}_{\mu,0,r}\right\|}{\mu_{0,r}\left\|\mathbf{R}_{\perp,0,r}\right\|}, \tag{19}$$

where $\left\|\mathbf{R}_{\mu,0,r}\right\|$ represents the net tangential force acting on the wheel contact interface as a result of the external forces, and $F_t = \mu_{0,r} F_{\perp,0,r}$ is the maximum traction force that can be transmitted through friction, represented by friction coefficient $\mu_{0,r}$. Index $T_\mu = 1$, when there is no limit to traction, and $T_\mu \leq 0$, when traction is completely restrained; in this case, slip happens when the wheel turns.

Therefore, in order for Conditions i–iii, described in Section 3 to be satisfied, the following must hold for every equilibrium configuration:

$$\begin{cases} T_\mu > 0 \\ R_{\perp,0,r} > 0 \\ R_{\perp,1,r} > 0 \end{cases}. \tag{20}$$

4. Numerical Implementation and Discussion

As was stated in Section 3, there are four main phases in the system operation. In order to show the behavior of the robot and to characterize the bounds of its geometrical design parameters, the model was implemented and the results are discussed. The fixed parameters are summarized in

Table 1; geometrical parameters are shown in Figure 10. The main condition for operation is expressed in Equation (20).

Table 1. Fixed design parameters.

Quantity	Symbol	Value	Unit
Magnetic permeability	η	1.257×10^{-6}	N/A^2
Magnetic moment (master)	$M_{(m,r)}$	0.9950	Am^2
Magnetic moment (slave)	$M_{(m,f)}$	0.9950	Am^2
Mass (master)	m_r	0.200	kg
Mass (slave)	m_f	0.045	kg
Frict. coeff. of wheel (master)	$\mu_{(0,r)}$	0.67	–
Frict. coeff. of skid (master)	$\mu_{(1,f)}$	0.05	–
Rolling resistance coeff. (slave)	$\mu_{(0,r)}$	0.01	–
Skid coefficient of friction (slave)	$\mu_{(1,f)}$	0.05	–

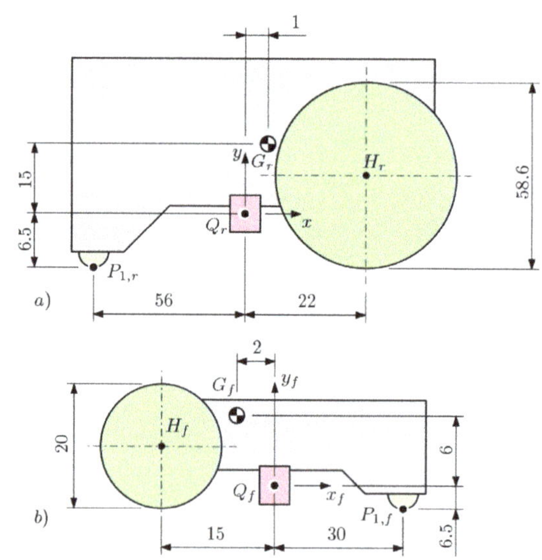

Figure 10. Geometry and dimensions of the robots. (**a**) Master and (**b**) slave.

The sheet that constitutes the main support surface is characterized mainly by its thickness t, curvature radius ρ, and the friction coefficient that emerges when coupled with the wheels and the skid of the master robot and slave. In the following, the results of traction margin T_μ are presented in different combinations of parameters, namely, t, the mass of the master robot m_m, the orientation of the gravity vector ϑ, and slope angle γ of the parking structure.

The first aspect that needs to be addressed is the one related to the curved portion of the support sheet. Indeed, as can be seen from Figure 2, the robot travels from an upright to an upside-down configuration by traveling along a cylindrical polycarbonate sheet. The chart in Figure 11 shows how the traction margin behaves for different sheet thicknesses t along the path around the cylinder, represented by gravity vector orientation ϑ. For design purposes, the selection of t can be done by making sure that Equation (20) is satisfied along the entire curved trajectory, i.e., when $-\pi/2 \leq \vartheta \leq \pi/2$; this guarantees that the robot operates correctly on the curved sheet.

Curvature ρ of the cylindrical surface being the second main parameter, we can see the behavior of T_μ against it in Figure 12 along with the contact forces. It is clear from the response surface, that the higher radius ρ is, the more diminished its influence becomes on the traction margin. Assuming that the robot has a larger footprint compared to that of the follower, the effect is explained by noting that, as ρ decreases, the distance between the magnets increases; the obvious consequence is a reduction in magnetic force.

Ceiling operation, i.e., when the robot is overturned, depends on same Conditions (20); however, in this case, the support surface is planar and horizontal: $\vartheta = \pi/2$. It is worth noting that the worst-case scenario is indeed when the surface is horizontal as opposed to otherwise oriented; in Figure 13, the traction margin is shown along with the contact forces: it can be seen that, for a fixed mass of $m_r = 0.200$ kg, contact forces are lower than those existing at $\vartheta = 0$ at the same thickness t, as seen in Figure 11. This observation is made assuming that the influence of curvature is negligible; in fact, as previously discussed, the curvature has a detrimental effect, which strengthens the thesis.

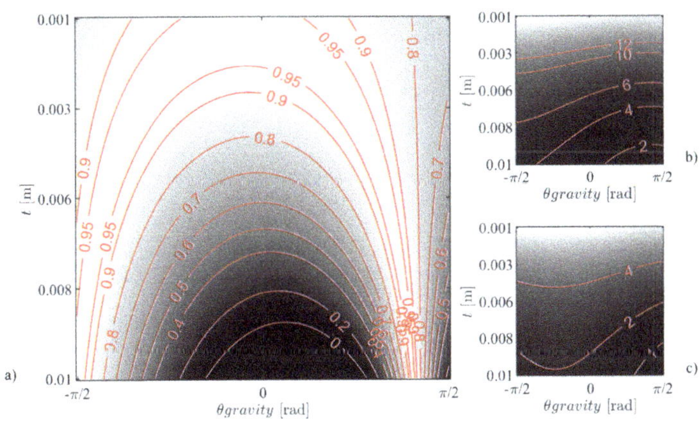

Figure 11. Traction margin T_μ and contact forces on a curved surface against thickness t and angular position $\vartheta_{gravity}$ along the curved path. (**a**) Traction margin; contact force for (**b**) the wheels in $\mathbf{P}_{0,r}$ and (**c**) the skid in $\mathbf{P}_{1,r}$.

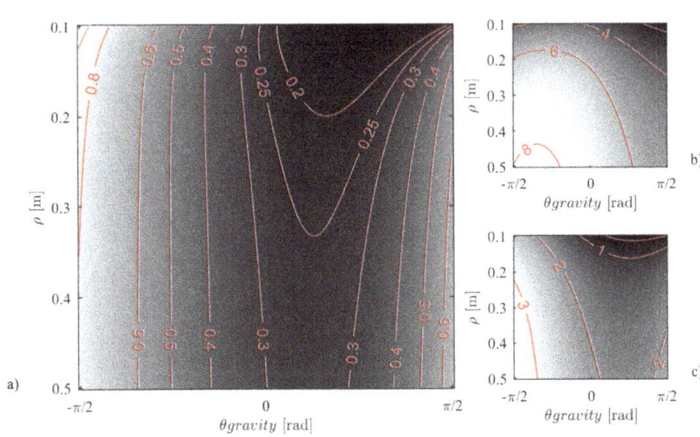

Figure 12. Traction margin T_μ and contact forces on a curved surface against curvature radius ρ and angular position $\vartheta_{gravity}$ along the curved path; (**a**) traction margin; contact force for (**b**) the wheels in $\mathbf{P}_{0,r}$ and (**c**) the skid in $\mathbf{P}_{1,r}$.

The last fundamental operating scenario is that of parking (Phases D and E). These phases are reversible, i.e., the system can transition $F \rightarrow E \rightarrow D \rightarrow C$ (engaging with the follower cart) similarly to $C \rightarrow D \rightarrow E \rightarrow F$ (parking); note that a certain hysteresis exists that stems from the friction of the follower cart, although this can safely be neglected. Based on the considerations put forth in Section 3.4, in order for coupling/decoupling to work correctly, Conditions i–ii in the same section should be satisfied. As the plot in Figure 14a shows, for small values of sheet thickness t that are otherwise either welcome or even necessary for the correct operation in a climbing phase or in upside-down situations (Figures 11 and 13, respectively), traction margin is null or negative past a certain traveling distance (here shown by misalignment s, which happens because the follower cart is prevented from moving by the bumper in the park area). The direct consequence would be that wheel slip would occur and the robot would not be able to drive away. This effect justifies the ramp structure (Phase E); in Figure 14b, the relation between traction margin, travel length L, and ramp angle γ is shown. It can be seen that lower angles allow a larger traction margin, as expected. Furthermore, once a good candidate for t_{park} has been selected using Figure 14a, a ramp must be designed that reaches said thickness starting from sheet thickness t; to this intent, the dashed line in Plot b can be used. This provides the travel $L = (t_{park} - t)/\sin\gamma$ required to reach parking thickness, and thus the value up to which the traction margin should be within limits to guarantee a successful operation.

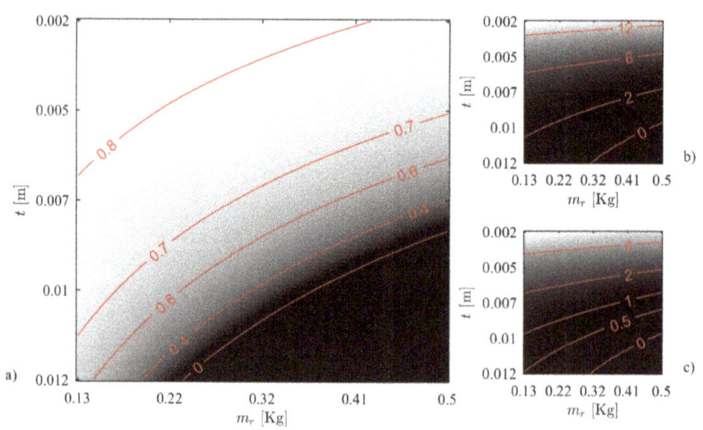

Figure 13. Traction margin T_μ and contact forces during ceiling operation in upside-down condition against mass m_r of the robot and thickness t of the support sheet. (**a**) Traction margin; contact force for (**b**) wheels in $\mathbf{P}_{0,r}$ and (**c**) the skid in $\mathbf{P}_{1,r}$.

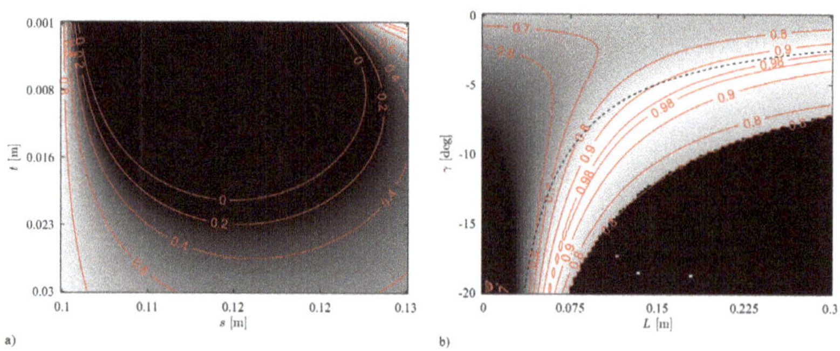

Figure 14. Traction margin during parking; (**a**) T_μ during proper parking phase (Phase E) against misalignment s and thickness t; (**b**) T_μ during ramp overcoming (Phase D) against travel L and ramp angle γ; dashed line: $L = L(t_{park} = 0.020)$, i.e., the value that L should have to reach $t_{park} = 0.020$ mm.

5. Prototype and Experiment Validation

In order to provide an early validation on the model, a functional prototype was built based on the design parameters shown in Table 1. The system is visible in Figure 15. A small 5V DC motor with a 900:1 reduction gearbox was used to drive the wheels forward at varying speed. A 3000 mAh battery was used to provide power on board. Given the simple setup, the robot is capable of driving in the forward direction at speeds between 0 and approximately 0.05 m/s. The maximum torque output of 1.2 Nm (at the wheel) is capable of overcoming friction force F_t even at $t = 0$, i.e., when magnetic force is at its highest.

Two experiments were defined to provide a comprehensive validation of the model: Experiment 1 was performed to validate the parking phase (Phase E), and Experiment 2 represents motion on a curved surface (Phase B). A comparison with numerical data from the numerical model illustrated in Section 3 is provided. The friction coefficient was lowered artificially from the rubber-PMMA value of $\mu_{r,0} \approx 0.90$ to a value of $\mu_{r,0} = 0.67$ with the addition of a thin layer of white adhesive tape, which can be seen in Figure 15a,b. The rationale for this is to reduce the forces required to induce slip, which has the added benefit of reducing stick-slip events.

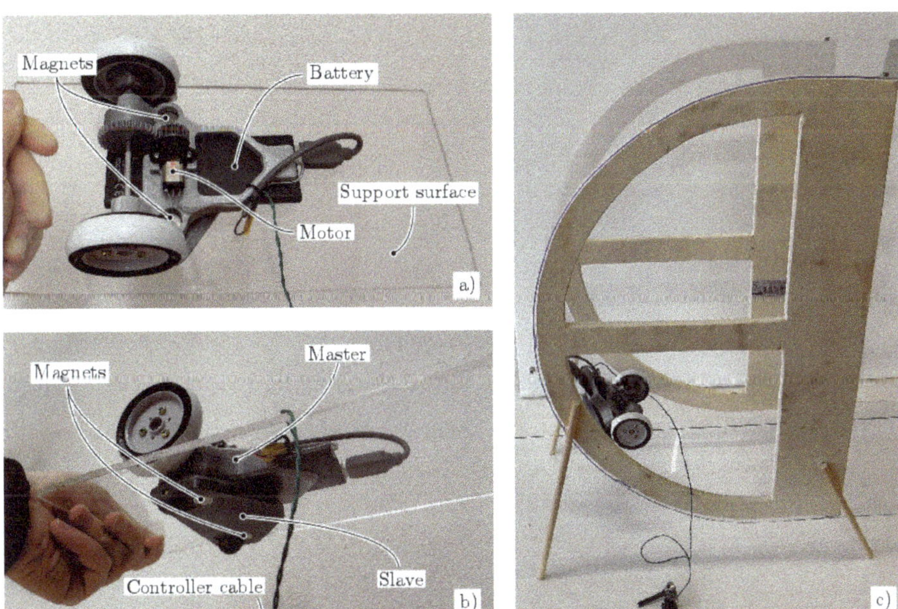

Figure 15. Functional prototype with nonmagnetic transparent support sheet. (**a**) Top view highlighting position of the master and its main component; (**b**) bottom view portraying the slave. The controller (a switch-potentiometer setup) is off-screen. (**c**) Experiment 2 setup, with robot located at approximately $\vartheta = -\pi/4$.

5.1. Experiment 1

The experiment consists of the measurement of minimum thickness t of the support sheet that allows the master robot to break free of the slave during the parking phase described in Section 3.4. Indeed, during this phase, the first prerequisite calls for magnetic force $F_{m,r}$ to be sufficiently small for the robot to drive away. This can indeed be seen in Figure 14, and was discussed in Section 4. The same measurement was repeated for several masses of the robot, namely, 0.2, 0.4, 0.6, and 1.2 kg, by adding commensurate additional weights to the robot close to its center of mass G_r.

This type of experiment was selected because the mechanics of the parking phase include the most important complex phenomena that influence the behavior of the system, which are

- friction of the master with respect to the support sheet and
- magnetic interaction between master and slave.

Being these inherently complex phenomena, described and modeled in detail in Sections 3.1, 3.2, and 3.4, the experimental validation of the parking phase was deemed the most adequate to provide insight on the adherence to reality of the model as a whole.

The results of the comparison between the numerical implementation of the model described in Section 3 and the experiment results are shown in Figure 16. It should be noted that a normalized thickness index is used in place of simple thickness value t of the support sheet; normalization allows, in principle, to compare systems with different friction coefficients, although in the present case, normalization is a simple scaling by value $1/\mu_{0,r}$.

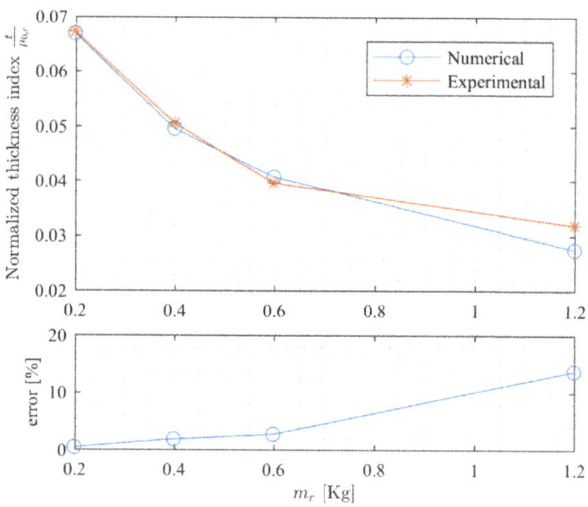

Figure 16. Comparison between numerical and experiment results (Experiment 1). Top plot: data points for different robot masses m_r; below plot: relative error.

It is apparent from the error plot shown on the bottom of the same figure that the error is comparatively small (from 0.5% to 3%) for large thicknesses and low masses of the master, while it becomes larger (approximately 14%) when thickness grows smaller following larger traction $\mathbf{R}_{\mu,0,r}$ associated with larger masses m_r. This is likely due to both the approximate model of the magnetic field that is described in Section 3.1 and the rigid-body approach used for the robot. In fact, large magnetic forces F_m arising from small thicknesses t tend to deform the rubber of the wheel, which decreases the distance of the master frame to the support sheet. The direct consequence of this is that the magnetic interaction deviates from its expected behavior. An additional possible source of error can be the model used for the friction; it assumes a linear relationship between force and frictional force. However, during deformation of the rubber due to increased mass m_r and magnetic force F_m, the footprint of the wheel increases in its area, which can have an unexpected influence on traction $\mathbf{R}_{\mu,0,r}$ and on traction margin T_μ.

5.2. Experiment 2

Phase B was experimentally investigated by measuring the maximum travel of the robot from the bottom across the curved portion of the surface, before the wheels lose traction. The angle at which slip occurs, ϑ_{slip}, is measured from the horizontal, coherently with the model presented in Section 3: the robot starts at $\vartheta = -\pi/2$ and travels towards the top of the surface, which is located at $\vartheta = +\pi/2$. Radius ρ of the curved sheet is 0.3 m. The test was performed for a set of four different masses m_r of the robot and for a set of four different support-sheet thicknesses t. In particular, m_r = [0.2, 0.3, 0.4, 0.5] kg, while t = [0.002, 0.003, 0.004, 0.005] m. This provides a total of 16 data points. The raw experimental measurements of ϑ_{slip}, which were taken for each data point, varied from 3 to 5.

Results shown in Figure 17 show a close similarity in the topology of the surface determined by the variation of parameters m_r and t. Results show that the larger the mass, the smaller the angle ϑ_{slip} of slip becomes. This means that the robot slips early in its travel along the curved surface. Similarly, the larger thickness t is, the smaller angle ϑ_{slip} is. This is expected, because thicker sheets increase the distance of the magnets, which then provide less adhesion force, which clearly translates in less of a traction margin. In order to better measure the error between numerical and experimental results, we define the following:

$$\varepsilon_{\vartheta,slip} = \frac{(\vartheta_{slip,num} - \vartheta_{slip,exp})}{\pi/2}. \tag{21}$$

This formulation measures the absolute error and is normalized against $\pi/2$, which is the travel from the horizontal to the vertical orientation of the robot. The comparison between numerical and experimental data, which is visible in both plots in Figure 17, shows that a comparatively small error is present in the measurements, with an average value of $\varepsilon_{\vartheta,slip} = 5\%$.

It is worth noting that the system appears nonfunctional in this configuration, although this is caused by the tape on the wheels, which considerably reduces traction. According to the now-validated model and to experimentation, the system is functional without the tape.

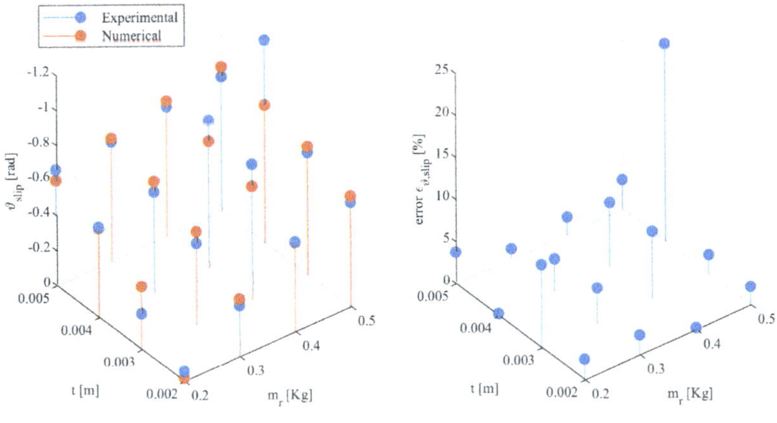

Figure 17. Comparison between numerical results of the model and results from Experiment 2. **Left** plot: data points for different robot masses m_r and for different thicknesses t for both the numerical and experimental campaigns; note that the z-axis is flipped. **Right** plot: relative error between numerical and experimental values, normalized with respect to $\pi/2$.

6. Conclusions

This paper discussed a type of climbing robot that takes advantage of a master–slave configuration that is connected through a magnetic interface. The master robot provides motion, while the slave component provides adhesion capabilities to the system even when the support sheet is nonmagnetic. Thanks to its design, once decoupled from the slave, the master robot is able to drive freely on horizontal ground.

In order to present a coherent framework, three conditions are defined that prove the feasibility of the specific design. The work envelope of the system is split in a set of phases, namely, driving on straight floor, vertical floor, curved surface, ceiling, and coupling and decoupling phases; each is characterized by its own challenges.

A Taylor expansion model was implemented to model the magnetic-field morphology, and a pseudostatic approach was used to model the master–slave robotic system to account for the inherently nontrivial interaction between the actors.

Comprehensive analysis was performed on the main variables of the master–slave system through numerical computation of the model in the operational envelope. Furthermore, experimental validation was performed both in the case of the parking phase, which is the most complex to model, and in the case of driving along a curved sheet. Both experiments showed that the numerical model had good adherence with reality, with errors around 3%–14% for the parking phase, and 5% for curved-surface driving.

Results clearly show the feasibility of the approach and provide some qualitative and quantitative insight. The experimental validation shows good accuracy of the model compared to reality. The design process of a working prototype can thus be successfully guided via the tools that are provided in the model and its analysis.

Author Contributions: Conceptualization, S.S., P.G.; methodology, S.S.; software, S.S.; validation, S.S.; formal analysis, P.G.; investigation, S.S., L.S.; resources, L.S.; data curating, S.S., L.S.; writing—original-draft preparation, S.S.; writing—review and editing, S.S., L.S., M.C., A.G.; visualization, S.S.; supervision, P.G., A.G.

Funding: This research received no external funding.

Conflicts of Interest: The authors declare no conflict of interest.

References

1. Cesetti, A.; Scotti, C.; Di Buo, G.; Longhi, S. A service oriented architecture supporting an autonomous mobile robot for industrial applications. In Proceedings of the 18th IEEE Mediterranean Conference on Control and Automation (MED'10), Marrakech, Morocco, 23–25 June 2010; pp. 604–609.
2. Bietresato, M.; Carabin, G.; D'Auria, D.; Gallo, R.; Ristorto, G.; Mazzetto, F.; Vidoni, R.; Gasparetto, A.; Scalera, L. A tracked mobile robotic lab for monitoring the plants volume and health. In Proceedings of the 12th IEEE/ASME International Conference on Mechatronic and Embedded Systems and Applications (MESA), Auckland, New Zealand, 29–31 August 2016.
3. Vidoni, R.; Gallo, R.; Ristorto, G.; Carabin, G.; Mazzetto, F.; Scalera, L.; Gasparetto, A. ByeLab: An agricultural mobile robot prototype for proximal sensing and precision farming. In Proceedings of the ASME 2017 International Mechanical Engineering Congress and Exposition. American Society of Mechanical Engineers, Tampa, FL, USA, 3–9 November 2017.
4. Palacín, J.; Palleja, T.; Valgañón, I.; Pernia, R.; Roca, J. Measuring coverage performances of a floor cleaning mobile robot using a vision system. In Proceedings of the 2005 IEEE International Conference on Robotics and Automation, Barcelona, Spain, 18–22 April 2005; pp. 4236–4241.
5. Quaglia, G.; Bruzzone, L.; Bozzini, G.; Oderio, R.; Razzoli, R.P. Epi. q-TG: Mobile robot for surveillance. *Ind. Robot Int. J.* **2011**, *38*, 282–291. [CrossRef]
6. Ristorto, G.; Gallo, R.; Gasparetto, A.; Scalera, L.; Vidoni, R.; Mazzetto, F. A Mobile Laboratory for Orchard Health Status Monitoring in Precision Farming. *Chem. Eng. Trans.* **2019**, *58*, 661–666.
7. Carbone, G.; Tedeschi, F.; Gallozzi, A.; Cigola, M. A robotic mobile platform for service tasks in cultural heritage. *Int. J. Adv. Robot. Syst.* **2015**, *12*, 88. [CrossRef]
8. Seriani, S.; Gallina, P.; Wedler, A. A modular cable robot for inspection and light manipulation on celestial bodies. *Acta Astronaut.* **2016**, *123*, 145–153. [CrossRef]
9. Seriani, S.; Gallina, P.; Wedler, A. Dynamics of a tethered rover on rough terrain. In *Advances in Italian Mechanism Science*; Springer: New York, NY, USA, 2017; pp. 355–361.
10. Seriani, S.; Scalera, L.; Gasparetto, A.; Gallina, P. Preloaded structures for space exploration vehicles. In *Mechanisms and Machine Science*; Springer: New York, NY, USA, 2018; pp. 129–137.

11. Silva, M.F.; Machado, J.T.; Tar, J.K. A survey of technologies for climbing robots adhesion to surfaces. In Proceedings of the 2008 IEEE International Conference on Computational Cybernetics, Stara Lesna, Slovakia, 27–29 November 2008; pp. 127–132.
12. Longo, D.; Muscato, G. Design of a single sliding suction cup robot for inspection of non porous vertical wall. In Proceedings of the (ISR 2004) 35th International Symposium on Robotics, Paris-Nord Villepinte, France, 23–26 March 2004.
13. Gao, X.; Kikuchi, K. Study on a kind of wall cleaning robot. In Proceedings of the 2004 IEEE International Conference on Robotics and Biomimetics, Shenyang, China, 22–26 August 2004; pp. 391–394.
14. Yoshida, Y.; Ma, S. Design of a wall-climbing robot with passive suction cups. In Proceedings of the 2010 IEEE International Conference on Robotics and Biomimetics, Tianjin, China, 14–18 December 2010; pp. 1513–1518.
15. Sintov, A.; Avramovich, T.; Shapiro, A. Design and motion planning of an autonomous climbing robot with claws. *Robot. Autom. Syst.* **2011**, *59*, 1008–1019. [CrossRef]
16. Balaguer, C.; Gimenez, A.; Jardón, A. Climbing robots' mobility for inspection and maintenance of 3D complex environments. *Autonom. Robot.* **2005**, *18*, 157–169. [CrossRef]
17. Seidl, T.; Vidoni, R. Adhesion to flat surfaces: From spiders to stickers. In *Spider Ecophysiology*; Springer: New York, NY, USA, 2013; pp. 463–473.
18. Gasparetto, A.; Seidl, T.; Vidoni, R. A mechanical model for the adhesion of spiders to nominally flat surfaces. *J. Bionic Eng.* **2009**, *6*, 135–142. [CrossRef]
19. Boscariol, P.; Henrey, M.A.; Li, Y.; Menon, C. Optimal gait for bioinspired climbing robots using dry adhesion: A quasi-static investigation. *J. Bionic Eng.* **2013**, *10*, 1–11. [CrossRef]
20. Henrey, M.; Ahmed, A.; Boscariol, P.; Shannon, L.; Menon, C. Abigaille-III: A versatile, bioinspired hexapod for scaling smooth vertical surfaces. *J. Bionic Eng.* **2014**, *11*, 1–17. [CrossRef]
21. Cai, J.; He, K.; Fang, H.; Chen, H.; Hu, S.; Zhou, W. The design of permanent-magnetic wheeled wall-climbing robot. In Proceedings of the 2017 IEEE International Conference on Information and Automation (ICIA), Macau, China, 18–20 July 2017; pp. 604–608.
22. Howlader, M.O.F.; Sattar, T.P. Development of magnetic adhesion based climbing robot for non-destructive testing. In Proceedings of the 2015 7th Computer Science and Electronic Engineering Conference (CEEC), Colchester, UK, 24–25 September 2015; pp. 105–110.
23. Howlader, M.O.F.; Sattar, T.P. Design and Optimization of Permanent Magnet Based Adhesion Module for Robots Climbing on Reinforced Concrete Surfaces. In *Intelligent Systems and Applications*; Springer: New York, NY, USA, 2016; pp. 153–171.
24. Zhang, Y.; Dodd, T.; Atallah, K.; Lyne, I. Design and optimization of magnetic wheel for wall and ceiling climbing robot. In Proceedings of the IEEE International Conference on Mechatronics and Automation, Xi'an, China, 4–7 August 2010; pp. 1393–1398.
25. Ishihara, H. Basic study on wall climbing root with magnetic passive wheels. In Proceedings of the IEEE International Conference on Mechatronics and Automation (ICMA), Takamatsu, Japan, 6–9 August 2017; pp. 1964–1969.
26. Hu, S.; Peng, R.; He, K.; Li, J.; Cai, J.; Zhou, W. Structural design and magnetic force analysis of a new crawler-type permanent magnetic adsorption wall—Climbing. In Proceedings of the 2017 IEEE International Conference on Information and Automation (ICIA), Macau, China, 18–20 July 2017; pp. 598–603.
27. Huang, H.; Li, D.; Xue, Z.; Chen, X.; Liu, S.; Leng, J.; Wei, Y. Design and performance analysis of a tracked wall-climbing robot for ship inspection in shipbuilding. *Ocean Eng.* **2017**, *131*, 224–230. [CrossRef]
28. Milella, A.; Maglietta, R.; Caccia, M.; Bruzzone, G. Robotic inspection of ship hull surfaces using a magnetic crawler and a monocular camera. *Sens. Rev.* **2017**, *37*, 425–435. [CrossRef]
29. Mao, J.; He, K.; Li, J.; Sun, X. Simulation and experimental verification of permanent magnet adsorption unit for wall-climbing robot. In Proceedings of the 2016 IEEE International Conference on Information and Automation (ICIA), Ningbo, China, 1–3 August 2016; pp. 1189–1194.
30. Tummala, R.L.; Mukherjee, R.; Xi, N.; Aslam, D.; Dulimarta, H.; Xiao, J.; Minor, M.; Dangi, G. Climbing the walls. *IEEE Robot. Autom Mag.* **2002**, *9*, 10–19. [CrossRef]
31. Seriani, S.; Scalera, L.; Gasparetto, A.; Gallina, P. A new family of magnetic adhesion based wall-climbing robots. In Proceedings of the 2nd International Conference of IFToMM ITALY, Cassino, Italy, 29–30 November 2018; pp. 223–230.

32. Belforte, G.; Gay, P.; Aimonino, D.R. Robotics for improving quality, safety and productivity in intensive agriculture: Challenges and opportunities. In *Industrial Robotics: Programming, Simulation and Applications*; IntechOpen: London, UK, 2006.
33. Fiorini, P.; Botturi, D. Introducing service robotics to the pharmaceutical industry. *Intell. Serv. Robot.* **2008**, *1*, 267–280. [CrossRef]
34. Griffiths, D.J. *Introduction to Electrodynamics*, 3rd ed.; Benjamin Cummings: San Francisco, CA, USA, 1998.
35. Landecker, P.B.; Villani, D.D.; Yung, K.W. Analytic solution for the torque between two magnetic dipoles. *Magn. Electr. Sep.* **1999**, *10*, 29–33. [CrossRef]

robotics

MDPI

Article

A Tactile-Based Wire Manipulation System for Manufacturing Applications

Gianluca Palli [1,†] and Salvatore Pirozzi [2,*,†]

1 Università di Bologna—DEI, Viale del Risorgimento 2, 40136 Bologna, Italy; gianluca.palli@unibo.it
2 Università degli Studi della Campania—DI, Via Roma 29, 81031 Aversa (CE), Italy
* Correspondence: salvatore.pirozzi@unicampania.it; Tel.: +39-081-5010-433
† The authors contributed equally to this work.

Received: 9 April 2019; Accepted: 10 June 2019; Published: 12 June 2019

Abstract: This paper presents experimental results developed within the WIRES experiment, whose main objective is the robotized cabling of switchgears. This task is currently executed by human operators; the WIRES Project tackles the development of a suitably designed sensorized end effector for the wire precise manipulation. In particular, the developed gripper with tactile sensors are shown and a procedure for the implementation of the insertion task is presented and discussed. Experimental results are reported both for quality of wire shape reconstruction and success rate of insertion task implementation.

Keywords: tactile sensors; manipulation task; assembly robot

1. Introduction

Robotic manipulation is a complex task especially when deformable and fragile objects have to be grasped. In these cases, the knowledge of geometrical and physical characteristics of the object to manipulate are fundamental for the successful implementation of the task. To this aim, specific sensing systems are developed to be integrated into robotic systems. This paper presents results of activities developed within the WIRES experiment (http://www-lar.deis.unibo.it/people/gpalli/WIRES/), where the main objective is the robotized cabling of switchgears. Switchgears are basic components in a wide range of applications. Currently, the switchgear wiring is executed by human operators due to the complex manipulation tasks. The WIRES Project tackles the development of a suitably designed end effector equipped with a vision system and a tactile sensor for wire-precise manipulation. Preliminary results have been presented in [1–3].

Standard approaches to this kind of problem use vision and/or tactile data. Often vision is used alone due to its efficiency in data collection ([4]). However, this solution may fail in the presence of varying lighting conditions and occlusions. The use of tactile sensors helps to improve the success rate by overcoming some environment limitations. As a consequence, there have been many papers where vision and tactile data are integrated in a single approach ([5–9]. The objective of these approaches is the estimation of object characteristics, such as pose, shape, surface features and so on. Among these, some researchers propose interesting algorithms for edge detection [10] that could be considered in future as alternative approaches with respect to the one proposed here in order to improve the estimation quality. At the moment, the estimation quality reached with the approach proposed here is sufficiently high for the task implementation, with a very simple formalization. Some researchers in recent papers [11] use vision systems directly integrated into fingers to evaluate both tactile and image data at the same time and with the same sensing system. Also, this approach demonstrates how the fusion among tactile and vision data can be a good approaches for manipulation tasks. However, none of these past papers tackle the estimation problem of shape and pose of flexible objects like wires.

In previous papers [2,12], the authors presented details of the tactile sensor design and a possible use of tactile data for the reconstruction of the grasped wire shape and the use of the estimated shape for the implementation of an insertion task. In those papers, the model used for the wire was constituted by a quadratic function for the grasped area and a straight line for the part outside the tactile sensor pad. The sensor was mounted on a commercial gripper and preliminary insertion tests have been carried out on a single hole of an electric component fixed on the workbench, with the same wire grasped from a single position.

This paper presents improvements with respect to the previous system in terms of mechatronic solutions that are integrated and tested in a new scenario much closer to real cases. In particular, for this paper, the tactile sensor has been integrated into the final end-effector designed for the WIRES Project, presented and equipped with an electric screwdriver used to automatically complete the connection task. A simplified solution for the wire shape estimation is considered in order to allow its integration directly into the on-board microcontroller. This solution is a subset of that proposed in [12], but is explicitly formalized to be used with the final end-effector in the current study. The quality of the reconstruction has been re-evaluated with the new model, by redefining the quality metric according to the different model, in order to check if the considered simplification does not strongly affect the expected results. Finally, unlike previous papers, the whole system has been tested in a more complex scenario, by grasping, inserting and connecting a sequence of wires in a testing switchgear as shown in the video as supplementary.

2. The Tactile Sensor and the Gripper

The tactile sensor working principle and its design is detailed in [12]. Here, a brief recalling is reported (related to the integration in the gripper). Figure 1 reports some pictures of the developed sensor where the main components are highlighted. The 16 taxels constituted by the optoelectronic components with the deformable layer bonded above represents the transduction part for the sensor. The optical signals are converted in electric signals by using simple resistors and the obtained voltage signals are acquired with a standard Analogue-to-Digital converter. All details about the components integrated in the PCB are reported in [12]. For the integration into the gripper finger, a second PCB with a microcontroller has been developed and connected to the first one. The second PCB is completed by a voltage regulator and a standard connector, which allows to interface the tactile sensor with a standard USB-TTL serial cable. A suitably designed finger case has been realized in aluminum via a 3D printing technique and the extended PCB has been integrated inside this case. The thickness of the designed case is the smallest in order to allow the insertion of the finger among the switchgear components and wires already connected. The case allows a mechanical connection to the gripper by using two screw.

The end effector developed in the WIRES experiment for the implementation of the whole cabling process can be seen in Figure 2. The end effector integrates a 2D camera providing top view of the scene, an computer-controlled screwdriver (to tight the terminal screws) and a 4-DOFs gripper equipped with the tactile sensor. The end effector is also equipped with an integrated torque/controlled screwdriver with remote PLC control and process data recording capabilities (Kolver PLUTO3CA electric screwdriver + EDU2AE/TOP/E control unit). In the final process implementation, the robot arm is used to position the screwdriver tip on the terminal screw, and the FT sensor will be used to control the contact with the screw during the tightening. Therefore, the end effector will be held in an almost fixed position, just the screw motion during the tightening will be compensated. Consequently, the wire insertion will be performed by using the gripper DOFs only. It results that the FT sensor can be used to estimate the interaction between the screwdriver and the terminal screw, but it cannot be used during the insertion and for the wire tightening check, because the magnitude of the force generated by the wire contact is much lower than the one generated by the contact between the screwdriver and the screw, making the former indistinguishable. For this reason, the use of the tactile sensor installed into the gripper fingertips is fundamental also during the insertion and for the wire tightening check, in order to reach a suitable success rate.

Figure 1. Some pictures of assembled tactile sensor: (**a**,**b**) report a front view and a rear view of the PCB components, respectively, while (**c**) reports the PCB integration into the gripper finger.

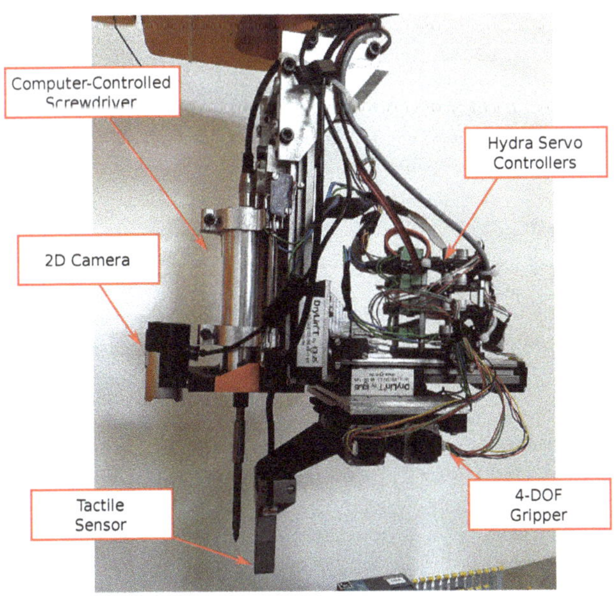

Figure 2. The end effector developed for the WIRES experiment. It is equipped with computer-controlled screwdriver, tactile sensor, 2D camera, Hydra servo controller boards, and a 4-DOF gripper.

Stepper motors with integrated encoder and lead screws have been adopted for the actuation of the end effector. This solution significantly simplifies the control and reduces the weight, the mechanical complexity and the cost of the end effector. Limit switches have been used for absolute position detection on both sides of all the end-effector movement axes. Each motor is driven by a Hydra servo drive control board, used as HW low-level motor controllers. These control boards are arranged on the end effector itself. The communication between the motor control boards and the high level WIRES controller is implemented through CAN bus. A ROS node has been developed to allow the control of the end effector and to ease the integration with other components of the WIRES system. At low level, the motors are controlled by means of the PLCOpen standard, allowing an easy implementation of the end effector controller. The tactile sensor has been integrated into the jaw tips (fingertips). Several versions of 3D printed fingers have been produced in order to evaluated different configurations during experiments.

3. Wire Shape Estimation

A specific sensor reference frame, $\Sigma_s(O_s, x_s, y_s)$, is defined at the center of the tactile sensor pad (see Figure 3) and the wire shape estimation problem is tackled with respect to this frame. The 16 taxels are organized as a matrix, where each cell can be identified by its row and column indices. Hence, for each c_{ij} cell it is possible to associate a couple of coordinates (x_i, y_j), corresponding to the physical distances of the cell mechanical center from the sensor frame origin. In particular, the x-coordinates of the columns are -4.5 mm, -1.5 mm, 1.5 mm and 4.5 mm, from left to right, while the y-coordinates of the rows, are 4.5 mm, 1.5 mm, -1.5 mm and -4.5 mm, from top to bottom. The measured voltage variation corresponding to the c_{ij} cell is indicated as Δv_{ij}.

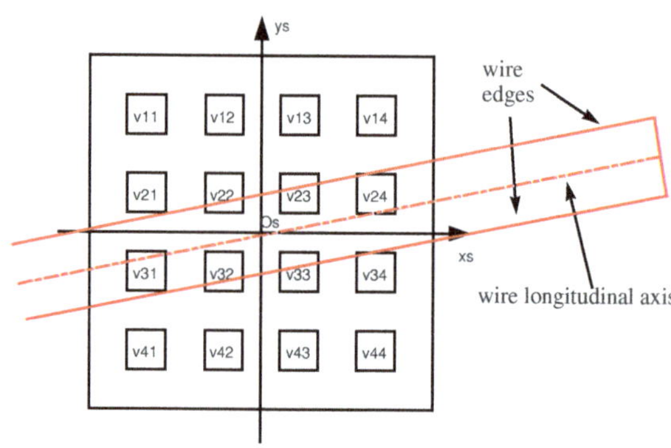

Figure 3. Scheme of the grasped wire with respect to the sensor frame Σ_s and taxels.

In this paper, in order to estimate the shape of the grasped wire, this is locally approximated as a straight line, coincident with the longitudinal axis of the wire (see Figure 3), modelled in the Σ_s frame as the line with equation

$$y_s = mx_s + n, \tag{1}$$

where m and n are the two parameters to be identified by using the tactile data. Since the section of the grasped wire is considered *a priori* known, estimating the grasped wire shape means to estimate the m and n parameters characterizing longitudinal axis of the wire. The initial position of the wire implies that the grasped wire has the main direction always mainly aligned with the x_s-axis (horizontal direction). In this hypothesis, the procedure for the wire shape estimation is constituted by a first step, where the centroid coordinates for each column are computed, and a second step, where the computation of the model parameters in (1) is implemented via a least squares method applied to the data set constituted by the coordinates of the column centroids. In detail, *step 1:*

the y coordinates y_j^c of the column centroids are computed from tactile data as

$$y_j^c = \frac{\sum_{i=1}^{4} y_i \Delta v_{ij}}{\sum_{i=1}^{4} \Delta v_{ij}} \quad j = 1, \ldots, 4, \tag{2}$$

where y_i is the mechanical y coordinate of the i-th row. Hence, the data set $\bar{\mathcal{D}}$ is constituted by the coordinates (x_j, y_j^c) of the 4 column centroids (where x_j is the mechanical x coordinate of the j-th column).

step 2:

the model (1) parameters, m, n, are estimated by using a least squares method applied to the data set $\bar{\mathcal{D}}$ resulting from step 1.

The presented procedure has been experimentally applied by grasping a wire in different configurations. Figure 4 reports a generic grasp: the tactile map shows how the cells on the second and the third rows present higher Δv_{ij} values. The column centroids (green stars) have been computed by using Equation (2) and than the wire shape has been computed via least squares method (straight line). To assess the accuracy of the algorithm, a comparison among the estimated shapes and the actual ones has been carried out, superimposing a picture of the corresponding grasp to the measured data and estimated shapes. Figure 5 shows the good matching between the estimated and the actual wire shapes. The quality of the shape reconstruction is fundamental to successfully complete the insertion task, as detailed below.

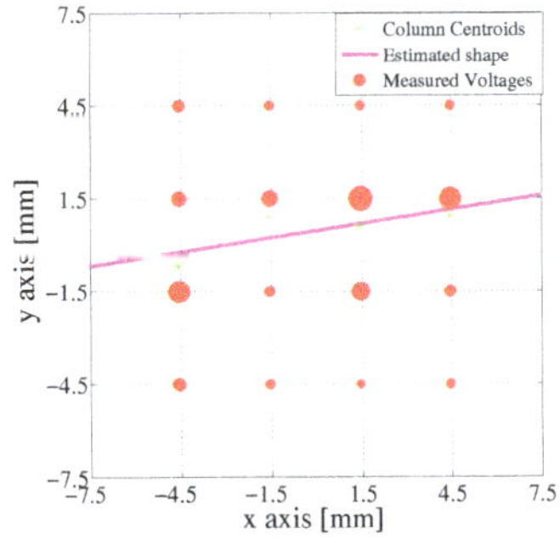

Figure 4. Tactile map and estimated shape for a grasped wire.

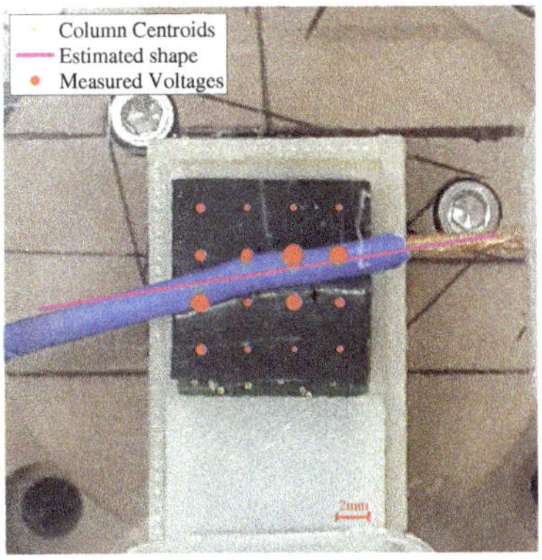

Figure 5. Comparison among estimated and actual wire shape for a grasped wire.

4. The Insertion Task

As discussed in Section 1, the main objective of the WIRES project is the robotic assembly of electric switchgears. To this aim, a fundamental subtask is represented by the insertion of the wire into the holes corresponding to the pins of the electrical components. The successful execution of the task allows the mechanical connection of the wire. The tactile sensor has been integrated into the gripper, suitably designed for the WIRES project in order to experimentally test its funcionalities during the insertion task. The proposed solution for the insertion task described in the following is based on the assumptions that the relative position between the robot system and the switchgear is known and, additionally, the length of the protruding part of the grasped wire is also known. Note that in real applications, a standard calibration procedure for the robot system allows us to obtain a precision that satisfies the first assumption. For the second assumption, the length of the protruding part of the wire can be estimated by using the camera integrated into the gripper as described in Section 2.

A human operator prepares the wire by placing it in a delimited area (based on the gripper stroke), with a random pose. The robotic system is used to grasp the wire and as a consequence, after grasping, the pose of the wire with respect to the tactile sensor is unknown. Then, the grasped wire shape is estimated by computing the model parameters m, n and applying the wire shape estimation algorithm. Figure 6a reports a sketch of a generic grasped wire, with the estimated shape. Let $\Sigma_h(O_h, x_h, y_h)$ be the hole frame, with the origin in the center of the hole and the x_h-axis aligned with the hole axis; let $\Sigma_w(O_w, x_w, y_w)$ be the wire end frame, with the origin in the end point of the wire actual axis and the x_w-axis aligned with the wire actual axis; let the frame $\hat{\Sigma}_w(\hat{O}_w, \hat{x}_w, \hat{y}_w)$, with the origin in the end point of the estimated wire axis and the $\hat{x}_w$-axis aligned with the estimated wire axis. On the basis of the assumption described above, the poses of Σ_s and Σ_h are perfectly known, while the pose of $\hat{\Sigma}_w$ can be computed from the estimated shape parameters m, n and the protruding part L value of the grasped wire. To this aim, the homogenous transformation matrix $\mathbf{T}^s_{\hat{w}}$ can be computed from Figure 6a with simple geometrical considerations

$$\mathbf{T}^s_{\hat{w}} = \begin{bmatrix} \cos\gamma & -\sin\gamma & 0 & l\cos\gamma \\ \sin\gamma & \cos\gamma & 0 & l\sin\gamma + n \\ 0 & 0 & 1 & 0 \\ 0 & 0 & 0 & 1 \end{bmatrix}, \tag{3}$$

where $\gamma = \arctan(m)$ and $l = L + a/\cos\gamma$ (a is the half side length of the sensor pad). After the computation of $\hat{\Sigma}_w$, a standard technique can be used to program the robotic system in order to align $\hat{\Sigma}_w$ with Σ_h. After that, the resulting configuration is sketched in Figure 6b, with $\hat{\Sigma}_w \equiv \Sigma_h$. From this point the insertion task can be easily completed with a linear movement along the x_h-axis. In real working conditions the hole diameter D is typically two times larger than the wire diameter d. Note that in this paper, the insertion has been tackled by considering the task a 2D problem; the z-axes are considered all aligned.

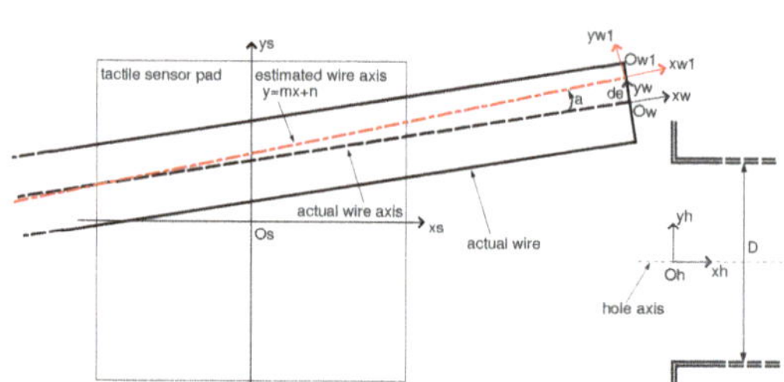

(a)

Figure 6. *Cont.*

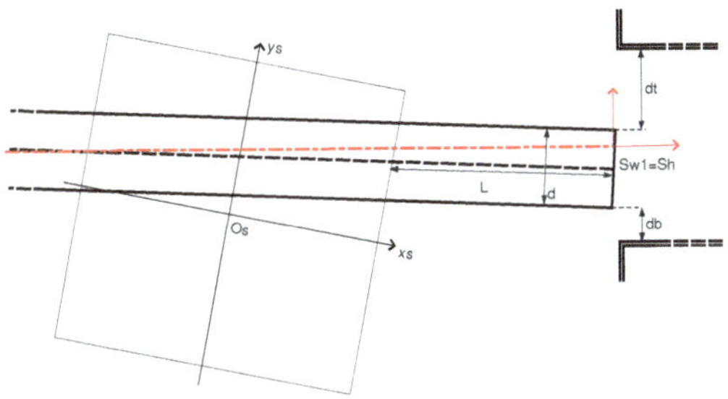

(**b**)

Figure 6. Sketch of the grasped wire with respect to the electric component before (**a**) and after (**b**) the alignment with the hole axis.

5. Assessment of Wire Shape Estimation and Expected Success Rate

In ideal conditions, when the estimation error of the wire shape is zero (i.e., $\hat{\Sigma}_w \equiv \Sigma_w$), under the described assumptions, the proposed procedure allows us to align the wire axis and the hole axis, by maintaining the distances between wire and hole edges (both above δ_a and below δ_b) equal to the maximum possible value $\bar{\delta} = (D - d)/2$. Obviously, in this case, the execution of the insertion task is guaranteed with a Success Rate $SR = 100\%$. In real working conditions, the estimation error of the wire shape implies $\hat{\Sigma}_w \neq \Sigma_w$, and since the alignment can be made only between $\hat{\Sigma}_w$ and Σ_h, when the estimation error increases the insertion task may fail. As a consequence, in real conditions the success rate of the insertion task is $SR < 100\%$.

The quality of the estimated grasped wire shape and the maximum SR reachable can be evaluated taking into account both the estimation error and the actual diameters of the hole and the wire. In particular, the estimation error can be quantified by considering the relative poses of Σ_w and $\hat{\Sigma}_w$. The relative pose of these two frames can be represented by the following homogeneous transformation

$$\mathbf{T}_w^{\hat{w}} = \begin{bmatrix} \cos\alpha & -\sin\alpha & 0 & -\Delta\sin\alpha \\ \sin\alpha & \cos\alpha & 0 & \Delta\cos\alpha \\ 0 & 0 & 1 & 0 \\ 0 & 0 & 0 & 1 \end{bmatrix}, \tag{4}$$

where α is the angle between the estimated wire axis and the actual one, while Δ is the distance between the origins of Σ_w and $\hat{\Sigma}_w$. In the ideal case, with a perfect shape estimation it is $\alpha = \Delta = 0$ and $\mathbf{T}_w^{\hat{w}} = \mathbf{I}$. In real working conditions ($\alpha \neq 0, \Delta \neq 0$), after the alignment $\hat{\Sigma}_w \equiv \Sigma_h$ (see Figure 6b), the distances between wire and hole edges depends on Δ and α. In particular, the estimation error implies that the actual position of the wire presents an offset along y_h-axis, which is responsible for any failure in the execution of the insertion task. This offset, computed from $\mathbf{T}_w^{\hat{w}}$, is equal to $\Delta\cos\alpha$ and it reduces the space between wire and hole edges. The maximum limit for this offset, in order to avoid the unsuccessful execution of the task, is represented by the value $\bar{\delta}$. As a consequence, the following metric

$$\delta = \bar{\delta} - \Delta\cos\alpha \tag{5}$$

can be computed to evaluate both the quality of the grasping and the expected result (success or not) of the insertion task execution. In conclusion, if $0 \leq \delta \leq \bar{\delta}$ the insertion can be successfully completed, while if $\delta < 0$ the insertion task cannot be correctly completed. Moreover, the more δ is close to $\bar{\delta}$, i.e., $\Delta\cos\alpha \to 0$, the better is the quality of the estimated wire shape.

6. Experiments

A number of experiments have been carried out to evaluate the proposed approach. For each experiment, the wire shape has been computed according to the procedure detailed in Section 3. Tens of static experiments, with the sensor fixed on the workbench, have been used to evaluate the shape estimation quality and expected success rate. Additional experiments have been carried out to evaluate the actual success rate of the insertion task in real working conditions, by using the sensorized gripper with standard wires and electrical component.

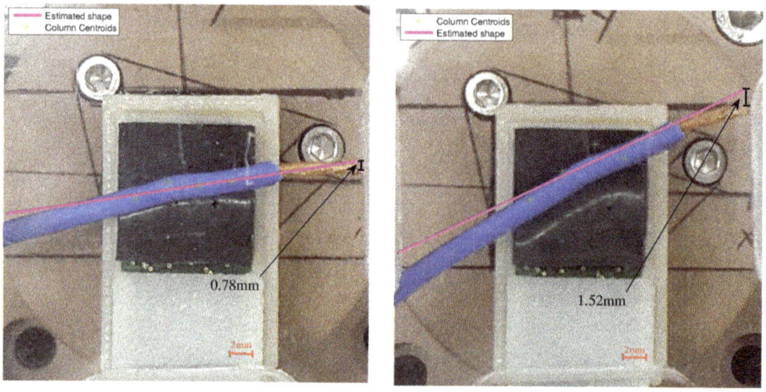

(a) (b)

Figure 7. Some pictures of grasped wires with the estimated shapes and the offset errors $\Delta \cos \alpha$: (**a**) reports a standard case, while (**b**) reports a borderline case.

6.1. Estimation Quality and Expected SR

For the first set of experiments, a standard wire with $d = 3$ mm has been grasped between the tactile sensor, fixed on the workbench, and a transparent methacrylate plate, in different poses. By considering the diameter of the hole of the electric component $D = 2d$, it is $\bar{\delta} = 1.5$ mm. A calibrated optical microscope has been used to take pictures from the transparent plate side. Hence, the offset errors $\Delta \cos \alpha$ between the estimated and the actual wire end points can be directly measured from the pictures. Figure 7 reports two sample cases, where the estimated wire shapes are compared to the actual ones. For each considered case the value of the offset error is reported. By using Equation (5) the metric can be computed, obtaining for the cases in Figure 7 the following values: $\delta = 0.72$ mm for case (a) and $\delta = -0.02$ mm for case (b). From these values it is evident that case (a) allows the correct execution of the insertion task, while case (b) does not guarantee a correct insertion phase ($\delta < 0$). Note that case (b) corresponds to a grasp configuration close to the diagonal of the sensor pad (that is quite unlikely). The same procedure has been applied to 20 considered experiments. Finally, all grasping cases have been divided into two sets: the first set corresponding to cases with a computed metric $\delta > 0$ (17 experiments) and a second set with $\delta < 0$ (3 experiments). The expected success rate for the insertion task has been computed, by relating the number of experiments within the first set with respect to the total number of experiments, by obtaining a $SR = 85\%$.

6.2. The Insertion Task Implementation

For the implementation of the insertion task, the sensorized gripper has been used. All measurements are reported with respect to the world reference frame $\Sigma(O, x, y)$, placed at the robot base. Position and orientation of Σ_s with respect to Σ is known in each time instant, by using the robot system kinematics. The pose of the electric component hole is defined by Σ_h, assumed known from the switchgear CAD. Figure 8 reports experimental results for the s pose during an insertion task. After the wire grasping, the x_s-axis is aligned to the x_h-axis during an approaching phase (see Figure 8a), leaving a distance between O_s and O_h equal to 22 mm in this specific case (it is the estimated length L plus the half side length a of the sensor pad). The reached configuration (at $t = 10$ s) is reported in Figure 9a, where the estimated wire shape and the frame poses $(\Sigma_s, \hat{\Sigma}_w, \Sigma_h)$ are reported with respect to Σ, together with the tactile sensor pad and the component hole. It is evident that without a correction the insertion cannot be completed correctly. The wire shape has been estimated by applying the wire shape estimation algorithm, and the parameters m and n have been used to compute the homogeneous transformation (3). For the experiment reported in the figures $m = -0.0694$ and $n = -3.4866$. By using the computed homogeneous transformation, $\hat{\Sigma}_w$ has been aligned with Σ_h during the correction phase. Figure 8b shows a zoom of the rotation and the translation applied during the correction. After this phase, the estimated wire axis is aligned with the component hole axis. The reached configuration (at $t = 22$ s) is reported in Figure 9b, where it is evident that the insertion can now be correctly completed with a simple translation along the x-axis. Figure 8a shows also the final insertion phase. Figure 10 reports the flowchart, where the connections among all subtasks of the whole insertion sequence are reported. Several checks are implemented by using the tactile sensor data during the insertion execution, in order to evaluate if the task is correctly completed or not. During experiments, the wire shape estimation error will affect the final success of the insertion phase. As discussed in Section 5, the actual wire end point is related to the Σ_w frame, while the estimated wire end point is identified by the $\hat{\Sigma}_w$ frame. To test how this estimation error affects the insertion phase during experiments in real working conditions, the insertion task has been repeated 40 times starting from different initial grasping conditions for the wire. The same experiment described above has been executed and, for each case, the final correct insertion has been evaluated. The number of successfully completed tasks was 33 with a success rate $SR = 82.5\%$. The obtained SR is slightly below the expected SR computed in static conditions (see Section 5), as was foreseeable, since during the experiments, additional errors (e.g., robotic system calibration, electric component position) appears together with the wire shape estimation error. Figure 11 reports a sequence of frames extracted from the video (https://youtu.be/oPxkeeQLKi8) related to the paper in order to show how the designed gripper with the proposed approach allow to correctly complete an insertion sequence. Each frame has been marked with the corresponding procedure subtask. The video shows the effectiveness of the proposed approach during a demo. In the video, the robot is used to fix the screwdriver position for the connection, while the insertion is completely implemented by the designed gripper.

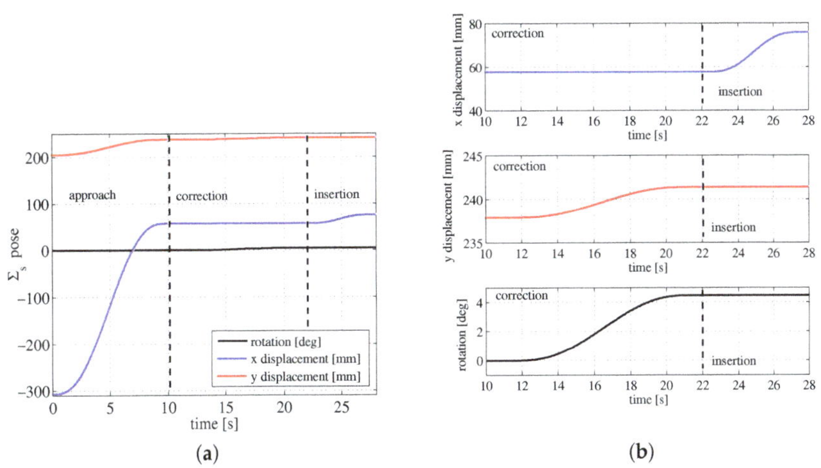

Figure 8. Experimental results: (**a**) Σ_s pose during the whole insertion subtask and (**b**) zoom of the correction and insertion phases.

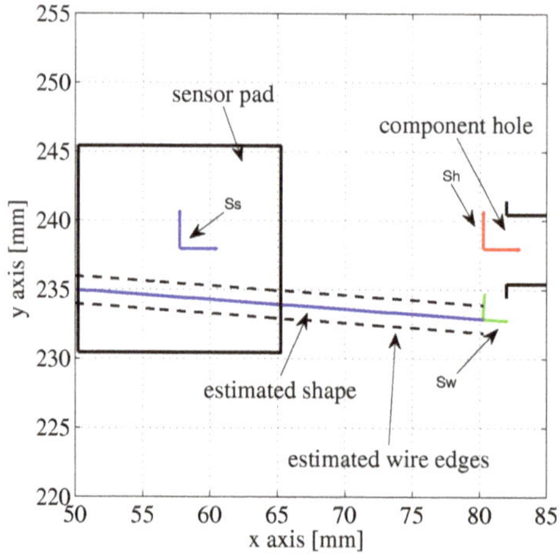

(**a**)

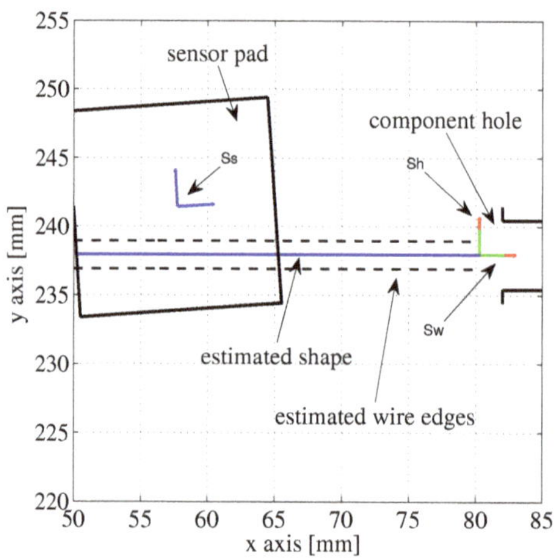

(**b**)

Figure 9. Positions of defined frames in the cartesian space for the experimental case (**a**) before ($t = 10\,\text{s}$) and (**b**) after ($t = 22\,\text{s}$) the correction phase.

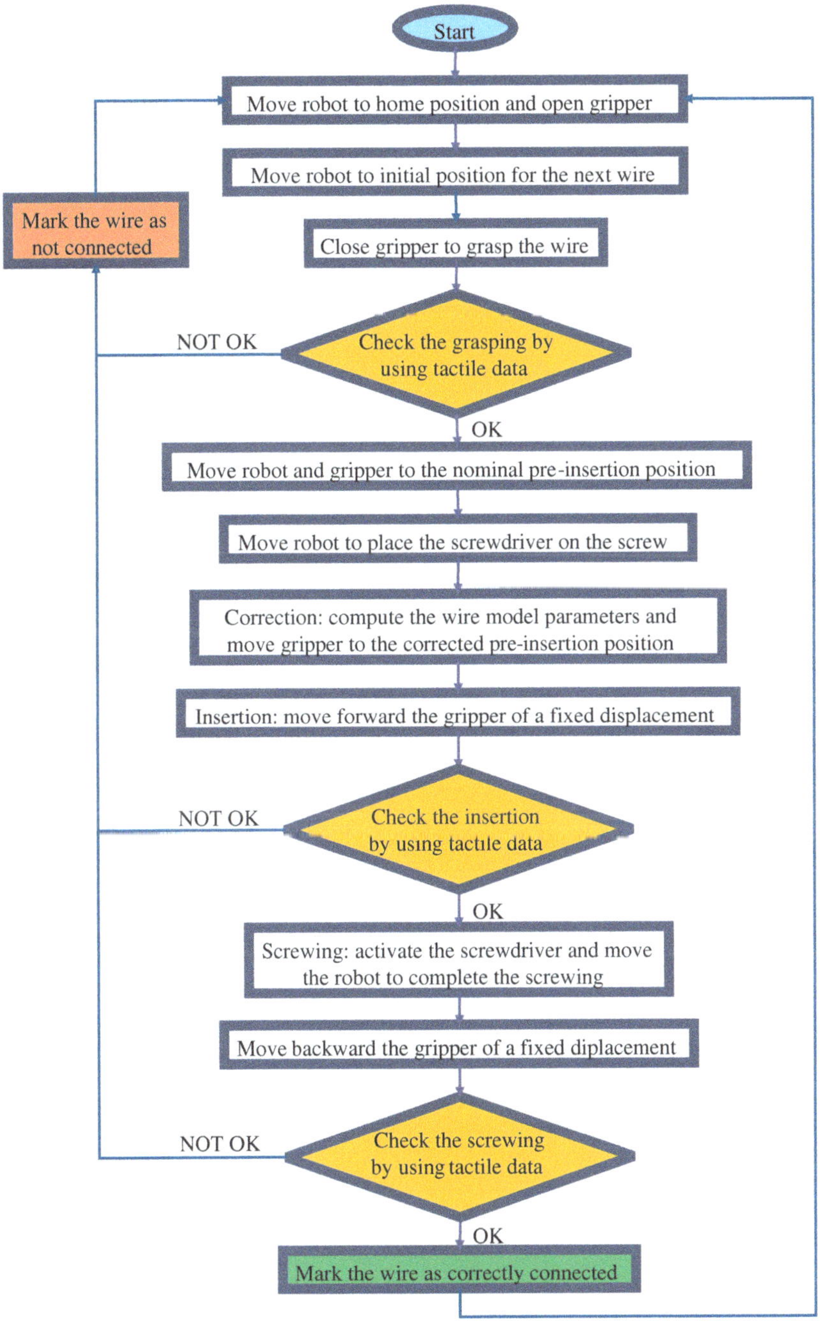

Figure 10. Flowchart of the whole insertion sequence.

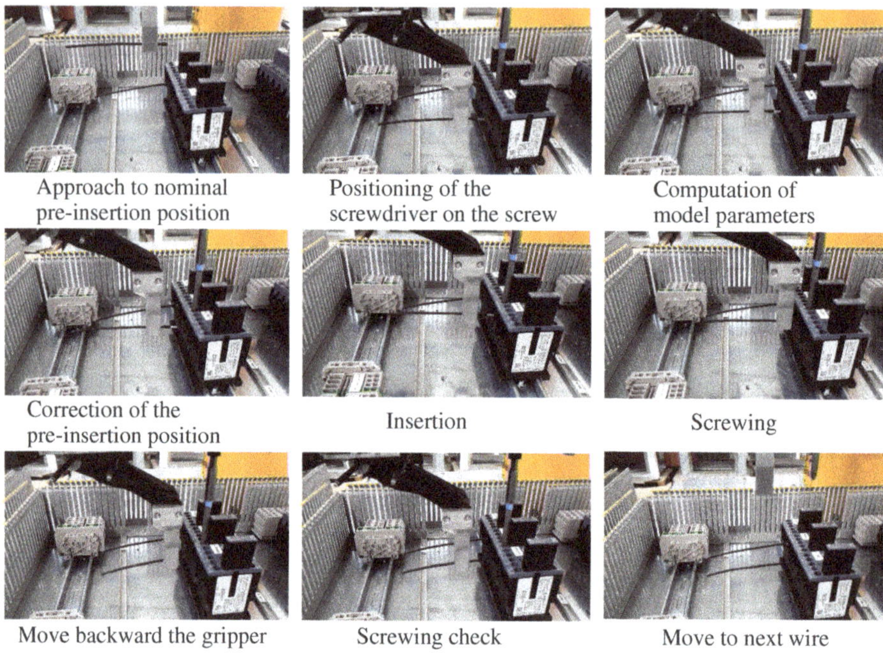

Figure 11. Sequence of frames during some detailed phases of an insertion task.

7. Conclusions

This paper presented a sensorized gripper for wire manipulation, and in particular, for their insertion into the electric components of a switchgear. The designed gripper integrates tactile sensors suitably optimized for this task. A specific procedure for the insertion task execution has been proposed and evaluated in terms of expected success rate. Experimental results have been reported to show the effectiveness of the proposed strategy. Future work will pprobably be devoted to using the sensor to estimate contact forces between the gripper and the manipulated wire during the whole assembly process of the switchgear.

Supplementary Materials: The following are available online at http://www.mdpi.com/2218-6581/8/2/46/s1.

Author Contributions: Design and development of the end-effector, G.P.; Design and development of the tactile sensor, S.P.; Integration, methodology and experimental validation, G.P. and S.P.

Funding: This work was supported by the European Commission's Seventh Framework Programme (FP7/2007-20013) under Grant agreement NO. 601116 (ECHORD++—WIRES Experiment).

Conflicts of Interest: The authors declare no conflict of interest. The founding sponsors had no role in the design of the study; in the collection, analyses, or interpretation of data; in the writing of the manuscript, and in the decision to publish the results.

References

1. Busi, M.; Cirillo, A.; De Gregorio, D.; Indovini, M.; De Maria, G.; Melchiorri, C.; Natale, C.; Palli, G.; Pirozzi, S. The WIRES Experiment: Tools and Strategies for Robotized Switchgear Cabling. *Procedia Manuf.* **2017**, *11*, 355–363. [CrossRef]
2. Cirillo, A.; De Maria, G.; Natale, C.; Pirozzi, S. Design and Evaluation of Tactile Sensors for the Estimation of Grasped Wire Shape. In Proceedings of the IEEE International Conference on Advanced Intelligent Mechatronics, Munich, Germany, 3–7 July 2017; pp. 490–496.

3. De Gregorio, D.; Zanella, R.; Palli, G.; Pirozzi, S.; Melchiorri, C. Integration of Robotic Vision and Tactile Sensing for Wire-Terminal Insertion Tasks. *IEEE Trans. Autom. Sci. Eng.* **2018**, in press. [CrossRef]
4. Saxena, A.; Driemeyer, J.; Ng, A.Y. Robotic Grasping of Novel Objects using Vision. *Int. J. Robot. Res.* **2008**, *27*, 157–173. [CrossRef]
5. Allen, P.K. Integrating Vision and Touch for Object Recognition Tasks. *Int. J. Robot. Res.* **1988**, *7*, 15–33. [CrossRef]
6. Björkman, M.; Bekiroglu, Y.; Högman, V.; Kragic, D. Enhancing visual perception of shape through tactile glances. In Proceedings of the IEEE/RSJ International Conference on Intelligent Robots and Systems, Tokyo, Japan, 3–7 November 2013; pp. 3180–3186.
7. Bimbo, J.; Seneviratne, L.D.; Althoefer, K.; Liu, H. Combining touch and vision for the estimation of an object's pose during manipulation. In Proceedings of the International Conference on Intelligent Robots and Systems, Tokyo, Japan, 3–7 November 2013; pp. 4021–4026.
8. Bhattacharjee, T.; Shenoi, A.A.; Park, D.; Rehg, J.M.; Kemp, C.C. Combining tactile sensing and vision for rapid haptic mapping. In Proceedings of the International Conference on Intelligent Robots and Systems, Hamburg, Germany, 28 September–2 October 2015; pp. 1200–1207.
9. Lepora, N.F.; Aquilina, K.; Cramphorn, L. Exploratory Tactile Servoing With Active Touch. *IEEE Robot. Autom. Lett.* **2017**, *2*, 1156–1163. [CrossRef]
10. Aggarwal, A.; Kirchner, F. Object Recognition and Localization: The Role of Tactile Sensors. *Sensors* **2014**, *14*, 3227–3266. [CrossRef] [PubMed]
11. Yamaguchi, A.; Atkeson, C.G. Combining finger vision and optical tactile sensing: Reducing and handling errors while cutting vegetables. In Proceedings of the 2016 IEEE-RAS 16th International Conference on Humanoid Robots (Humanoids), Cancun, Mexico, 15–17 November 2016; pp. 1045–1051.
12. Pirozzi, S.; Natale, C. Tactile-based manipulation of wires for switchgear assembly. *IEEE/ASME Trans. Mechatron.* **2018**, *23*, 2650–2661. [CrossRef]

robotics

MDPI

Review

A Survey of Behavioral Models for Social Robots

Olivia Nocentini, Laura Fiorini, Giorgia Acerbi, Alessandra Sorrentino, Gianmaria Mancioppi and Filippo Cavallo *

The BioRobotics Institute, Scuola Superiore Sant'Anna, 56025 Pontedera (PI), Italy
* Correspondence: filippo.cavallo@santannapisa.it

Received: 17 May 2019; Accepted: 5 July 2019; Published: 9 July 2019

Abstract: The cooperation between humans and robots is becoming increasingly important in our society. Consequently, there is a growing interest in the development of models that can enhance and enrich the interaction between humans and robots. A key challenge in the Human-Robot Interaction (HRI) field is to provide robots with cognitive and affective capabilities, by developing architectures that let them establish empathetic relationships with users. Over the last several years, multiple models were proposed to face this open-challenge. This work provides a survey of the most relevant attempts/works. In details, it offers an overview of the architectures present in literature focusing on three specific aspects of HRI: the development of adaptive behavioral models, the design of cognitive architectures, and the ability to establish empathy with the user. The research was conducted within two databases: Scopus and Web of Science. Accurate exclusion criteria were applied to screen the 4916 articles found. At the end, 56 articles were selected. For each work, an evaluation of the model is made. Pros and cons of each work are detailed by analyzing the aspects that can be improved to establish an enjoyable interaction between robots and users.

Keywords: social robots; behavioral models; assistive robotics; cognitive architectures; empathy; human-robot interaction

1. Introduction

Social Robotics is commonly defined as the research field dedicated to the socially skillful robots [1]. The main ability of social robots is to establish a natural interaction with humans. The Human-Robot Interaction (HRI) field of study tries to shape the interactions between one or more humans and one or more robots. Over the latest several years, there is an increasing interest in HRI due to the increasing usage of robots not only in industrial fields, but also in other areas as schools [2], homes [3], hospitals [4], and rehabilitation centers [5].

Consequently, in the near future, robots will concretely share environments with human beings to actively collaborate with them in specific daily tasks. The presence of a robot, in fact, could be a useful support during the management of daily activities [6,7], the promotion of social inclusion [8,9], and the suggestion of healthy activities [10,11]. Particularly, recent literature findings underline that robots could help users in their daily life by bringing them objects that they need (i.e., a bottle of water, a specific drug) [12], which helps them in dressing tasks [13,14] or in getting in contact with their families or authorities in dangerous situations [9]. An easy and continuous connection with other people (i.e., relatives, friends, or doctors), could promote social inclusion of people with disabilities or elderly people and increase the quality of their life [15]. Therefore, in this context, there is a growing necessity for developing behavioral models for social robots to have a high quality interaction and level of acceptability in providing useful and efficient services [16,17]. Remarkably, how people accept, perceive, interact, and cooperate with this intelligent machine in their life is still somewhat unknown. However, researchers with different backgrounds are trying to meet this challenge [18].

First, to achieve fluent and effective human-like communication, robots must seamlessly integrate the necessary social behaviors for a given situation using a large number of patterned behaviors that people employ to achieve particular communicative goals. Furthermore, robots should be endowed with the capability to understand feelings, intentions, and beliefs of the user, which are not only directly expressed by the user, but that are also shaped by bodily cues (i.e., gaze, posture, facial expressions) and vocal cues (i.e., vocal tones and expressions) [19]. The non-verbal immediacy, which characterizes communications between humans, should be conveyed in Human-Robot Interaction (HRI). Moreover, the ability to replicate human non-verbal immediacy in artificial agents is twofold. On one side, it allows the detection of emotional and cognitive state of the user, which is useful to develop proactive robots. On the other side, it allows us to shape the behavior of the robot in order to encode behavior capabilities in the interaction as those of humans. The latter case leads to the possibility to automatically generate new robotic behaviors that the robot learns directly from the user.

The first attempts to solve this challenge have been performed by developing intelligent systems able to detect user's emotions [20] and by identifying the key factors that should be adjusted to make the interaction smoother (i.e., interpersonal distance, mental state, user's feedback, and user's profile) [21]. More advanced steps should be performed so that robots are endowed with cognitive and affective capabilities that could provide them with tools to establish empathetic relationships with users and to gain social cognitive mechanisms that are necessary to be perceived as a teammate [22,23].

Second, it is important to remark that the robot's ability to establish empathic relationships has a key role in HRI since it indicates the degree of perceived bodily and psychological closeness between people. Over the last several years, researchers put a lot of effort in understanding how psychology and cognitive neuroscience could be integrated in the design process of artificial cognitive architectures to achieve this target. The field of brain-inspired technologies has become a hot topic in the last several years.

In this context, this paper aims to analyze the current state of the art of behavioral models to find barriers and limitations to provide guidelines for future research studies in this area. Particularly, two databases (namely Scopus and Web of Science) were analyzed to retrieve papers linked with cognitive robotics architecture and model robot empathy, affordance, facial expression, cultural adaptation, and the social robot. In effect, this survey expresses this growing interest and the need to support the research studies in this field and to organize the large amount of work, which is loosely related to the topic underling the scientific challenges. The main issues of this area is related to the fact that several models are too often described from a theoretical point of view without being tested on a real robot and the ones that are tested on a real robot are often tried in a single environment with people belonging to a specific culture [24]. Specifically, researchers are working on the development of cognitive architectures approaching a fully cognitive state, embedding mechanisms of perception, adaptation, and motivation [25]. Particularly, from the analysis of the state of the art, the papers of this survey are grouped according to three main application areas: cognitive architectures, behavioral adaptation, and empathy.

- **Cognitive architectures**—This term refers to research works where both abstract models of cognition and software instantiations of such models, employed in the field of artificial intelligence, are described [26]. Cognitive architectures have the fundamental role to enable artificial intelligence in robotic agents, in order to exhibit intelligent behaviors.
- **Behavioral adaptation**—Behavioral adaptation is defined as "learning new tasks and to adapt to changes in environmental conditions, or to failures in sensors and/or actuators" [27]. Thus, the papers included in this group describe robot's social abilities enhanced by the robot's capability of adapting its behavior to the user's need and habits [28].
- **Empathy**—Empathy is defined as "The act of perceiving, understanding, experiencing, and responding to the emotional state and ideas of another person" [29]. In human-human relationships, this term explains the capacity to take the role of the other to adopt alternative perspectives [28]. Works clustered in this category present a particular emphasis on the attempts

to reproduce this ability in robotic agents to establish an empathetic connection with the user, which improves Human-Robot Interaction (HRI). Empathy is a sub-category of the behavioral adaptation. However, we decide to make separate categories to be aligned with some recent papers [17,30,31].

In this review, several models and architectures used in social robots are presented to evaluate how these attempts fare in achieving an efficient robot-human interaction. A comparison with works presenting experimentation to demonstrate the persuasiveness of robots is also provided to highlight limitations and future trends. In details, the paper is organized as follows: in Section 2, the research methodology for the review is explained. In Sections 3 and 4, the results and the discussions regarding the papers are shown. In Section 5, a summary of the review and its conclusions are presented.

2. Materials and Methods

This section presents the methodology used in the paper to select the most appropriate recent developments as published in the literature, covering the topics of behavioral models for robots.

Study Selection Procedures

This paper reviews empirical studies published between 2010 and 2018 since most of the advances in this area have occurred within that timeframe. A bibliography was developed upon research in Scopus and Web of Science electronic databases. Reference lists of included articles and significant review papers were examined to include other relevant studies. The search queries contained the following terms and were summarized in Table 1.

Table 1. List of keywords used in this review work.

Research Keywords
(cult*) AND (adapt*) AND (behavio*) AND (model* OR system*) AND (robot*)
(cult*) AND (adapt*) AND (cognitive) AND (model* OR architecture*) AND (robot*)
affordance* AND (behavio*) AND (adapt*) AND (robot*)
affordance* AND (cognitive) AND (model* OR architecture*) AND (robot*)
fac* AND expression AND cognitive AND (model* OR architecture*) AND (robot*)
fac* AND expression AND (behavio*) AND (model* OR system*) AND (robot*)
cognitive AND robot* AND architecture*
learning AND assistive AND robot*
affective AND robot* AND behavio*
empathy AND social AND robot*

Application of these search keys provided a total of 4916 hits with 1520 hits in Web of Science in the field "Topic" and 3396 hits in Scopus in "Article title, abstract, keywords" fields.

After deletion of duplicates, the titles and abstracts retrieved by the electronic search were read first, to identify articles deserving a full review. Papers about users' emotion recognition and about emotions as unique input for HRI were excluded. Additionally, papers not written in English were excluded. A total of 1297 works was selected at this stage.

Then, a full-text assessment was carried out. The reading process led to the exclusion of 1241 papers that were out of topic, papers focusing only on definitions and taxonomy, papers for missing the model's evaluation, and papers focusing more on users' perception about robot's abilities without behavioral adaptation.

The final list of papers includes 56 studies, which satisfy all the following selection criteria: (i) employment of cognitive architectures and/or behavioral models, (ii) explanation of cognitive architectures and/or behavioral models, (iii) research focus on robotic agent's capabilities; (iv) behavioral adaptation according to different strategies, and (v) analysis conducted on social or assistive robots. The studies' selection process is shown in Figure 1.

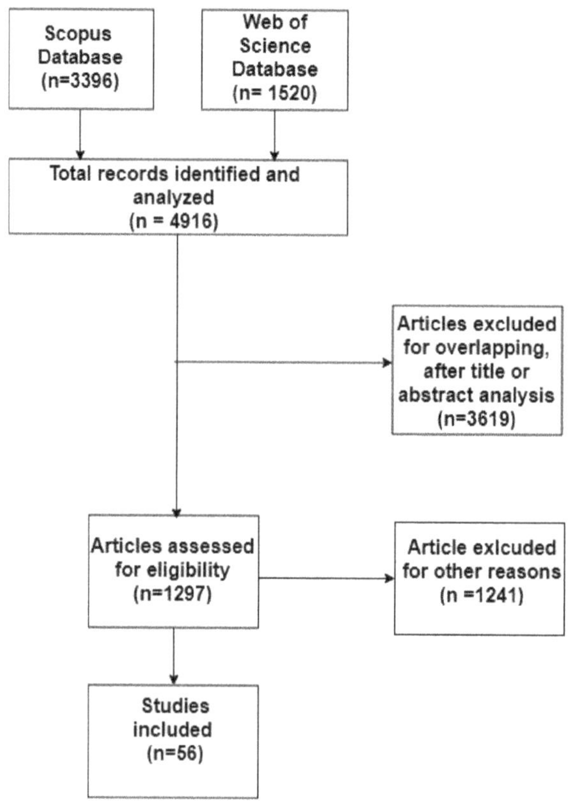

Figure 1. Selection process of relevant papers.

3. Results

3.1. Application Overview

The interest toward behavioral architectures has grown, as shown in Figure 2. Particularly, of the fully evaluated papers, 7 papers (12.5%) were published before 2014 and 49 papers (87.5%) were published within the past five years.

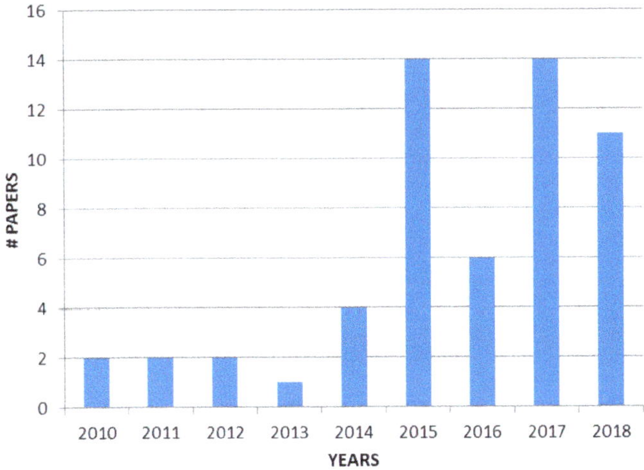

Figure 2. Number of papers analyzed in this review, categorized by the year of publication.

The selected papers can be divided into two big groups: the works describing cognitive architectures, behavioral adaptation models, and empathy models from a conceptual point of view (seven papers as summarized in Table 2) and the works presenting experimental studies (twenty-three papers summarized in Table 3). Additionally, the papers can be grouped subsequently on the basis of the three areas described in the introduction and Figure 3 shows that most of the papers included in this review focus on behavioral adaptation strategies (41.07%) together with cognitive architectures (46.43%), and empathy (15.22%).

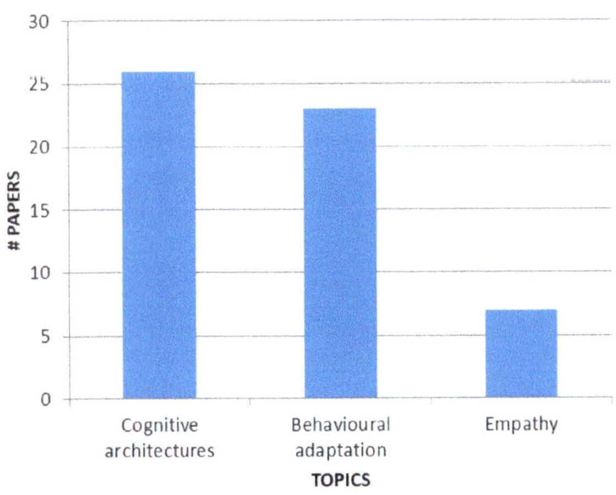

Figure 3. Number of papers, categorized by the main application covered.

Table 2. Summary of recently published works on behavioral models without an experimentation loop.

Ref.	Title and Year	Aim	Robot	Social Cues	Models	Area
[32]	Persuasive AI Technologies for Healthcare Systems (2016)	Development of an IoT toolbox toward AI-based persuasive technologies for healthcare systems	NAO	Gaze behavior, facial behavior, speech cues	A system that contains: • a sensor network architecture used to observe the environment and provide real-time access to the data for interpretation • a block containing context-aware applications (object detection, activity recognition) • a prediction block of future states	Cognitive architectures
[23]	Enabling robotic social intelligence by engineering human social-cognitive mechanisms (2017)	Suggestions and overviews on social cognitive mechanisms	-	-	A model that takes into consideration social-cognitive mechanisms to facilitate the design of robots	Cognitive architectures
[33]	An Architecture for Emotional and Context-Aware Associative Learning for Robot Companions (2015)	Theoretical architectural model based on the brain's fear learning system	-	Environmental cues	Theoretical architectural model that uses artificial neural networks (ANNs) representing sensory thalamus, sensory cortex, amygdala, and orbitofrontal cortex.	Cognitive architectures
[34]	From learning to new goal generation in a bio-inspired robotic setup (2016)	Imitation of the neural plasticity, the property of the cerebral cortex supporting learning	NAO	Eye behavior	A model called Intentional Distributed Robotic Architecture (IDRA) that takes inspiration from the amygdala-thalamo-cortical circuit in the brain at its functional level	Cognitive architectures
[35]	Human-aware interaction: A memory-inspired artificial cognitive architecture (2017)	Human-aware cognitive architecture to support HRI	Care-O-Bot 4	Eyes and vocal behaviors	Partially observable Markov decision process (POMDP) model	Cognitive architectures
[31]	Artificial Empathy: An Interdisciplinary Investigation (2015)	Overview the research field aimed at building emotional and empathic robots focusing on its main characteristics and ongoing transformations	-	Gestures and posture cues	-	Empathy

Table 2. *Cont.*

Ref.	Title and Year	Aim	Robot	Social Cues	Models	Area
[17]	How can artificial empathy follow the developmental pathway of natural empathy? (2015)	A conceptual model of artificial empathy is proposed and discussed with respect to several existing studies	Emotional communication robot, WAMOEBA - WE humanoid robot	-	Conceptual model of artificial empathy	Empathy
[36]	A Cognitive Control Architecture for the Perception–Action Cycle in Robots and Agents (2013)	Visual perception, recognition, attention, cognitive control, value attribution, decision-making, affordances, and action can be melded together in a coherent manner in a cognitive control architecture of the perception–action cycle for visually guided reaching and grasping of objects by a robot or an agent.	-	-	Model composed of four modules: object localization and recognition, cognitive control, decision-making, value attribution affordances, motion, and planning	Cognitive architectures
[37]	A computational model of perception and action for cognitive robotics (2011)	A novel computational cognitive model that allows for direct interaction between perception and action as well as for cognitive control, demonstrated by task-related attentional influences.	-	-	HiTEC architecture composed of a task level, a feature level, and a sensory motor level	Cognitive architectures
[38]	Interaction of culture-based learning and cooperative co-evolution and its application to automatic behavior-based system design (2010)	A bio-inspired hybridization of reinforcement learning, cooperative co-evolution, and a cultural-inspired memetic algorithm for the automatic development of behavior-based agents		-	Reinforcement learning for structure learning that finds the organization of behavior modules during the agent's lifetime	Behavioral adaptation
[39]	Affective Facial Expression Processing via Simulation: A Probabilistic Model (2014)	Simulation Theory and neuroscience findings on Mirror-Neuron is used as the basis for a novel computational model, as a way to handle affective facial expressions.	-	postural and vocal cues	Simulation Theory and neuroscience findings on Mirror-Neuron Systems as the basis for a novel computational model,	Cognitive architectures
[40]	Multi-robot behavior adaptation to local and global communication atmosphere in humans-robots interaction (2014)	A multi-robot behavior adaptation mechanism based on cooperative–neutral–competitive fuzzy-Q learning is developed in a robot	-	Eyes cues	Cooperative–neutral–competitive fuzzy-Q learning	Behavioral adaptation

Table 2. *Cont.*

Ref.	Title and Year	Aim	Robot	Social Cues	Models	Area
[41]	An Ontology Design Pattern for supporting behavior arbitration in cognitive agents (2017)	An Ontology Design Pattern for the definition of situation-driven behavior selection and arbitration models for cognitive agents.	MARIO robot-	affordances cues	Affordance Ontology Design Pattern (ODP)	Cognitive architectures
[42]	Context-Aware Cloud Robotics for Material Handling in Cognitive Industrial Internet of Things (2018)	In this paper, a cognitive industrial entity called context-aware cloud robotics (CACR) for advanced material handling is introduced and analyzed.	-	-	Energy-efficient and cost-saving algorithms for material handling	Cognitive architectures

Table 3. Summary of recently published works on behavioral models with an experimentation loop.

Ref.	Title and Year	Aim	Robot	Social Cues	Participants	Models	Area
[30]	Compassion, empathy, and sympathy expression features in affective robotics (2017)	Comparison between expression features of compassion, sympathy, and empathy in British English and Polish that need to be tuned in social robots to enable them to operate successfully	Generic assistive robots	Linguistic cues, facial cues, movement cues, and physiological features	British English-speaking participants (mean age 23.2 years, 21 females) and 29 Polish-speaking subjects (mean age 25.6 years, 26 females)	Creating culture-specific emotion (compassion, sympathy, and empathy) models	Empathy
[43]	Persuasive robotic assistant for health self-management of older adults: Design and evaluation of social behaviors (2010)	Comparison of a text-based interface with a character persuading user	I-Cat	Gaze and posture cues	24 middle age adults age 45 to 65 years old	The model is based on persuasion and uses Big five questionnaire on personality	Empathy
[16]	Inclusion of service robots in the daily lives of frail older users: A step-by-step definition procedure on users' requirements (2018)	Definition of metrics to build an empathic robot that helps elderly people	Pepper	-	42 participants (elderly people)	The model is based on an emotion recognition algorithm	Empathy
[44]	Promoting interactions between humans and robots using robotic emotional behavior (2015)	Emotion-based assistive behavior for a socially assistive robot	Brian	Gestures and emotional state cues	34 subjects aged 17 to 68 years old	Markov model that uses a human affective state classifier and a non-verbal interaction and states analysis (NISA)	Empathy

Table 3. *Cont.*

Ref.	Title and Year	Aim	Robot	Social Cues	Participants	Models	Area
[45]	A personalized and platform-independent behavior control system for social robots in therapy: development and applications (2018)	A behavior control system for social robots in therapies is presented with a focus on personalization and platform-independence	NAO Pepper	-	Children and elderly people	Model based on a Body Action Coding System (BACS)	Behavioral adaptation
[46]	Adaptive artificial companions learning from users' feedback (2017)	Capacity of an intelligent system to learn and adapt its behavior/actions	EMOX	Current user(s) attributes and current environmental attribute cues	Children, teenagers, and adults	Markov Decision Processes (MDPs) model	Behavioral adaptation
[47]	Why robots should be social: Enhancing machine learning through a social human-robot interaction (2015)	How additional social cues can improve learning performance	Robot head mounted on an articulated robot arm	Gaze behavior	41 healthy and young participants with an average age of 24 years old.	The interaction model of the robot is based on language games	Behavioral adaptation
[25]	The affective loop: A tool for autonomous and adaptive emotional human-robot interaction (2015)	Affective model for social robotics	NAO	-	Children age 5 to 7 years old.	Flutchik emotional model; uses Fuzzy rules and learning by demonstration	Empathy
[48]	Puffy: A Mobile Inflatable Interactive Companion for Children with Neurodevelopmental Disorder (2017)	Robot's ability in gestures' interpretation	Puffy	Visual, auditory, and Tactile cues	19 children with an average age of 6 years old	Interactional Spatial Model which considers: interpersonal distance - relative (child-robot) bodily orientations - child's and robot's movements in space - child's emotional state - child's eye contact - robot's emotional state	Behavioral adaptation

Table 3. *Cont.*

Ref.	Title and Year	Aim	Robot	Social Cues	Participants	Models	Area
[49]	Social intelligence for a robot engaging people in cognitive training activities (2012)	Definition of user state and task performance, adjusting robot's behavior	Brian 2.0	-	10 healthy and young participants from 20 to 35 years old.	A hierarchical reinforcement learning approach is used to create a model that allow the robot to learn appropriate assistive behaviors based on the structure of the activity	Behavioral adaptation
[50]	Empathizing with emotional robot based on cognition reappraisal (2017)	Continuous cognitive emotional regulation model for robot	-	-	10 subjects	Hidden Markov Model; uses cognitive reappraisal strategy	Behavioral adaptation
[51]	Interpretation of interaction demanding of a user based on nonverbal behavior in a domestic environment (2017)	Model for decision making on a user's non-verbal interaction demand	MIRob	Eyes behavior	Eight subjects from 25 to 58 years old.	A model to decide when to interact with the user. It observes movements and behavior of the patient put them through a module called Interaction Demanding Pose Identifier. The data obtained are fed into the Fuzzy Interaction Decision Making Module in order to interpret the degree of interaction demanding of the user.	Behavioral adaptation
[52]	Understanding human intention by connecting perception and action learning in artificial agents (2017)	System inspired by humans' psychological and neurological phenomena	-	Eyes behavior	-	Generic model; uses supervised multiple timescale recurrent neural networks	Cognitive architectures
[53]	Improving human-robot interaction based on joint attention (2017)	A novel cognitive architecture for a computational model of the limbic system is proposed, inspired by human brain activity, which improves interactions between a humanoid robot and preschool children	Robotis Bioloid	Eyes, auditory, and sensory cues	16 pre-school children from 4 to 6 years old.	Dynamic neural fields (DNFs) model; used with reinforcement and unsupervised learning-based adaptation processes	Cognitive architecture

Table 3. *Cont.*

Ref.	Title and Year	Aim	Robot	Social Cues	Participants	Models	Area
[54]	A Robot Learns the Facial Expressions Recognition and Face/Non-face Discrimination Through an Imitation Game (2014)	A robotic system that can learn online is shown to recognize facial expressions without having a teaching signal associating a facial expression	-	Gestures, gaze direction, vocalization cues	20 persons	Theoretical model for online learning of facial expression recognition	Cognitive architecture
[55]	Development of first social referencing skills: Emotional interaction as a way to regulate robot behavior (2014)	Studying how emotional interactions with a social partner can bootstrap increasingly complex behaviors such as social referencing	-	Facial expression cues	20 persons	Model that uses the child's affective states and adapts its affective and social behavior in response to the affective states of the child	Behavioral adaptation
[24]	May I help you? (2015)	Model of adaptive behaviors to pedestrians' intentions	Robovie	Body and facial expression cues	The participants were visitors of the shopping mall where the robot was placed.	State transition model; used with an Intention-Estimation Algorithm	Behavioral adaptation
[56]	A framework for the design of person following behaviors for social mobile robots (2012)	Framework for people detection, state estimation, and trajectories generation for an interactive social behavior	Kompai	Body and facial expression cues	-	The model combines perception, decision, and action and uses fuzzy logic and SLAM algorithms	Behavioral adaptation
[57]	People-aware navigation for goal-oriented behavior involving a human partner (2011)	Person-aware navigation system	-	Body and facial expression cues	8 healthy and young subjects (7 male and 1 female) with an average age of 20.8 years.	The model weights trajectories of a robot's navigation system for autonomous movement using Algorithms in navigation-stack of ROS	Behavioral adaptation
[58]	Promoting interactions between humans and robots using robotic emotional behavior (2015)	Map of human psychological traits to make robots emotive and sociable	NAO	Body and facial expression cues	10 subjects	Emotional Intelligence (EI) model; uses Meyer Briggs theory (MBT) and fuzzy logic	Behavioral adaptation

Table 3. *Cont.*

Ref.	Title and Year	Aim	Robot	Social Cues	Participants	Models	Area
[59]	Moving toward an intelligent interactive social engagement framework for information gathering (2017)	To develop an integrated robotic framework including a novel architecture and an interactive user interface to gather information	NAO, Milo, and Alice	Body and facial expression cues	186 children from 8 to 12 years old	Interactive Social Engagement Architecture (ISEA) is designed to integrate behavior-based robotics, human behavior models, cognitive architectures, and expert user input to increase social engagement between a human and system	Cognitive architectures
[60]	Moving toward an intelligent interactive social engagement framework for information gathering (2018)	Software architecture allowing a robot to socially interact with human beings, sharing with them some basilar cognitive mechanisms	NAO	non- verbal cues such as social signals	Children	Hidden Markov Model (HMM) used as a reasoning approach The model encoded social interaction and uses natural and intuitive communication channels, both to interpret the human behavioral and to transfer knowledge to the human	Cognitive architectures
[61]	Persuasive robotic assistant for health self-management of older adults: Design and evaluation of social behaviors (2010)	To implement models obtained from biological systems to humanoid robots	Robothespian	Body and facial expression cues	237 subjects of different age, gender, and education	Model that uses a facial action coding system	Behavioral adaptation
[62]	A Logic-Based Computational Framework for Inferring Cognitive Affordances (2018)	A framework that reasons about affordances in a more general manner than described in the existing literature	-	Eye, gaze body orientation; verbal request	The robot is tested in different scenarios with the following domains as, for example, at an elder care facility	A DS theory framework often interpreted as a generalization of the Bayesian framework	Cognitive architectures
[63]	The Role of Functional Affordances in Socializing Robots (2015)	A paper where affordances of objects were used as a starting point for the socialization of robots	Jenny Robot	gesture, gaze, head movements, vocal features, posture, proxemics, and touch and affordances cues	Jenny is implemented for the tea-making task under various scenarios	OWL-DL model	Cognitive architectures

Table 3. *Cont.*

Ref.	Title and Year	Aim	Robot	Social Cues	Participants	Models	Area
[64]	Intelligent emotion and behavior based on topological consciousness and adaptive resonance theory in a companion robot (2016)	An artificial topological consciousness that uses a synthetic neurotransmitter and motivation, including a logically inspired emotion system is proposed	CONBE robot	Gaze cues	Experiments done for the object recognition and face recognition	Behavioral model for recognition of objects and faces	Cognitive architectures
[65]	Practical aspects of deploying Robotherapy systems (2018)	Our robotic system helps therapists in sessions of cognitive stimulation. Without	NAO	Orientation of the gaze	Robot tested with different kind of cultures	A new module capable of communicating implemented with a behavioral architecture BICA	Behavioral adaptation
[66]	Staged Development of Robot Skills: Behavior Formation, Affordance Learning and Imitation with Motions (2015)	Realization of an integrated developmental system where the structures emerging from the sensorimotor experience of an inter- acting real robot are used as the sole building blocks of the subsequent stages that generate increasingly more complex cognitive capabilities	7 DOF Motoman robot arm	Motionese cues robot	The robot performed 64 swipe action executions towards a graspable object that is placed in a reachable random position	Model that includes three stages: discovering behavior primitives, learning to detect affordances, learning to predict effects	Behavioral adaptation
[67]	Personality affected robotic emotional model with associative memory for human-robot interaction (2016)	This paper discusses human psychological phenomena during communication from the point of view of internal and external factors, such as perception, memory, and emotional	Iphonod robot	Facial, gesture, voice cues	The experimental part is divided into two parts; first, the processing of multi-modal information into emotional information in the emotion model is simulated. Next, based on multi-modal information and emotional information, which came from the first one, the association process will be performed to determine the robot behaviors.	Model for object, facial and gesture, voice, and biometric recognition	Behavioral adaptation

Table 3. *Cont.*

Ref.	Title and Year	Aim	Robot	Social Cues	Participants	Models	Area
[68]	Multimodal emotional state recognition using sequence-dependent deep hierarchical features (2015)	This model uses a hierarchical feature representation to deal with spontaneous emotions, and learns how to integrate multiple modalities for non-verbal emotion recognition, which makes it suitable for use in an HRI scenario	-	Facial expressions	Three experiments are executed and evaluated. The first one uses information of the face expression to determine emotional estates. The second one extracts information from the body motion, composed by arms, torso, and head movements, and the third one uses both types of information.	Multichannel Convolutional Neural Network (MCCNN) to extract hierarchical features	Behavioural adaptation
[69]	Artificial cognition for social human–robot interaction: An implementation (2015)	This article is an attempt to characterize these challenges and to exhibit a set of key decisional issues that need to be addressed for a cognitive robot to successfully share space and tasks with a human	Manipulator	Verbal communication, gestures, and social gaze	A scenario involving multi-modal, interactive grounding: the humans can refer to invisible or ambiguous objects that the robots anchor to physical objects through multi-modal interactions with the user. A second task that the robot has to achieve is cleaning a table	Model that has collaborative cognitive skills: geometric reasoning and situation assessment based on perspective-taking and affordance analysis; acquisition and representation of knowledge models for multiple agents (humans and robots, with their specificities); natural and multi-modal dialogue; human- aware task planning; human–robot joint task achievement.	Cognitive architectures

Table 3. *Cont.*

Ref.	Title and Year	Aim	Robot	Social Cues	Participants	Models	Area
[70]	Collaborative Autonomy between High-level Behaviours and Human Operators for Remote Manipulation Tasks using Different Humanoid Robots (2017)	This article discusses the technical challenges that two teams face and overcame during a DARPA completion to allow the human operators to interact with a robotic system with a higher level of abstraction and share control authority with it	THORMANG "Johnny" and Atlas "Florian"	Affordances Visual cues	The two robots have to complete the two tasks of opening a door and a valve	Model for a humanoid robot that is composed by a remote manipulation control approach, a high-level behavior control approach, an overarching principle, and a collaborative autonomy, which brings together the remote manipulation and high-level control approaches	Behavioral adaptation
[71]	Probabilistic Movement Primitives for Coordination of Multiple Human-Robot Collaborative Tasks (2017)	This paper proposes an interaction learning method for collaborative and assistive robots based on movement primitives	Dual arm manipulator	Social-cognitive cues	A robot co-worker must recognize the intention of the human to decide some actions: if it should hand over a screwdriver or hold the box or coordinate the location of the handover of a bottle with respect to the location of the hand of the human.	Imitation learning to construct a mixture model of human-robot interaction primitives. This probabilistic model allows the assistive trajectory of the robot to be inferred from human observations	Cognitive architectures
[2]	Robots in Education and Care of Children with Developmental Disabilities: A Study on Acceptance by Experienced and Future Professionals (2017)	A study on the acceptance of robots by experienced practitioners and university students in psychology and education sciences is presented	NAO	Verbal cues	Demonstration of the capabilities of the robot in front of participants that had to filled a questionnaire at the end according to robot's behavior	The aim is to examine the factors, through the Unified Theory of Acceptance and Use of Technology (UTAUT) model	Cognitive architectures
[72]	Optimized Assistive Human–Robot Interaction Using Reinforcement Learning (2016)	The proposed HRI system assists the human operator to perform a given task with minimum workload demands and optimizes the overall human–robot system performance.	PR2	-	The robot has to draw a prescribed trajectory	Reinforcement Learning LQR Method	Behavioral adaptation

Table 3. *Cont.*

Ref.	Title and Year	Aim	Robot	Social Cues	Participants	Models	Area
[73]	The Impact of Social Robotics on L2 Learners' Anxiety and Attitude in English Vocabulary Acquisition (2015)	This study aimed to examine the effect of robot assisted language learning (RALL) on the anxiety level and attitude in English vocabulary acquisition amongst Iranian EFL junior high school students	NAO		Iranian EFL junior high school students. Forty-six female students, who were beginners at the age of 12, participated in this study.	FLCAS questionnaire that evaluates anxiety	Cognitive Architectures
[74]	The multi-modal interface of Robot-Era multi-robot services tailored for the elderly (2018)	ROBOTERA project has the objective of implementing easy and acceptable service robotic system for the elderly.	Three Robot: Coro, Doro, Oro	Gaze, vocal, facial cues	Elderly people	Architectures based on emotion recognitions and object's recognition	Cognitive Architectures
[75]	Human–Robot Facial Expression Reciprocal Interaction Platform: Case Studies on Children with Autism (2018)	In this research, a robotic platform has been developed for reciprocal interaction consisting of two main phases, namely as Non-structured and Structured interaction modes	Mina Robot	Facial expressions	Children	The model is composed of two modules: Non-structured and Structured interaction modes. In the Non-structured interaction mode, a vision system recognizes the facial expressions of the user through a fuzzy clustering method. In the Structured interaction mode, a set of imitation scenarios with eight different posed facial behaviors were designed for the robot	Cognitive Architectures
[76]	Adaptive Robotic Tutors that Support Self-Regulated Learning: A Longer-Term Investigation with Primary School Children (2018)	This paper explores how personalized tutoring by a robot, achieved using an open learner model (OLM), promotes self-regulated learning (SRL) processes and how this can impact learning and SRL skills compared to personalized domain support alone	NAO	Gaze	Children	Using an open learner model (OLM) to learn SRL processes	Behavioral adaptation

Table 3. *Cont.*

Ref.	Title and Year	Aim	Robot	Social Cues	Participants	Models	Area
[77]	DAC-h3: A Proactive Robot Cognitive Architecture to Acquire and Express Knowledge About the World and the Self (2018)	This paper introduces a cognitive architecture for a humanoid robot to engage in a proactive, mixed-initiative explo- ration and manipulation of its environment, where the initiative can originate from both humans and robots.	i-Cub	Gaze	Picking objects	The framework, based on a biologically grounded theory of the brain and mind, integrates a reactive interaction engine, a number of state-of-the-art perceptual and motor learning algorithms, as well as planning abilities and an autobiographical memory	Cognitive Architectures
[78]	Using a Humanoid Robot to Elicit Body Awareness and Appropriate Physical Interaction in Children with Autism (2015)	A human–robot interac- tion study, focusing on tactile aspects of interaction, in which children with autism interacted with the child-like humanoid robot KASPAR	KASPAR	Tactile, gaze cues	Children	Model for the behavior analysis	Cognitive Architectures

3.2. Data Abstraction

Data were abstracted from each selected article, as reported in Table 4 The tables give the main purpose of each work, the robot used, the extracted features, and a short description of the implemented model/algorithms. The last column reports the area to which they belong (cognitive architectures, behavioral adaptation criterion, and empathy). In addition, for those papers which describe an experimental protocol, the number and the type of participants involved in the experimental session are also reported. The objective of the abstraction is to provide an overview of the papers included in this survey and to facilitate their comparison.

3.3. Theoretical Works on the Development of Robotics Behavioral Models

In this section, published works on theoretical studies of robotic models are described. Occurrences of theoretical studies belong to three areas (Table 2).

3.3.1. Concepts for the Cognitive Application Area

Human cognitive systems are often adopted as an inspiration to develop a cognitive architecture for robots. In the last several years, in fact, assistive and companion robots have accomplished advanced social proficiency whenever they were equipped with cognitive architectures. Relevant examples of this trend are listed below in this section.

Reference [23] described cognitive architectures citing the Learning Intelligent Distribution Agent, Soar, and the Adaptive Control of Thought-Rationale architecture with the aim to provide a set of commitments useful to develop intelligent machines. In this work are presented the Theory of Mind (ToM) and the "perceptual-motor simulation routines," which are two of the fundamental theories of social cognition. Particularly, the ToM would represent the inherent human ability in attributed mental states to other social agents. That is possible through the application of theoretical inference mechanisms on cues gathered by people during social interactions (e.g., facial expression could be used in order to probabilistically determine the person's emotional state). On the other hand, the paradigm of "perceptual-motor simulation routines" state that people would be able to understand others' mental state by the use of simulation mechanisms, which would help the subject attribute a mental state to his\her interlocutor. The authors suggested their approach, Engineering Human Social Cognition (EHSC), which incorporates social signal processing mechanisms to allow a more natural HRI and focus on verbal and non-verbal cues to support interaction. Social signal processing is able to interpret social cues and then individual mental states. The authors underlined that modelling recommendations have centered primarily on the perceptual, motor, and cognitive modelling of a robotic system that spans disciplinary perspectives. This is the area that will require extensive work in the future. As such, the next steps in this area must include both research and modelling efforts that assess the issues and challenges of integrating the proposed types of models and formalisms. That effort can aid in the development of an integrated and working system based on these recommendations. These recommendations, if instantiated, would provide some basic perceptual, motor, and cognitive abilities, but future efforts should address whether these would also support more complex forms of social interaction. Such a capability would permit an artificial system to better express or perceive emotions while interacting and communicating with humans in even more complex social scenarios that would require shared decision-making and problem-solving.

Among cognitive architectures to be implemented into social robots to improve HRI, Pieters et al. [35] presented a work with the aim to develop a human-aware cognitive architecture. This system is conceived to provide robots with the ability to understand the human state, physical and affective, and then to interact in a suitable manner. Starting from cognitive models, the authors organized the architecture by considering a cognitive model that represents how memory is organized: a declarative memory for semantic and episodic facts, and procedural memory. According to this organization, the robot's tasks are encoded as a sequence of actions and events, thanks to a symbolic task planner, with the aim to verify if, and in what way, the task has already been executed. In Reference [77], the authors proposed an architecture that drives the robot behavior to acquire language capabilities, execute goal-oriented behavior, and express a verbal narrative of its own experience in the world.

To provide robots with believable social responses and to have a more natural interaction, a theoretical model was developed by Reference [33]. The proposed architecture is human brain-inspired and it is structured into four principal modules, which encompasses anatomic structures and cognitive functions, such as the sensory system, the amygdala system, the hippocampal system, and the working memory. This brain-inspired system provides robots with emotional memory, which is fundamental to be able to learn and adapt to dynamic environments. In particular, the authors focus on artificial emotional memory, which lets robots remember emotions, associate them with stimuli, and react in an appropriate way if unpleasant stimuli occur. External stimuli are pre-processed by the sensory system, which is composed of the sensory cortex and thalamus. Prediction and association between stimuli and emotions are conducted via the amygdala system that provides emotional feedback to the hippocampal system.

Lastly, another brain-inspired architecture was developed in Reference [34]. It focuses on the autonomous development of new goals in robotic agents. Starting from neural plasticity, the Intentional Distributed Robotic Architecture (IDRA) is an attempt to simulate a brain circuit composed of the amygdala, the thalamus, and the cortex. The cortex is responsible for receiving signals from sensory organs. The thalamus develops new motivations in mammals, while the amygdala manages the generation of somatosensory responses. Elementary units, called Deliberative Modules (DM), enable a learning process that lets the robot learn and improve its skills during the execution of a task. This process is known as Intentional Distributed Robotic Architecture (IDRA). Working memory (acting as the cerebral cortex) and goal generator (acting as the thalamus) modules compose each DM. Amygdala is represented by instincts modules. Experiments were made to verify the ability of a NAO robot (*https://www.softbankrobotics.com/emea/en/robots/nao/find-out-more-about-nao*. Retrieved July 2018) in learning to distinguish particular object shapes and in exploring in an autonomous way and learning new movements. Sensing and actuation as main activities required for learning and cognitive development were tested: NAO was able to learn new shapes taking sensorial inputs and to compose new behaviors, which are consistent with these goals. The authors underlined their choice to opt for directly using the high-level representation of the neural function, even though a system that uses neural coding as basic representation could be integrated into IDRA. NAO is often used to implement cognitive architectures. It was used in Reference [79] to evaluate a robot-assisted therapy for children with autism and intellectual disability (the same was done in Reference [78] with a robot named Kaspar) and in Reference [73] to examine the effect of robot-assisted language learning (RALL) on the anxiety level and attitude in English vocabulary acquisition among Iranian EFL junior high school students.

Another important element that should be considered in the field of behavioral models is the mechanism of affordances. The concept of affordance refers to the relationship between human perceivers and aspects of their environment. Being able to infer affordances is central to common sense reasoning, tool use, and creative problem solving in artificial agents.

Cutsuridis et al. [36] created a cognitive control architecture of the perception–action cycle for visually guided reaching and grasping of objects by a robot or an agent melded perception, recognition, attention, cognitive control, value attribution, decision-making, affordances, and action. The suggested visual apparatus allows the robot/agent to recognize both the object's shape and location, extract affordances, and formulate motor plans for reaching and grasping.

Haazebroek et al. [37] presented HiTEC, a novel computational (cognitive) model that allows for direct interaction between perception and action as well as for cognitive control, demonstrated by task-related attentional influences. In their model, the notion of affordance is effectively realized by allowing for automatic translation of perceptual object features (e.g., object shape) to action by means of overlap with anticipated action effect features (e.g., hand shape). Reference [39] proposed a Simulation

Theory and neuroscience findings on Mirror-Neuron Systems as the basis for a novel computational model, as a way to handle affective facial expressions. The model is based on a probabilistic mapping of observations from multiple identities onto a single fixed identity ('internal transcoding of external stimuli'), and then onto a latent space ('phenomenological response'). Asprino et al. [41] presented in this paper an Ontology Design Pattern for the definition of situation-driven behavior selection and arbitration models for cognitive agents. The proposed pattern relies on the descriptions and situations ontology pattern, combined with a frame-based representation scheme. Inspired by the affordance theory and behavior-based robotics principles, their reference model enables the definition of weighted relationships, or affordances, between situations (representing agent's perception of the environmental and social context) and agent's functional and behavioral abilities. These weighted links serve as a basis for supporting runtime task selection and arbitration policies, to dynamically and contextually select agent's behavior.

Lastly, a different use of a cognitive industrial entity called context-aware cloud robotics (CACR) is used for advanced material handling. Compared with the one-time on-demand delivery, CACR is characterized by two features: (1) context-aware services and (2) effective load balancing. The CACR case study is performed to highlight its energy-efficient and cost-saving material handling capabilities.

3.3.2. Concepts for the Empathy Area

Empathy is becoming an important field of social robotics and several behavioral models take this aspect into consideration.

Reference [31] showed how different models based on emotions were created to build empathetic and emotional robots. The main cues used in these models are movements, gestures, and postures. In another paper, the same authors explored different dimensions of artificial empathy and revealed different empathy models: a conceptual model of artificial empathy that was structured on the developmental axis of self-other cognition, statistical models based on battery level or temperature, and a four-dimension empathy model were presented and described. The cues used in this article were unimodal and multimodal communication cues as opposed to the previous one that used movements.

Reference [80] discussed a conceptual model of artificial empathy with respect to several existing studies. This model is based on affective developmental robotics, which provide more authentic artificial empathy based on the concept of cognitive developmental robotics. The authors showed how the model worked using two different robots: an emotional communication robot called WAMOEBA and a humanoid robot called WE.

3.3.3. Concepts for Behavioral Adaptation Area

Designing an intelligent agent is a difficult task because the designer must see the problem from the agent's viewpoint, considering all its sensors, actuators, and computation systems. Farahmand et al. [38] introduced a bio-inspired hybridization of reinforcement learning, cooperative co-evolution, and a cultural-inspired memetic algorithm for the automatic development of behavior-based agents. Reinforcement learning is responsible for the individual-level adaptation. Cooperative co-evolution performs at the population level and provides basic decision-making modules for the reinforcement-learning procedure. The culture-based memetic algorithm, which is a new computational interpretation of the meme metaphor, increases the lifetime performance of agents by sharing learning experiences between all agents in the society. To accelerate the learning process, the authors introduced a cultural-based method based on their new interpretation of the meme metaphor. Their proposed memetic algorithm is a mechanism for sharing learned structures among agents in society and lifetime performance of the agent, which is quite important for real-world applications, increases considerably when the memetic algorithm is in action.

3.4. Experimental Works on the Development and Implementation of the Behavioral Model

In this section, published works on behavioral models with the experimental loop are shown and are divided into sub-categories, according to the application area (Table 3).

3.4.1. Experimental Works for Cognitive Architectures

Concerning cognitive architectures, Reference [53] proposed a cognitive framework inspired by the human limbic system to improve HRI between humanoid robots and children during a game session. The robot's emotional activity was modelled with computational modules representing amygdala, hippocampus, hypothalamus, and basal ganglia and used to suggest users' optimal game actions. The results showed that this cognitive architecture provided an efficient mechanism for representing cognitive activity in humanoid robots. The children's attention level was higher when compared to those of a game session without the use of the robot.

Reference [59] aimed to use an Interactive Social Engagement Architecture (ISEA) and an interactive user interface to gather information from children. The authors tested the developed architecture with an NAO robot and two other humanoids with 186 children. The ISEA is able to integrate and combine human behavior models, behavior-based robotics, cognitive architectures, and expert user input to improve social HRI. Eight modules compose the framework presented: knowledge, user input, sensor processing, perceptual, memory, behavior generation, behavior arbitration, and behavior execution modules. The knowledge module models human behaviors, while the perceptual module manages external sensor data from the environment, and processes and interprets data, sending results to the memory module. The behavioral generation module calculates which behavior and communication strategies must be used and sends data to the behavioral generation module. Novel emergent behaviors can be obtained by combining newly generated behaviors with the stored behaviors in memory modules. Every time that behavior is displayed, the robot's internal state is updated to keep track of the new data storage. Preliminary results showed that children seemed to find it more comfortable to establish an engagement with a robot, rather than with humans, in sharing information about their bullying experiences at school. Although this research is only midway through the grant award period, the developments and results are promising. Moreover, the authors said that slow and steady progress is occurring with the development of this Integrated Robotic Toolkit, but there is still significant and ongoing work to be explored with this approach.

Reference [52] proposed an intention understanding system that consists of perception and action modules. It is an object-augmented model, composed of two neural network models able to integrate perception and action information to allow the robot to better predict the user's intention. The model was tested in a cafeteria with customers and clerks. The action module was able to understand the human intention and associate a meaning to predict an object related to that action. The combination of these modules resulted in an improved human intention detection.

As explained in the theory section of the cognitive area, affordances are important elements for building a behavioral model for social robots. Those ones encode relationships between actions, objects, and effects and play an important role in basic cognitive capabilities such as prediction and planning [62], which also developed a computational framework based on the Dempster-Shafer (DS) theory for inferring cognitive affordances. They explained that this, much richer level of affordance representation is needed to allow artificial agents to be adaptable to novel open-world scenarios. Reference [63] also underlined the fact that affordances play an important role on basic cognitive capabilities such as prediction and planning. The authors said that the problem of learning affordances is a key step toward understanding the world properties and developing social skills.

Reference [69] also proposed a model that has collaborative cognitive skills such as geometric reasoning and situation assessment based on perspective-taking and affordance analysis. Another important element to be taken into consideration in the implementation of a behavioral model are facial expressions. Those ones are often based on an inner model that is related to the emotional state and are not only based on categorical choice. Chumkamon et al. [64] proposed a framework that focuses on three main topics including the relation between facial expressions and emotions. The first point of their model is the organization of the behavior including inside-state emotion regarding the consciousness-based architecture. The second one presents a method whereby the robot can have empathy toward its human user's expressions of emotion. The last point shows the method that enables the robot to select a facial expression in response to the human user, which provides instant human-like 'emotion' and is based on emotional intelligence (EI) that uses a biologically inspired topological online method to express, for example, encouragement or being delighted. Another application of facial expressions in a cognitive architecture is shown in Reference [39] and in Reference [75]. Reference [55] proposed a robotic system that could learn online to recognize facial expressions without having a teaching signal associated with a facial expression. Reference [74] also created a system composed of three robots that helped elderly people during their daily works such as reminding them of taking drugs or bringing them the objects that they desired and analyzing their facial expressions to recognize them.

Lastly, learning from demonstration is used in Reference [71]. The authors proposed a learning method for collaborative and assistive robots based on movement primitives. The method allows for both action recognition and human-robot movement coordination.

3.4.2. Experimental Works on Empathy

When a social robot interacts with human users, empathy represents one of the key factors to increase natural HRI. Emotional models are fundamental for social abilities to reach empathy with users.

Reference [81], for example, evaluated and compared the emotion recognition algorithm in two different robots (NAO and Pepper) and created metrics to evaluate the empathy of these social robots.

Reference [44] developed emotion-based assistive behavior to be implemented in social assistive robots. According to the user's state, the model is able to provide abilities to the robot to show appropriate emotions, which elicits suitable actions in humans. The robot's environmental and internal information plus user affective state represent the inputs for the Brian robot (Brownsell, Alex (29 May 2013). *"Confused.com overhauls brand in search of 'expert' positioning"*. Marketing Magazine. http://www.marketingmagazine.co.uk/article/1183890/confusedcom-overhauls-brand-search-expert-positioning. Retrieved July 2018) to alter its emotional state according to the well-being of a participant and to the assistant in executing tasks. In this work, the robot emotional module is employed not to provoke emotional feelings, but rather in terms of assistive tasks that the robot should perform to satisfy the user's well-being.

The experiments show the potential of integrating the proposed online updating Markov chain module into a socially assistive robot to obtain compliance from individuals to engage in activities. Using robotic behavior that focuses on the well-being of the person could be beneficial to the person's health. Moving to a fully encompassing target user group is needed to test the overall robot in its intended assistive applications.

Reference [43] implemented an experiment with the I-Cat robot (http://www.hitech-projects.com/icat/. Retrieved July 2018), which aims to provide a computer-based assistant that could persuade and guide elderly people to behave in a healthy way. Previous works demonstrated that combining the robot's empathy with the user's state contributed to a better appreciation of a personal assistant [82]. I-Cat features an emotional model that makes it able to smile and express sadness. Authors implemented natural cues such as understanding, listening, and looking, to perform different roles for the robot (educator, buddy, and motivator). The analysis was conducted by considering participants' personalities. The percentage of the total time that participants talked, laughed, and looked at the robot, and how many times the participants said "goodbye," as a sign of interpretation of the robot as a social entity. The aim of the work was to establish behaviors for an electronic personal assistant with a high level of dialogue, emotions, and social competencies. The findings showed that natural cues used by I-Cat provoked more empathy and social involvement with users. When non-social cues were used, users perceived the robot as less trustworthy and less persuasive, while avoiding its suggestions.

During experiments, the physical characters were found to be more trustworthy but less empathetic than the virtual character, which was not expected. This negative outcome on empathy might be due to specific constraints of the iCat: it makes a relatively high amount of noise when it moves, and the head and body movements may not be fluent enough. Another technical constraint was the (occasional) appearance of errors in the movements and speech, such as skipping choices of the multiple-choice questions. Furthermore, it may be that the three-character roles did not capture important advantages of a physical character that can act in the real environment. For instance, more positive outcomes might show up with a character that helps to attend to a medicine box with a specific location in the house, compared to a virtual character that is not a real actor in the house.

Reference [25] provided their contribution to social pervasive robotics by proposing an affective model for social robots, empathizing the concept of empathy. Behavioral adaptation according to users' needs and preferences resulted in preliminary tests that achieved a better social inclusion in a learning scenario. The first part of the model, called "Affective loop," was a module for the perception of humans, characterized by body-based emotion recognition that can recognize human emotions. According to the perception for human module's outputs, the internal state of the robot changed, which generates a complex emotional spectrum using a psycho-evolutionary theory of emotions. The user was able to visualize the robot's internal state and adjust some system parameters for the duration and intensity of each emotion. The user's interest in interaction was then monitored by the visual system: when it decreased, the robot changed its behavior to socially involve the user and selected its emotion according to the user's state. Affective behaviors were also adapted to the goal of interaction in a cooperative task between the robot and users.

Lastly, a comparison between two different cultures was made in Reference [30]. They made, in fact, a comparison between expression features of compassion, sympathy, and empathy in British English and Polish using emotion models that had sensory cues as inputs.

3.4.3. Experimental Works on Behavioral Adaptation

An attempt to develop robots to be emotive and sociable like humans, showing a capability to adapt behavior in a social manner, is presented in Reference [58]. Starting from the Meyer-Briggs Theory on human personality, the authors mapped human psychological traits to develop an artificial emotional intelligence controller for the NAO robot. The proposed model was modelled as a biological system, and as a structure of emotionally driven and social behavior represented by three fuzzy logic blocks. Three variables were used as system input: "trigger event" that incites different psychological reactions, "behavior profiler" that models event-driven behavior to fit profiles of individuals whose behavior needs to be modelled, and "behavior booster/inhibitor" that augments or decreases the affective expressiveness. Social behavior attributes were implemented in the NAO robot controller according to this model. The robot interacted with young researchers, recognizing calls and gestures, and locating people in the environment, and showing personality traits of joy, sociability, and temperament. The model considers personality traits, social factors, and external/internal stimuli as human psychology does when interacting with others. In Reference [76], Nao was also used to assist children in developing self-regulated learning (SRL) skills. Combining the knowledge about personality traits discovered with Meyer-Briggs Theory and validated by Reference [58] and experimental measurements of affective reactions from a live model performed by an actor, Reference [61] developed a cognitive model of human psychological behavior. This model includes personality types and human temperaments to be implemented into the Robothespian humanoid robot. The authors tuned the block scheme developed in Reference [58] according to measurements from an actor performing as a behavioral live model. Different affective behaviors were played to create affective reactions to be added to the previous model.

Studies on proxemics, speed, and velocity provided unique suggestions to improve HRI, especially in behavior adaptation according to the user's movements and position. In Reference [56], the authors investigated a robot's trajectories and speed when it follows a user in a real domestic environment to provide a comfortable social interactive behavior. The authors presented a framework for people detection, state estimation, and trajectory generation that can regulate robotic behavior. To select the appropriate behavior, the robot used the state of the user and his/her localization as input, considering movements and the context. Trajectories and velocity were considered in Reference [57], with a robot moving with a social partner toward the same goal. The authors developed and tested a person-aware navigation system modifying a trajectory planner. The criterion to change the planner was the distance between the robot and the user, according to which the robot's behavior adapted its velocity and trajectory to reach the goal, but remained close to the user at the same time. The approach described in this paper is limited because it only considers distance to the goal while ignoring the available free space. This model could be augmented to consider free-space features, such as free space in front of each social agent, distances to walls, and distances to other obstacles, to be more informed.

A similar work is presented by Reference [51] with a model to interpret the user's behavior and inclination toward interaction with an assistant robot. The robot was able to determine the user's behavior through body movements and extraction of posture features. According to its interpretation, the robot decided if it should move closer or should wait for a better inclination from the user to interact. The major benefit of this model is that it does not use verbal instruction from the user, which allows the robot to assess the suitability of starting a conversation by using posture and movement analysis.

Behavioral adaptation according to users' preferences and feedback on robot's actions is presented in Reference [46]. Two learning algorithms were applied to an internally developed adaptive robot, known as the EMOX (EMOtioneXchange) robot. After having identified the user's profile, the robot proposed a personalized activity, while assisting and interacting with the user after the activity selection. The user's feedback after each activity was traced, letting the robot have a memory about the user's preferences to aid in suggesting a more appreciated activity later. The robot's architecture has observations of user behavior, feedback, and environment to use as input. The robot's actions are the system output, which are determined through knowledge rules as interaction traces, users' profiles, decision process, and learning from the feedback process. The results showed that, even if the interaction modality, with hand gestures, was found difficult, most participants found the robot behavior adaptable and pertinent to their preferences.

Reference [49] presented a novel control architecture for the internally developed Brian 2.0 robot. The aim was to adapt the robot's behaviors according to the user state, which is a social motivator and assists if needed. To be effectively integrated into society, robots should be provided with social intelligence to interact with humans. This architecture promoted the robot's abilities to support and motivate users during a game memory session to stimulate humans cognitively. Encouragement and assistance were provided through a modular learning architecture that determined the user's state and performances, which modified the robot's behavior according to these inputs, recorded through sensors, cameras, and modules. The combination of the robot's emotional state module and intelligence layer led establishment of the current robot's assistive action related to the user's state and adapts the robot's behavior to the interactive scenario, using non-verbal modalities of communication.

Reference [24] investigated a robot's behavior by proposing a model that adapted to the visitor's intention. In a shopping mall, a humanoid robot was tested during approaching and interaction tasks. The robot was provided with two interaction strategies depending on users' behaviors: when visitors showed uncertain intentions, the "proactively waiting" strategy was used and the robot went toward them. The "collaboratively initiating" strategy, instead, was used when visitors' willingness to interact was seen and the robot started a conversation and moved closer to them. To reach a more natural context in interacting with robots, Reference [47] presented an experiment with a social robot learning to perform word-meaning associations. The authors hypothesized that a different human attitude in approaching the robot could be obtained. The robot's design had the aim to evoke a strong social response from humans. The social cues used influenced the tutoring of the human teacher and his behavior. An HRI interaction was measured through a language game, during which the learner assimilated a lexicon and associated meanings. Based on the teacher's feedback, the learner modified the word-meaning association. It could be considered as a sort of behavioral adaptation, applied in a different context that could improve the robot's social abilities. Through users' facial tracking, the robot was able to address participants during the interaction, which emphasized the social involvement. Additional multi-modal social cues (gaze and verbal statement) to express its learning preference was used by the robot, which modulates the interaction and positively influences it. Reference [48] developed a spatial relationship model that considers interpersonal distance, body orientations, emotional state, and movements. On the basis of these inputs, the robot decides how to proceed, which sets its voice and moves toward the user or not. As the robot comes close to the child, entering the "personal" distance zone, the current status of the user is re-evaluated to adapt better to the robot's actions. Children with cognitive disabilities interacted with the robot, executing free and structured game sessions. Robot tactile sensors led us to understand tangible interaction, as an expression of touch-interaction through physical contact with the children. Depending on the touch-contact typology, the robot was able to select an appropriate behavior using multimodal emotional expressions. The robot's behavior can be adapted depending on the user's emotion, seen as an emotional stimulus for the robot's cognitive architecture. Reference [50] proposed a cognitive-emotional interactive model for interactive and communication tasks between young users and a robot. During the interaction, the emotional robot acted its emotions using facial expression, movements, and gesture as a consequence of the user's emotion, according to the Hidden Markov Model. The use of this model allowed the robot to regulate emotions as humans do, which provides a better interaction. The model starts from the hypothesis that robots might know a human's cognitive process, in order to understand human's behaviors. To do that, an object-functional role perspective method allowed robots to understand humans' behaviors: objects are interpreted as object-functional roles and role interactions. An activity is interpreted as an integration of object role interactions, so the robot is able to predict and understand a human activity. Because this model is only involved in emotional intensity attenuation, the continuous prediction of spontaneous affect still needs to be improved in the future, and the authors are considering expanding the experimental sample size and seeking more effective evaluation approaches for affective computing. Reference [54] also proposed a model that used a child's affective states and adapted its affective and social behavior in response to the affective states of the child.

In Reference [83], the authors also try to adapt robots' behavior to human emotional intention and an information-driven multi-robot behavior adaptation mechanism is proposed for human–robot interaction (HRI). In the mechanism, the optimal policy of behavior is selected by information-driven fuzzy friend-Q learning (IDFFQ), and facial expression with identification information are used to understand human emotional intention. It aims to make robots become capable of understanding and adapting their behaviors to human emotional intention, in such a way that HRI runs smoothly. The importance of facial expressions for the implementation of social robots is shown in the other two works. Reference [67] created a model for object, facial, gesture, voice, and biometric recognition and Reference [68] used a Multi-channel Convolutional Neural Network (MCCNN) to extract emotions from facial expressions.

Affordances are also used to implement a behavioral model that can adapt to users' needs. Another important aspect related to behavioral models is the cultural adaptation of the robot. Reference [40] proposed a multi-robot behavior adaptation mechanism based on cooperative-neutral-competitive fuzzy Q learning for coordinating local communication atmospheres in human-robots interaction. The Fuzzy Q learning is an approach that fuses fuzzy logic with the discrete Q-learning method and the authors called communication atmospheres significant information were introduced for human-robot interactions. This approach was tested with people from different countries and with different backgrounds to overcome the problem of cultural adaptation. Reference [65] also proposed a robotic system that helps therapists in sessions of cognitive stimulation. Without taking into account aspects such as the patient's perception of the robot, or the impact of the cultural environment, the application of these systems may be doomed to failure. The authors showed pieces of evidence of how the cultural adaptation of the robots has been considered decisive in their success.

Inspired by infant development, Reference [66] proposed a three-staged developmental framework for an anthropomorphic robot manipulator. In the first stage, the robot is initialized with a basic reach-and-enclose-on-contact movement capability and discovers a set of behavior primitives by exploring its movement parameter space. In the next stage, the robot exercises the discovered behaviors on different objects and learns the caused effects. This effectively builds a library of affordances and associated predictors. In the third stage, the learned structures and predictors are used to bootstrap complex imitation and action learning with the help of a cooperative tutor. Reference [70] developed an innovative approach that allows one or more human operators to share control authority with a high-level behavior controller on the basis of previous work on operator-centric manipulation control at the level of affordances. In their work, the affordances of the object template can be requested from the Object Template Server (OTS) and can be executed so that the robot performs the required arm motions to achieve the manipulation task.

Lastly, Reinforcement Learning techniques were also used to create an HRI system (robot that assists the human operator) to perform a given task with minimum workload demands and optimizes the overall human–robot system performance [72].

4. Discussion

The aim of this work is to analyze the state of the art and, thus, to provide a list of hints regarding cognitive architectures, behavioral adaptation, and empathy. Future research efforts should lead to overcoming the limitation of the current state of the art, as summarized in Table 4.

Table 4. Challenges and opportunity.

Keywords	Barriers/Limitations	Challenges and Opportunities	Research Topics
Sensors Technology	Multimodal sensors [32]	A multisensory system should be implemented in the model of a robot to create an improved architecture	• Development of a multisensory system that could be used to detect different social cues at the same time (i.e., vocal, facial, and gaze cues).
	Reliable and usable sensor technology [32]	Sensors should be designed to be reliable and acceptable in a real-life situation to reduce the time-to-market	• Design new sensors that can be used for a long time by the robot without being damaged. • Design and test sensors to be resistant to possible impacts that the robot can have during its work. • Design sensors resistant to external agents (i.e., water)
Perception	Real-time learning [28,84]	Real-time learning should be developed to adapt the behavior of the robot, according to the changing needs of the user	• Develop a robot capable of adapting in real time to the changing needs of users. • Analyze "social cues" that the robot should have according to the person it is approaching (i.e., children, aged people).
	Emotional state transitions [44]	Research on the emotional state module should be done more deeply	• Investigate a larger variety of emotional states including the emotion transitions.

Table 4. *Cont.*

Keywords	Barriers/Limitations	Challenges and Opportunities	Research Topics
	Improving object detections [52]	Different approaches in the area of object detection should be investigated to obtain a strong model of the robot	• Development of real-time object detection and recognition using geometrical model or simple CNN, deep and sophisticated CNNs
	Learning from the user [46,71]	The robot should be able to learn from the user in order to accomplish complex tasks	• A model based on learning from demonstration methods could be very useful to let the robot achieve better and more complex skills.
	Affordances [41,62,63]	Affordances are important elements to be analyzed in the process of the implementation of a behavioral model for a cognitive robot	• Develop a behavioral model that includes action selection of cognitive agents, following the notion of affordance. • Create a network that covers some areas as a personal sphere (e.g., people information), life events (e.g., information about memories, scheduling, plans, etc.), environment sphere (e.g., information about rooms, furniture, objects, etc.), the health sphere (e.g., living patterns, health patterns, vital signs, etc.), and the emotional sphere (e.g., emotions, sentiments, opinions, etc.). • Investigate affordance strategies related to deformable objects.
Experimental	Experimental session [35,61,85]	The model should be implemented on a real robot and tested to evaluate the proposed artificial cognitive architecture in dynamical environments	• Test of the model of a robot in dynamics environments (i.e., outdoor and indoor, in crowded or not crowded places) and with people from different ages.
Architecture Design	Brain-inspired architecture [34]	Research in robot's behavioral model should be conceived with a multidisciplinary approach to be able to adapt to the user's needs	• Implement a multidisciplinary approach to create better models for social factors (i.e., engineering, neuroscience, and psychology).
	Modular and flexible architecture [19]	Robot should adapt to different context and different preferences, which could change over time. Therefore, the architecture of a robot should be modular and flexible.	• Develop modular and adaptable model of a robot to operate in different contexts (schools, hospitals, and industries). • Implement a cloud architecture to offload intensive tasks to the cloud, to access a vast amount of data, and to share knowledge and new skills.
	Behavioral consistency, predictability, and repeatability [48]	These requirements should be investigated to obtain a complex model	• Integrating those fields in the model of a robot could be obtained by creating an algorithm that represents the episodic memory. This could bring the creation of an artificial intelligent agent that can act more independently.

Table 4. *Cont.*

Keywords	Barriers/Limitations	Challenges and Opportunities	Research Topics
	Standardization	Having a high level of interoperability	• Investigate the system interoperability to create a bridge among the different components of the system.
Ethical, legal, and social	Ethical and social aspects [19]	Ethical implications also should be investigated when creating a new model for a robot	• Investigate the ethical and social implications of designing social robots with advanced HRI abilities.
	Legal aspect	No rules can be found in the legal field of social robotics	• Investigate the regulation for social robots to overcome the problem that social robots cannot operate in environments with people without a supervision of an operator.
	Cultural adaption [24,40,65]	The robot should be able to adjust parameters for different cultures	• Test new robots with different cultures to obtain a generic model for a robot. Many models work well when tested with people born in the country where they were developed.

It shows several areas that have to be analyzed in future works as sensors technology, perception, architecture design, and the presence/lack of an experimental phase.

Moreover, ethical, legal, and social aspects should be taken into consideration to build an efficient behavioral model for future robots.

4.1. Sensors Technology

A crucial aspect in HRI is how robots manage to understand intentions and emotions of the users using social cues (i.e., posture and body movements, facial expression, head and gaze orientation, and voice quality). Sensors play a fundamental role because they are used to detect these cues, which are then processed in the robot model. An issue related to sensors is the data acquisition that can face delays, so an effort for the future could be to design sensors that are reliable and usable in real-life situations. Moreover, the robot should have a multisensory to acquire different types of signals [32]. To achieve this goal, microphones, 2D and 3D vision sensors, thermal cameras, leap motion, Myo, and face-trackers could be collected to create a system that gives a complete sensor coverage to the robot. Each device could cover a different area. Microphones could be used for speech, Myo to collect IMU and EMG data, face-trackers to find the head pose, gaze, and FACS, vision sensors to acquire point cloud data, thermal cameras to detect objects in dark environments, and a leap motion to track and estimate the position of the hand.

4.2. Perception and Learning from the User

A second ability that should be deeply analyzed is the area of perception. The main problem of perception is to have a reliable real-time sensing and learning system, as expressed in Reference [84]. In this paper, the authors show that people's preferences and knowledge change over time and a good system should be capable of adapting in real time to these changes and should be able to learn from the user. To achieve the latter competence, advanced learning-based methods [86] should be used to satisfy user needs, while increasing the performance of the robot [87]. Additionally, future works should detect and handle the emotion transition since humans change their emotions steadily [44]. This topic is one of the main issues of HRI because the robot must have a real time data acquisition to handle the emotional transitions and this is quite difficult to gain due to delays during the data acquisition.

Another limitation of this area is the disconnection between perception and actions [52] and this problem should be overcome to have a reactive robot. Reference [28] underlined this concept, saying that changes in the user profile should be promptly detected to adapt behavior automatically.

An effort that could be interesting to study more deeply in the future is the possibility for the robot to achieve complex skills learned from the user (i.e., for cleaning a table). Reference [46] proposed that the robot should be able to change the internal model on the basis of what it learns from human beings, using the learning from the demonstration approach.

Moreover, it is worth underlining the importance to increase the number of non-verbal input parameters considered in the analysis, to make the robot more compliant and adaptable to the user's state and preferences [51].

4.3. Architecture Design

Concerning the architecture design, more modular and more flexible architectures should be created, in the future: robots should be able to, for example, autonomously react to an unplanned event and a complete risk analysis procedure should be performed in the design phase to correctly handle the unplanned situations [19,24].

Behavioral consistency, predictability, and repeatability should be investigated since they are fundamental requirements in the design of socially assistive robots in different contexts, such as for children with autism as underlined in Reference [48]. Addressing them requires an accurate case analysis, grounded on the current practice and on extensive experimentation. A possible approach could be the use of learning from demonstration to teach the robot some skills to achieve a better performance [87].

Moreover, a multidisciplinary approach should be pursued to design and develop reliable and acceptable behavioral models [34]. Psychology, biology, and physiology, among others, are areas of expertise, which should be part of the development process since they can help improve the HRI experience [88]. Innovative behavioral models for assistive robots could be developed by taking inspiration from the studies of human social models or from the study of a specific anatomical apparatus. In this manner, future research will not only consider the physical safety (of the robot and the human beings) but also psychological, anatomical, and social spheres of humans [48].

It is important to underline that the robot might be appropriate not only in the context for which it was originally conceived (i.e., private home, hospital, and residential facility) but also for people with different levels of residual abilities. In other words, the robot should be able to adapt to the variability and different cultural and social contexts [89].

Lastly, a model of a robot should have a cloud architecture to have the ability to offload intensive tasks to the cloud, to access a vast amount of data, to access shared knowledge, and not to lose information in case of connection problems.

4.4. Experimental Phase

The advantage to test robotic solutions with real users is absolutely remarkable, as underlined by the comparison between papers with or without the experimental sessions. Particularly, some works [33,35,61] remark on the importance of testing the proposed model as a fundamental step for future research. The main issue is that, in some scenarios, robots that completed a task during the simulation phase, do not succeed in the experimental phase. This is the reason why testing the behavior of the robot in a real environment is of extreme importance in order to find good parameters that work for the experimental phase. In addition, the architecture should be validated using physical robots that interact with users in dynamic environments (i.e., schools, industries, and hospitals).

Lastly, another limitation that can be found in several papers is the absence of a database (i.e., physical forces or emotional states of the user). A database could be useful to collect information achieved from the experimental phase and could be useful not only for the researchers that are working on the project, but also for other researchers that want to use the data for their own projects.

4.5. Ethical, Legal, and Social Aspects

Future research effort should include the ethical implications of designing robots that interact with people and the data that should be acquired and correctly stored to guarantee privacy [19].

Regarding legal and social aspects, Reference [24] underlined that the robot should be able to adjust its parameters for different cultures that have different needs, in order to be able to satisfy real user requests. Lastly, the birth of a regulation for social robots, like the one created for drones during recent years, could be an important step in order to have the possibility to use a robot in crowded environments [90].

5. Conclusions

This paper focuses on behavioral adaptation, cognitive architectures, and the establishment of empathy between social robots and users. The current state-of-the-art of existing systems used in this field is presented to identify the pros and cons of each work with the aim to provide recommendations for future improvements.

To establish a set of benchmarks to define an HRI similar to human-human interaction is an enormous challenge because of the complexity of non-verbal phenomena in social interactions. Its interpretation needs the support of psychological processes and neural mechanisms. The topic of the behavioral model is huge, and several factors contribute to making robots more accepted, perceived as friends, and empathetic with users. A common limitation of the works presented is that, often, authors focused on a particular aspect of HRI emphasizing a communication strategy or a particular behavior as a reaction to the user's action. Since it is not easy to include behavioral adaptation techniques, cognitive architectures, persuasive communication strategies, and empathy in a unique solution, researchers are often limited to organize experimental studies, which include only some of these factors. This, unfortunately, provides useful information only to a limited part of persuasive robotics. To maintain the importance of each contribution, it is fundamental to include, in a whole vision, all the suggestions provided by each work. Although many improvements remain to be accomplished, the already satisfying results from the authors have achieved an optimum starting point to develop a better solution using knowledge of human cognitive and psychological structures.

Author Contributions: O.N. and G.A. were responsible for the literature research and methodology definition and search strategies for synthesizing the information from the papers into text and tables. O.N., L.F., A.S., and G.M. collaborated on the discussion. O.N. and L.F. were responsible for the paper structure. They contributed to the methodology definition and search strategies and wrote the discussion. F.C. was the scientific supervisor, guarantor for the review, and contributed in methodology definition, paper writing, discussion, and conclusion. All authors were involved in paper screening and selection. All authors read, provided feedback, and approved the final manuscript.

Funding: Research supported by "SocIalROBOTics for active and healthy ageing" (SI-ROBOTICS) project founded by the Italian "Ministero dell'Istruzione, dell' Università e della Ricerca" under the framework "PON—Ricerca e Innovazione 2014–2020", Grant Agreement ARS01_01120.

Conflicts of Interest: The authors declare no conflict of interest.

References

1. Goodrich, M.A.; Schultz, A.C. Human-Robot Interaction: A Survey. *Found. Trends®. Hum. Comput. Interact.* **2008**, *1*, 203–275. [CrossRef]
2. Conti, D.; Trubia, G.; Buono, S.; Di Nuovo, S.; Di Nuovo, A. Evaluation of a robot-assisted therapy for children with autism and intellectual disability. In Proceedings of the Annual Conference towards Autonomous Robotic Systems, Bristol, UK, 21 July 2018.

3. Cavallo, F.; Aquilano, M.; Bonaccorsi, M.; Limosani, R.; Manzi, A.; Carrozza, M.C.; Dario, P. *Improving Domiciliary Robotic Services by Integrating the ASTRO Robot in an AmI Infrastructure*; Springer International Publishing: Cham, Switzerland, 2014; pp. 267–282.
4. Sancarlo, D.; D'Onofrio, G.; Oscar, J.; Ricciardi, F.; Casey, D.; Murphy, K.; Giuliani, F.; Greco, A. *MARIO Project: A Multicenter Survey About Companion Robot Acceptability in Caregivers of Patients with Dementia*; Springer International Publishing: Cham, Switzerland, 2017; pp. 311–336.
5. Loi, S.M.; Bennett, A.; Pearce, M.; Nguyen, K.; Lautenschlager, N.T.; Khosla, R.; Velakoulis, D. A pilot study exploring staff acceptability of a socially assistive robot in a residential care facility that accommodates people under 65 years old. *Int. Psychogeriatr.* **2018**, *30*, 1075–1080. [CrossRef] [PubMed]
6. Limosani, R.; Manzi, A.; Fiorini, L.; Cavallo, F.; Dario, P. Enabling Global Robot Navigation Based on a Cloud Robotics Approach. *Int. J. Soc. Robot.* **2016**, *8*, 371–380. [CrossRef]
7. Gerłowska, J.; Skrobas, U.; Grabowska-Aleksandrowicz, K.; Korchut, A.; Szklener, S.; Szczęśniak-Stańczyk, D., Tzovaras, D.; Rejdak, K. Assessment of perceived attractiveness, usability, and societal impact of a multimodal Robotic Assistant for aging patients with memory impairments. *Front. Neurol.* **2018**, *9*, 392. [CrossRef] [PubMed]
8. Reppou, S.; Karagiannis, G. *Progress in Automation, Robotics and Measuring Techniques*; Springer International Publishing: Cham, Switzerland, 2015; Volume 352, pp. 233–234.
9. Cesta, A.; Cortellessa, G.; Orlandini, A.; Tiberio, L. Long-Term Evaluation of a Telepresence Robot for the Elderly: Methodology and Ecological Case Study. *Int. J. Soc. Robot.* **2016**, *8*, 421–441. [CrossRef]
10. Fasola, J.; Mataric, M. A Socially Assistive Robot Exercise Coach for the Elderly. *J. Hum. Robot. Interact.* **2013**, *2*, 3–32. [CrossRef]
11. Fiorini, L.; Esposito, R.; Bonaccorsi, M.; Petrazzuolo, C.; Saponara, F.; Giannantonio, R.; Petris, G.D.; Dario, P.; Cavallo, F. Enabling personalised medical support for chronic disease management through a hybrid robot-cloud approach. *Auton. Robot.* **2017**, *41*, 1263–1276. [CrossRef]
12. Burke, N.; Dautenhahn, K.; Saunders, J.; Koay, K.L.; Syrdal, D.S. "Teach Me–Show Me"—End-User Personalization of a Smart Home and Companion Robot. *IEEE Trans. Hum. Mach. Syst.* **2015**, *46*, 27–40.
13. Chance, G.; Jevtic, A.; Caleb-Solly, P.; Alenya, G.; Torras, C.; Dogramadzi, S. "Elbows Out" - Predictive Tracking of Partially Occluded Pose for Robot-Assisted Dressing. *IEEE –. Autom. Lett.* **2018**, *3*, 3598–3605. [CrossRef]
14. Woiceshyn, L.; Wang, Y.; Nejat, G. A Socially Assistive Robot to Help with Getting Dressed 2 Clothing Recommendation System. *IEEE Robot. Autom. Lett.* **2018**, *4*, 13–14.
15. Turchetti, G.; Micera, S.; Cavallo, F.; Odetti, L.; Dario, P. Technology and innovative services. *IEEE Pulse* **2011**, *2*, 27–35. [CrossRef] [PubMed]
16. García-Soler, Á.; Facal, D.; Díaz-Orueta, U.; Pigini, L.; Blasi, L.; Qiu, R. Inclusion of service robots in the daily lives of frail older users: A step-by-step definition procedure on users' requirements. *Arch. Gerontol. Geriatr.* **2018**, *74*, 191–196. [CrossRef] [PubMed]
17. Asada, M. Towards Artificial Empathy: How Can Artificial Empathy Follow the Developmental Pathway of Natural Empathy? *Int. J. Soc. Robot.* **2015**, *7*, 19–33. [CrossRef]
18. Cross, E.S.; Hortensius, R.; Wykowska, A. From social brains to social robots: Applying neurocognitive insights to human-robot interaction. *Philos. Trans. R. Soc. B Biol. Sci.* **2019**, *374*, 5–8. [CrossRef] [PubMed]
19. Chidambaram, V.; Chiang, Y.-H.; Mutlu, B. Designing Persuasive Robots: How Robots Might Persuade People Using Vocal and Nonverbal Cues. In Proceedings of the 7th Annual. ACM/IEEE International Conference Human-Robot Interaction. (HRI '12), New York, NY, USA, 5–8 March 2012; pp. 293–300.
20. Cavallo, F.; Semeraro, F.; Fiorini, L.; Magyar, G.; Sinčák, P.; Dario, P. Emotion Modelling for Social Robotics Applications: A Review. *J. Bionic Eng.* **2018**, *15*, 185–203. [CrossRef]
21. Hall, E.T. *The Hidden Dimension*; Doubleday: Garden City, NY, USA, 1966.
22. Wiltshire, T.J.; Smith, D.C.; Keebler, J.R. *Cybernetic Teams: Towards the Implementation of Team Heuristics in HRI*; Springer International Publishing: Berlin, Gemany, 2013; pp. 321–330.
23. Wiltshire, T.J.; Warta, S.F.; Barber, D.; Fiore, S.M. Enabling robotic social intelligence by engineering human social-cognitive mechanisms. *Cogn. Syst. Res.* **2017**, *43*, 190–207. [CrossRef]
24. Kato, Y.; Kanda, T.; Ishiguro, H. May I help you? In Proceedings of the 10th Annual ACM/IEEE International Conference Human-Robot Interaction (HRI 2015), New York, NY, USA, 3–5 March 2015; pp. 35–42.

25. Vircikova, M.; Magyar, G.; Sincak, P. The affective loop: A tool for autonomous and adaptive emotional human-robot interaction. *Adv. Intell. Syst. Comput.* **2015**, *345*, 247–254.
26. Lieto, A.; Bhatt, M.; Oltramari, A.; Vernon, D. The role of cognitive architectures in general artificial intelligence. *Cogn. Syst. Res.* **2018**, *48*, 1–3. [CrossRef]
27. Silva, F.; Correia, L.; Christensen, A.L. Evolutionary online behaviour learning and adaptation in real robots. *R. Soc. Open Sci.* **2017**, *4*, 160938. [CrossRef]
28. Tapus, A.; Aly, A. User adaptable robot behavior. In Proceedings of the 2011 International Conference on Collaboration Technologies and Systems (CTS 2011), Philadelphia, PA, USA, 23–27 May 2011; pp. 165–167.
29. Cuff, B.M.P.; Brown, S.J.; Taylor, L.; Howat, D.J. Empathy: A review of the concept. *Emot. Rev.* **2014**, *8*, 144–153. [CrossRef]
30. Lewandowska-Tomaszczyk, B.; Wilson, P.A. Compassion, empathy and sympathy expression features in affective robotics. In Proceedings of the 2016 7th IEEE International Conference on Cognitive Infocommunications (CogInfoCom), Wroclaw, Poland, 16–18 October 2017; pp. 65–70.
31. Damiano, L.; Dumouchel, P.; Lehmann, H. Artificial Empathy: An Interdisciplinary Investigation. *Int. J. Soc. Robot.* **2015**, *7*, 3–5. [CrossRef]
32. Sonntag, D. Persuasive AI Technologies for Healthcare Systems. In Proceedings of the 2016 AAAI Fall Symposium Series, The Westin Arlington Gateway, Arlington, Virginia, 17–19 November 2016; pp. 165–168.
33. Raymundo, C.R.; Johnson, C.G.; Vargas, P.A. An architecture for emotional and context-aware associative learning for robot companions. In Proceedings of the 2015 24th IEEE International Symposium on Robot and Human Interactive Communication (RO-MAN), Kobe, Japan, 31 August–4 September 2015; pp. 31–36.
34. Franchi, A.M.; Mutti, F.; Gini, G. From learning to new goal generation in a bioinspired robotic setup. *Adv. Robot.* **2016**, *1864*, 795–805. [CrossRef]
35. Pieters, R.; Racca, M.; Veronese, A.; Kyrki, V. Human-aware interaction: A memory-inspired artificial cognitive architecture. *Cogn. Robte Archit.* **2017**, *1855*, 38–39.
36. Cutsuridis, V.; Taylor, J.G.; Asprino, L.; Nuzzolese, A.G.; Russo, A.; Gangemi, A.; Presutti, V.; Nolfi, S. A Cognitive Control Architecture for the Perception-Act1. *Cognit. Comput.* **2013**, *5*, 383–395. [CrossRef]
37. Haazebroek, P.; Van Dantzig, S.; Hommel, B. A computational model of perception and action for cognitive robotics. *Cogn. Process.* **2011**, *12*, 355–365. [CrossRef] [PubMed]
38. Farahmand, A.M.; Ahmadabadi, M.N.; Lucas, C.; Araabi, B.N. Interaction of culture-based learning and cooperative co-evolution and its application to automatic behavior-based system design. *IEEE Trans. Evol. Comput.* **2010**, *14*, 23–57. [CrossRef]
39. Vitale, J.; Williams, M.A.; Johnston, B.; Boccignone, G. Affective facial expression processing via simulation: A probabilistic model. *Biol. Inspired Cogn. Archit.* **2014**, *10*, 30–41. [CrossRef]
40. Chen, L.F.; Liu, Z.T.; Wu, M.; Dong, F.Y.; Yamazaki, Y.; Hirota, K. Multi-robot behavior adaptation to local and global communication atmosphere in humans-robots interaction. *J. Multimodal User Interfaces* **2014**, *8*, 289–303. [CrossRef]
41. Asprino, L.; Nuzzolese, A.G.; Russo, A.; Gangemi, A.; Presutti, V.; Nolfi, S. An ontology design pattern for supporting behaviour arbitration in cognitive agents. *Adv. Ontol. Des. Patterns* **2017**, *32*, 85–95.
42. Wan, J.; Tang, S.; Hua, Q.; Li, D.; Liu, C.; Lloret, J. Context-aware cloud robotics for material handling in cognitive industrial Internet of Things. *IEEE Internet Things J.* **2018**, *5*, 2272–2281. [CrossRef]
43. Looije, R.; Neerincx, M.A.; Cnossen, F. Persuasive robotic assistant for health self-management of older adults: Design and evaluation of social behaviors. *Int. J. Hum. Comput. Stud.* **2010**, *68*, 386–397. [CrossRef]
44. Ficocelli, M.; Terao, J.; Nejat, G. Promoting interactions between humans and robots using robotic emotional behavior. *IEEE Trans. Cybern.* **2015**, *46*, 2911–2923. [CrossRef] [PubMed]
45. Cao, H.L.; Van de Perre, G.; Kennedy, J.; Senft, E.; Esteban, P.G.; De Beir, A.; Simut, R.; Belpaeme, T.; Lefeber, D.; Vanderborght, B. A personalized and platform-independent behavior control system for social robots in therapy: Development and applications. *IEEE Trans. Cogn. Dev. Syst.* **2018**, *8920*, 1–13. [CrossRef]
46. Karami, A.B.; Sehaba, K.; Encelle, B. Adaptive artificial companions learning from users' feedback. *Adapt. Behav.* **2016**, *24*, 69–86. [CrossRef]
47. De Greeff, J.; Belpaeme, T.; Bongard, J. Why robots should be social: Enhancing machine learning through social human-robot interaction. *PLoS ONE* **2015**, *10*, e0138061. [CrossRef] [PubMed]

48. Garzotto, F.; Gelsomini, M.; Kinoe, Y. Puffy: A Mobile Inflatable Interactive Companion for Children with Neurodevelopmental Disorder. In Proceedings of the IFIP Conference on Human-Computer Interaction, Bombay, India, 25–29 September 2017; pp. 467–492.
49. Chan, J.; Nejat, G. Social intelligence for a robot engaging people in cognitive training activities. *Int. J. Adv. Robot. Syst.* **2012**, *9*, 113. [CrossRef]
50. Liu, X.; Xie, L.; Wang, Z. Empathizing with emotional robot based on cognition reappraisal. *China Commun.* **2017**, *14*, 100–113. [CrossRef]
51. Sirithunge, H.P.C.; Viraj, M.A.; Muthugala, J.; Buddhika, A.G.; Jayasekara, P.; Chandima, D.P. Interpretation of interaction demanding of a user based on nonverbal behavior in a domestic environment. In Proceedings of the 2017 IEEE International Conference on Fuzzy Systems (FUZZ-IEEE), Naples, Italy, 9–12 July 2017.
52. Kim, S.; Yu, Z.; Lee, M. Understanding human intention by connecting perception and action learning in artificial agents. *Neural Netw.* **2017**, *92*, 29–38. [CrossRef]
53. Dağlarlı, E.; Dağlarlı, S.F.; Günel, G.Ö.; Köse, H. Improving human-robot interaction based on joint attention. *Appl. Intell.* **2017**, *47*, 62–82. [CrossRef]
54. Boucenna, S.; Gaussier, P.; Andry, P.; Hafemeister, L. A Robot Learns the Facial Expressions Recognition and Face/Non-face Discrimination Through an Imitation Game. *Int. J. Soc. Robot.* **2014**, *6*, 633–652. [CrossRef]
55. Boucenna, S.; Gaussier, P.; Hafemeister, L. Development of first social referencing skills: Emotional interaction as a way to regulate robot behavior. *IEEE Trans. Auton. Ment. Dev.* **2014**, *6*, 42–55. [CrossRef]
56. Granata, C.; Bidaud, P. A framework for the design of person following behaviors for social mobile robots. In Proceedings of the 2012 IEEE/RSJ International Conference on Intelligent Robots and Systems, Vilamoura, Portugal, 7–12 October 2012; pp. 4652–4659.
57. Feil-Seifer, D.; Matarić, M. People-aware navigation for goal-oriented behavior involving a human partner. In Proceedings of the 2011 IEEE International Conference on Development and Learning (ICDL), Frankfurt am Main, Germany, 24–27 August 2011.
58. Rodić, A.D.; Jovanović, M.D. How to Make Robots Feel and Social as Humans Building attributes of artificial emotional intelligence with robots of human-like behavior. In Proceedings of the 6th IARIA International Conference on Advanced Cognitive Technologies and Applications, Venice, Italy, 25–29 May 2014; pp. 133–139.
59. Bethel, C.L.; Henkel, Z.; Eakin, D.K.; May, D.C.; Pilkinton, M. Moving toward an intelligent interactive social engagement framework for information gathering. In Proceedings of the 2017 IEEE 15th International Symposium on Applied Machine Intelligence and Informatics (SAMI), Herl'any, Slovakia, 26–28 January 2017; pp. 21–26.
60. Infantino, I.; Augello, A.; Maniscalto, U.; Pilato, G.; Vella, F.; Prestazioni, A.; Nazionale, C.; Malfa, V.U. La A Cognitive Architecture for Social Robots. In Proceedings of the 2018 IEEE 4th Internation Forum Research Technology Society India, Palermo, Italy, 10–13 September 2018.
61. Rodić, A.; Urukalo, D.; Vujović, M.; Spasojević, S.; Tomić, M.; Berns, K.; Al-Darraji, S.; Zafar, Z. Embodiment of human personality with EI-Robots by mapping behaviour traits from live-model. *Adv. Intell. Syst. Comput.* **2017**, *540*, 438–448.
62. Sarathy, V.; Scheutz, M. A Logic-Based Computational Framework for Inferring Cognitive Affordances. *IEEE Trans. Cogn. Dev. Syst.* **2018**, *10*, 26–43. [CrossRef]
63. Awaad, I.; Kraetzschmar, G.K.; Hertzberg, J. The Role of Functional Affordances in Socializing Robots. *Int. J. Soc. Robot.* **2015**, *7*, 421–438. [CrossRef]
64. Chumkamon, S.; Hayashi, E.; Koike, M. Intelligent emotion and behavior based on topological consciousness and adaptive resonance theory in a companion robot. *Biol. Inspired Cogn. Archit.* **2016**, *18*, 51–67. [CrossRef]
65. Mart, F. Practical aspects of deploying Robotherapy systems. In Proceedings of the ROBOT 2017: Third Iberian Robotics Conference, Sevilla, Spain, 22–24 November 2018.
66. Ugur, E.; Yukie, N.; Erol, S.; Erhan, O. Staged development of robot skills: Behavior formation, affordance learning and imitation with motionese. *IEEE Trans. Auton. Ment. Dev.* **2015**, *7*, 119–139. [CrossRef]
67. Masuyama, N.; Loo, C.K.; Seera, M. Personality affected robotic emotional model with associative memory for human-robot interaction. *Neurocomputing* **2018**, *272*, 213–225. [CrossRef]
68. Barros, P.; Jirak, D.; Weber, C.; Wermter, S. Multimodal emotional state recognition using sequence-dependent deep hierarchical features. *Neural Netw.* **2015**, *72*, 140–151. [CrossRef]

69. Lemaignan, S.; Warnier, M.; Sisbot, E.A.; Clodic, A.; Alami, R. Artificial cognition for social human–robot interaction: An implementation. *Artif. Intell.* **2017**, *247*, 45–69. [CrossRef]
70. Romay, A.; Kohlbrecher, S.; Stumpf, A.; von Stryk, O.; Maniatopoulos, S.; Kress-Gazit, H.; Schillinger, P.; Conner, D.C. Collaborative Autonomy between High-level Behaviors and Human Operators for Remote Manipulation Tasks using Different Humanoid Robots. *J. Field Robot.* **2017**, *34*, 333–358. [CrossRef]
71. Maeda, G.J.; Neumann, G.; Ewerton, M.; Lioutikov, R.; Kroemer, O.; Peters, J. Probabilistic movement primitives for coordination of multiple human–robot collaborative tasks. *Auton. Robot.* **2017**, *41*, 593–612. [CrossRef]
72. Modares, H.; Ranatunga, I.; Lewis, F.L.; Popa, D.O. Optimized Assistive Human-Robot Interaction Using Reinforcement Learning. *IEEE Trans. Cybern.* **2016**, *46*, 655–667. [CrossRef] [PubMed]
73. Alemi, M.; Meghdari, A.; Ghazisaedy, M. The Impact of Social Robotics on L2 Learners' Anxiety and Attitude in English Vocabulary Acquisition. *Int. J. Soc. Robot.* **2015**, *7*, 523–535. [CrossRef]
74. Di Nuovo, A.; Broz, F.; Wang, N.; Belpaeme, T.; Cangelosi, A.; Jones, R.; Esposito, R.; Cavallo, F.; Dario, P. The multi-modal interface of Robot-Era multi-robot services tailored for the elderly. *Intell. Serv. Robot.* **2018**, *11*, 109–126. [CrossRef]
75. Ghorbandaei Pour, A.; Taheri, A.; Alemi, M.; Meghdari, A. Human–Robot Facial Expression Reciprocal Interaction Platform: Case Studies on Children with Autism. *Int. J. Soc. Robot.* **2018**, *10*, 179–198. [CrossRef]
76. Jones, A.; Castellano, G. Adaptive Robotic Tutors that Support Self-Regulated Learning: A Longer-Term Investigation with Primary School Children. *Int. J. Soc. Robot.* **2018**, *10*, 357–370. [CrossRef]
77. Moulin-Frier, C.; Fischer, T.; Petit, M.; Pointeau, G.; Puigbo, J.Y.; Pattacini, U.; Low, S.C.; Camilleri, D.; Nguyen, P.; Hoffmann, M.; et al. DAC-h3: A Proactive Robot Cognitive Architecture to Acquire and Express Knowledge about the World and the Self. *IEEE Trans. Cogn. Dev. Syst.* **2018**, *10*, 1005–1022. [CrossRef]
78. Costa, S.; Lehmann, H.; Dautenhahn, K.; Robins, B.; Soares, F. Using a Humanoid Robot to Elicit Body Awareness and Appropriate Physical Interaction in Children with Autism. *Int. J. Soc. Robot.* **2015**, *7*, 265–278. [CrossRef]
79. Conti, D.; Di Nuovo, S.; Buono, S.; Di Nuovo, A. Robots in Education and Care of Children with Developmental Disabilities: A Study on Acceptance by Experienced and Future Professionals. *Int. J. Soc. Robot.* **2017**, *9*, 51–62. [CrossRef]
80. Horii, T.; Nagai, Y.; Asada, M.; Access, O. Imitation of human expressions based on emotion estimation by mental simulation. *J. Behav. Robot.* **2016**, *7*, 40–54. [CrossRef]
81. Bechade, L.; Dubuisson-Duplessis, G.; Pittaro, G.; Garcia, M.; Devillers, L. Towards metrics of evaluation of pepper robot as a social companion for the elderly. *Lect. Notes Electr. Eng.* **2019**, *510*, 89–101.
82. Liu, K.; Picard, R. Embedded empathy in continuous, interactive health assessment. In Proceedings of the CHI Workshop on HCI Challenges in Health Assessment, Portland, Oregon, 2–7 April 2005.
83. Chen, L.; Wu, M.; Zhou, M.; She, J.; Dong, F.; Hirota, K. Information-Driven Multirobot Behavior Adaptation to Emotional Intention in Human-Robot Interaction. *IEEE Trans. Cogn. Dev. Syst.* **2018**, *10*, 647–658. [CrossRef]
84. Admoni, H.; Series, B.S. Nonverbal Behavior Modeling for Socially Assistive Robots. In Proceedings of the 2014 AAAI Fall Symposium Series, Arlington, VA, USA, 24 September 2014.
85. Ham, J.; Cuijpers, R.H.; Cabibihan, J.J. Combining Robotic Persuasive Strategies: The Persuasive Power of a Storytelling Robot that Uses Gazing and Gestures. *Int. J. Soc. Robot.* **2015**, *7*, 479–487. [CrossRef]
86. Martins, G.S.; Santos, L.; Dias, J. BUM: Bayesian User Model for Distributed Learning of User Characteristics from Heterogeneous Information. *IEEE Trans. Cogn. Dev. Syst.* **2018**. [CrossRef]
87. Billard, A.; Siegwart, R. Robot learning from demonstration. *Robot. Auton. Syst.* **2004**, *47*, 65–67. [CrossRef]
88. Di Nuovo, A.; Jay, T. Development of numerical cognition in children and artificial systems: A review of the current knowledge and proposals for multi-disciplinary research. *Cogn. Comput. Syst.* **2019**, *1*, 2–11. [CrossRef]

89. Martins, G.S.; Santos, L.; Dias, J. User-Adaptive Interaction in Social Robots: A Survey Focusing on Non-physical Interaction. *Int. J. Soc. Robot.* **2019**, *11*, 185–205. [CrossRef]
90. Clarke, R. The regulation of civilian drones' impacts on behavioural privacy. *Comput. Law Secur. Rev.* **2014**, *30*, 286–305. [CrossRef]

robotics

MDPI

Article

Mechanical and Control Design of an Industrial Exoskeleton for Advanced Human Empowering in Heavy Parts Manipulation Tasks

Alessandro Mauri [1,2], Jacopo Lettori [1,3], Giovanni Fusi [4], Davide Fausti [4], Maurizio Mor [4], Francesco Braghin [2], Giovanni Legnani [1,3] and Loris Roveda [1,5,*]

1 Consiglio Nazionale delle Ricerche (CNR), Istituto di Sistemi e Tecnologie Industriali per il Manifatturiero Avanzato (STIIMA), 20133 Milano, Italy; alessandro7.mauri@mail.polimi.it (A.M.); j.lettori@studenti.unibs.it (J.L.); giovanni.legnani@unibs.it (G.L.)

2 Department of Mechanical Engineering, Politecnico di Milano, 20156 Milano, Italy; francesco.braghin@polimi.it

3 Department of Mechanical and Industrial Engineering, University of Brescia, 25121 Brescia, Italy

4 Polibrixia, 25123 Brescia, Italy; giovanni.fusi@polibrixia.it (G.F.); davide.fausti@polibrixia.it (D.F.); maurizio.mor@polibrixia.it (M.M.)

5 Istituto Dalle Molle di Studi sull'Intelligenza Artificiale (IDSIA), Scuola Universitaria Professionale della Svizzera Italiana (SUPSI), Università della Svizzera Italiana (USI), IDSIA-SUPSI, 6928 Manno, Switzerland

* Correspondence: loris.roveda@idsia.ch

Received: 4 May 2019; Accepted: 31 July 2019; Published: 2 August 2019

Abstract: Exoskeleton robots are a rising technology in industrial contexts to assist humans in onerous applications. Mechanical and control design solutions are intensively investigated to achieve a high performance human-robot collaboration (e.g., transparency, ergonomics, safety, etc.). However, the most of the investigated solutions involve high-cost hardware, complex design solutions and standard actuation. Moreover, state-of-the-art empowering controllers do not allow for online assistance regulation and do not embed advanced safety rules. In the presented work, an industrial exoskeleton with high payload ratio for lifting and transportation of heavy parts is proposed. A low-cost mechanical design solution is described, exploiting compliant actuation at the shoulder joint to increase safety in human-robot cooperation. A hierarchic model-based controller with embedded safety rules is then proposed (including the modeling of the compliant actuator) to actively assist the human while executing the task. An inner optimal controller is proposed for trajectory tracking, while an outer safety-based fuzzy logic controller is proposed to online deform the task trajectory on the basis of the human's intention of motion. A gain scheduler is also designed to calculate the inner optimal control gains on the basis of the performed trajectory. Simulations have been performed in order to validate the performance of the proposed device, showing promising results. The prototype is under realization.

Keywords: industrial exoskeleton design; industrial exoskeleton control; human-robot collaboration; optimal control; empowering fuzzy control

1. Introduction

Exoskeletons are one of the key technologies to assist humans in a wide range of applications, such as rehabilitation, daily activities and so forth [1–3]. In particular, the adoption of exoskeletons in industrial applications is nowadays a *hot-topic* [4], since their capabilities to assist humans executing onerous tasks [5,6].

1.1. Industrial Exoskeletons Design Solutions

Industrial exoskeletons can be classified as passive and active. Passive solutions are not provided by actuation, indeed, they use springs and/or dampers to store energy from human's motion and releasing it when required [7,8]. Passive exoskeletons are commonly supporting the human operator in order to relieve him/her from repetitive tasks, while improving ergonomics [9]. Considering upper limbs solutions, different devices are available on the market [10–13]. Such exoskeletons assist humans in specific tasks, e.g., in over-the-head tasks, supporting the arm to reduce muscular stress. The main advantages of these solutions are the reduced weight and size, do not requiring motors and batteries. However, the comfortable range of postures for the human worker is restricted to specific configurations (such as in the over-the-head tasks assistance), making passive exoskeleton typically tailored to specific applications. Proposed solutions also consider a limited payload, since active assistance cannot be generated. In addition, common solutions does not support the forearm, therefore not guaranteeing the elbow support in case, e.g., of heavy parts transportation.

Active exoskeletons [14] are instead provided by actuation, allowing to empower the human worker. Different solutions have been developed in order to face different tasks, adopting different kinematics and hardware solutions [15–17]. Many solutions are available on the market, facing different tasks and scenarios. The *Panasonic Corporation* has developed an active device, called *AWN03*, that supports the operator's back when lifting heavy loads [18] thanks to electric motors. Another active human's back support has been developed in Reference [19]. *AUTON* proposes a back-exoskeleton to support workers in the transportation of heavy parts [20]. *Muscle Suit* by *Innophys* is an active back-support for lifting tasks [21]. *HAL* exoskeleton developed by *Tsukaba University* [22] is a full body exoskeleton for medical and industrial applications. *Sarcos Robotics* proposes the full-body exoskeleton *Guardian XO* for heavy tasks execution [23]. The proposed state-of-the-art solutions are still high-cost, characterized by a complex design, having a limited payload ratio (i.e., the ratio between the exoskeleton payload and the weight of the device) and not involving compliant actuation. Compliant actuation, such as series elastic actuation and variable stiffness actuation [24–26] are increasingly implemented in human-robot interaction tasks [27–29]. A compliant actuation, in fact, can intrinsically increase the safety in human-robot collaboration at the hardware level, having a deformable structure. Some upper limbs exoskeletons exploiting compliant actuation can be found for medical/rehabilitation applications [30,31]. However, due to the higher target payloads and increased design and implementation complexity, available state-of-the-art active industrial upper limbs exoskeletons are not equipped with compliant actuation. The safety issues are therefore tackled only at the software side (i.e., saturating velocities/forces). In addition, low-cost solutions can only be found in rehabilitation/medical domain, where limited power is required [32,33].

1.2. Exoskeletons Control Solutions

Exoskeleton control is widely investigated in order to assist humans in different applications [34]. Many control approaches have been developed, integrating different sensors and control techniques. Brain-control schemes have been developed exploiting a electroencephalogram signals [35]. Surface electromyograpy measurements have been exploited in order to control the exoskeleton on the basis of the human's muscles activation [36,37], also exploiting variable impedance control [38]. Admittance force control has been also exploited in order to control the exoskeleton on the basis of the measured interaction between the human and the robot [39]. External devices and measurement systems have also been used to control the exoskeleton on the basis of muscular activation, such as the Myo armband [40] or IMU sensors [41]. Common state-of-the-art approaches, however, show difficulties in the estimation of the human intention, especially while manipulating (partially) unknown payloads. Moreover, common approaches does not allow to online regulate the assistance given to the human during the task on the basis of the human-robot interaction. In addition, safety is commonly tackled in the controller only as a pre-defined saturation on the control action. Considering the empowering scenario and the manipulation of a (heavy) part, safety-based rules modulating the assistance to the human on the basis of the current interaction state (i.e., velocity, interaction force and derivative of the interaction force) should be included in the controller.

1.3. Paper Contribution

The aim of this paper is to fill the gap above described in the industrial exoskeleton field. More in details, the paper proposes (i) the mechanical design of a low-cost exoskeleton (hardware costs <10,000 Euro) for industrial applications with (ii) high payload ration (>0.8), (iii) involving compliant actuation (to achieve intrinsic safety in human-robot interaction), together with (iv) the design of an empowering safety-based control framework.

A lifting and transportation task of a heavy component has been considered as an objective for the exoskeleton design specifications definition (case study: car bumper part with weight of 10 kg). On the basis of such task, the kinematics of the exoskeleton has been defined, together with the performance required by the exoskeleton—objective (i) and (ii). A series elastic actuator (SEA) has been designed for the shoulder joint to embed compliance into the device—objective (iii). The SEA has been designed exploiting a compliant transmission (i.e., a compliant belt) between the shoulder joint motor and the link side. The target belt compliance has been calculated in order to achieve a target equivalent shoulder joint compliance. On the basis of such specifications, components from the market (e.g., motors, elastic belt, etc.) has been selected to implement the designed solution.

The proposed empowering controller has been designed in order to actively assist the human during the task execution. Intrinsic safety rules have been embedded into the control design in order to modulate the assistance on the basis of the current interaction state (i.e., velocity, interaction force and derivative of the interaction force) —objective (iv). Furthermore, the controller has been designed to be robust to (partially) unknown payloads (i.e., the weight of the part). A hierarchic controller has been designed, composed by an inner optimal controller (for trajectory tracking purposes) and by an outer safety-based fuzzy logic controller (for human empowering purposes), online modulating the assistance. The inner model-based controller includes the compliant modeling of the shoulder joint. The outer controller (on the basis of the proposed membership functions) is capable to identify the intention of motion of the human, reacting consequently. Moreover, a gain scheduler has been designed in order to store inner optimal control gains as a function of the performed task trajectory (i.e., control gains are a function of the executed trajectory).

Simulations studies have been performed in order to validate the proposed approach, simulating different task scenarios. Simulation results are promising and the proposed methodology will be applied to the real exoskeleton for final evaluation. The proposed exoskeleton is under realization.

2. Task Specifications & Exoskeleton's Design Guidelines

2.1. Task Description

Considering the industrial context, the main purpose of the proposed exoskeleton is to relieve the human from lifting efforts of heavy parts, while actively assisting him/her during the task.

In order to design the proposed exoskeleton, a bumper lifting task in a car assembly line (Figure 1a) has been considered as the reference task. Such task implies a target payload for the upper limbs exoskeleton equal to $P_e = 10$ kg (i.e., each exoskeleton arm has to lift—in ideal conditions—$P_a = 5$ kg).

In the following, the task is analyzed in order to define the exoskeleton configurations assumed during the task execution. Two main configurations will be assumed by the human wearing the exoskeleton during reference task (Figure 1b):

- **Configuration** 1: The arm is extended along the operator's trunk and the elbow configuration is 90° flexed;
- **Configuration** 2: The shoulder is flexed of 90° and the upper limb is completely extended. This is the most critical configuration considering the required motor torques.

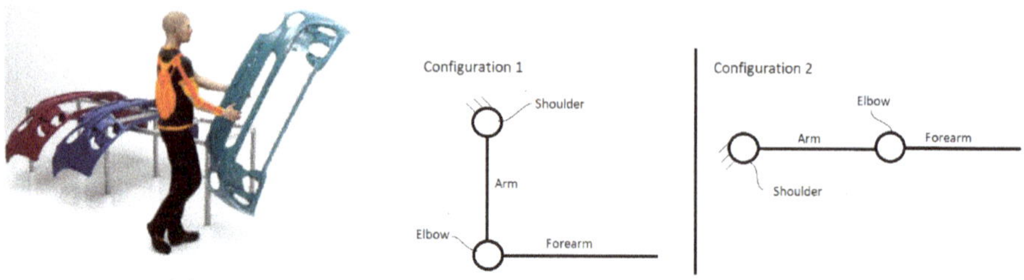

(a) (b)

Figure 1. (**a**) Reference task: car bumper lifting and transportation. (**b**) Human arm + exoskeleton configurations during task execution.

More in details, the task is composed of nine phases (Figure 2):

- *phase 0:* the operator configures the upper limbs in configuration 1 and the exoskeleton is turned on;
- *phase 1:* the operator moves the upper limbs without the external load, in order to grasp the object (critical configuration 2);
- *phase 2:* the operator grasps the object (critical configuration 2);
- *phase 3:* the operator moves back to configuration 1 carrying the part;
- *phase 4:* the operator transports the object maintaining upper limbs in configuration 1;
- *phase 5:* the operator lifts the external object up to the critical configuration 2 to release the part;
- *phase 6:* the operator releases the object staying in the critical configuration 2;
- *phase 7:* the operator moves back in configuration 1 without the part;
- *phase 8:* exoskeleton assistance is turned off (configuration 1).

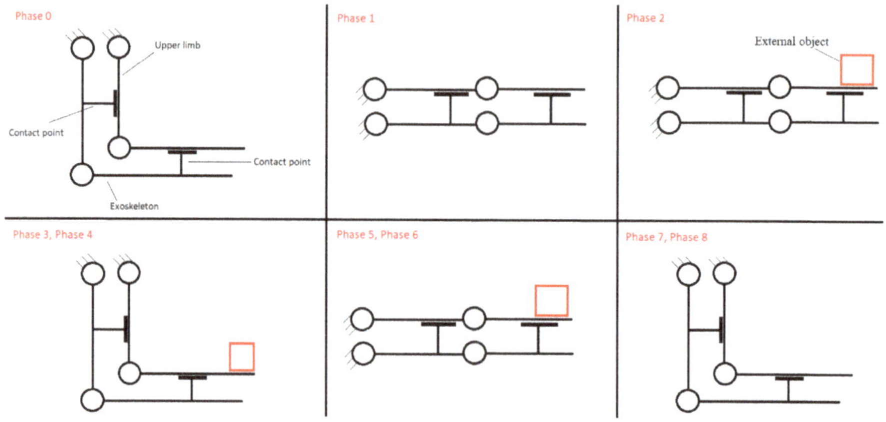

Figure 2. Lifting task phases.

2.2. Exoskeleton Specifications

In order to design the exoskeleton to face the proposed task, the following requirements have to be considered:

- kinematics;
- torque requirements;

- weight and size of the designed device;
- human safety;
- economic affordability.

2.2.1. Kinematics

Considering the proposed task and its phases, the kinematics shown in Figure 3 has been proposed for the exoskeleton. In particular, the proposed kinematics is composed by 2 degrees of freedom (DoFs), allowing to perform the complete sequence of task phases while minimizing the DoFs of the device and, therefore, its complexity (i.e., minimizing size, weight and cost of the solution). The *Denavit-Hartenberg parameters* of the proposed solution are summarized in Table 1.

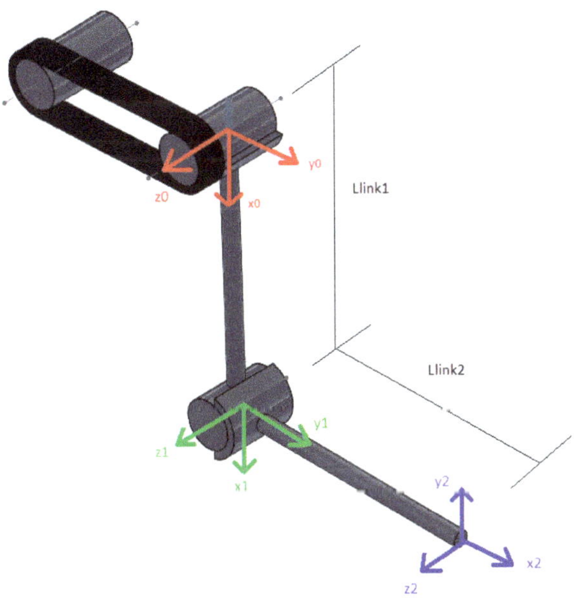

Figure 3. Kinematics of the proposed device.

Table 1. Denavit Hartenberg parameters of the proposed device.

θ	α	r	d
θ_1	0	L_{linka}	0
θ_2	0	L_{linkf}	0

2.2.2. Torque Requirements

In order to define the torque requirements associated to the reference task, the most critical configuration assumed by the exoskeleton has to be considered. Therefore, configuration 2 will be used in order to calculate the required motor torques. The applied motor torque has to compensate for the lifted load P_{load} and the human arm weight (composed by the human arm weight P_1 and by the human forearm and hand weight P_2), while applying the required assistance to the human. The exoskeleton will, therefore, apply two forces on the upper limbs due to arm support and forearm support, assisting the worker during the target task execution. These forces compensate the external torques applied by the load, therefore, having the human generating no torques.

Figure 4 shows the exoskeleton-upper limb model, highlighting all the acting forces/torques, where:

- R_a: interaction force between the forearm and the exoskeleton;
- R_b: interaction force between the arm and the exoskeleton;
- R_s: vertical reaction force of the shoulder;
- a_1: application point of the R_b force;
- a_2: application point of the R_a force;
- L_{arm}: human's arm length;
- L_{f+h}: human's forearm and hand;
- P_{link_2}: weight of the forearm exoskeleton link applied in the middle of the link;
- P_{motor_2}: weight of the elbow motor;
- P_{link_1}: weight of the arm link applied in the middle of the link;
- τ_B: torque of the elbow motor to support the joint;
- τ_{mot}: torque of the shoulder motor to support the joint;
- T_s: vertical reaction force of the support connecting the exoskeleton link to the back-plate.

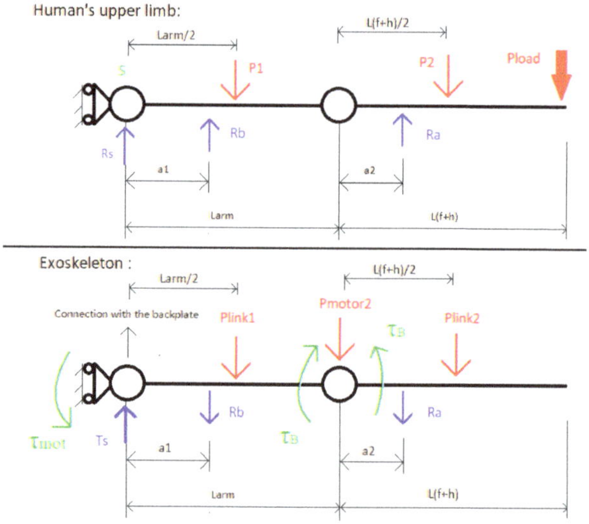

Figure 4. Loads acting on the exoskeleton in the critical configuration 2.

On the basis of Reference [42] (detailing the average height of European adults) and on the basis of Reference [43] (detailing the average weight of European adults), the following parameters have been chosen in order to define the human arm weight and length to be used in the above modeling:

- $Height = H = 180$ cm.
- $Weight = W = 80$ kg.

From the references above, it is therefore possible to calculate the modeling parameters from anthropometric tables [44]:

- $L_{arm} = 0.2898$ m.
- $L_{forearm+hand} = 0.4536$ m.
- $w_{arm} = 2.24$ kg ($P_1 = 22$ N).

- $w_{forearm+hand} = 1.76$ kg ($P_2 = 17.27$ N).

Assuming that (from a preliminary design) the weight of exoskeleton links is 0.5 kg (i.e., $P_{link_1} = P_{link_2} = 4.905$ N), the weight of the elbow motor is 1 kg ($P_{motor_2} = 9.81$ N), $a_1 = \frac{L_{arm}}{2}$ and $a_2 = L_{f+h}$, it is possible to calculate reaction forces and the maximum static torques to be applied by the shoulder and elbow motors considering configuration 2 (Table 2). On the basis of such torque requirements, motors can be selected.

Table 2. Reaction forces and motor torques calculated to compensate for human arm weight and component weight in the critical configuration 2.

R_a N	R_b N	T_s N	R_s N	τ_{mot} Nm	τ_B Nm
57.68	39.29	116.56	−8.63	54.7	27.3

2.2.3. Weight and Size of the Designed Device

Considering that the upper limbs exoskeleton has to be transported by the human worker, its weight and size have to be reduced as much as possible. Therefore, the selection of the components (e.g., motors) has to consider also such goal. Moreover, in order to limit the size of the designed device (in particular lateral dimension of the device) while ensuring the torque requirements, the shoulder motor has been placed on the back support of the exoskeleton. Mechanical design of the links, etc, also considers both weight and size requirements.

2.2.4. Human Safety

In order to improve the transparency and the safety of the device while collaborating with the human, compliance has been embedded into the mechanical structure of the exoskeleton. In particular, a compliant transmission has been designed to connect the shoulder motor to the shoulder link. Such transmission has to be capable to produce an equivalent shoulder stiffness $K_{eq} \in [150, 200]$ Nm/rad. Such equivalent stiffness is a medium-level value capable to give a degree of compliance to the exoskeleton while avoiding too high deformations.

2.2.5. Economic Affordability

Components from the market have to be selected both satisfying the previous design requirements and the costs constraints). In fact, one of the main goal of the proposed design is to have hardware costs <10,000 Euro. Therefore, a balance between design specifications and hardware costs has to be found.

3. Exoskeleton Modeling & Design

3.1. Compliant Shoulder Joint Actuation Modeling

As mentioned in Section 2.2, the proposed exoskeleton design includes a compliant actuation for the shoulder joint. Such actuation system is composed by a compliant belt as a transmission between the motor and the shoulder joint. The concept of the compliant belt actuator is shown in Figure 5.

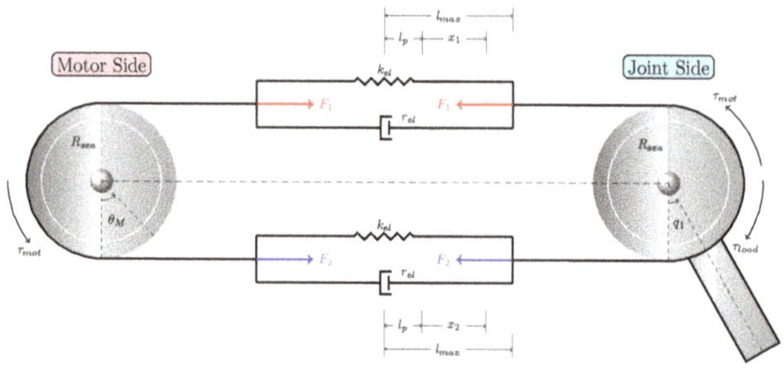

Figure 5. Mechanical model of the compliant belt shoulder joint actuation.

The aforementioned actuation system is modeled as the parallel of two mass-less spring-damper elements (with same stiffness k_{el} and damping r_{el} parameters) as in Reference [45]. Being the radius of the pulleys both equal to $R_{sea} = 0.04$ m, the deformation of each spring-damper element x_i is given by: $x_1 = -R_{sea}(q_1 - \theta_M)$ and $x_2 = R_{sea}(q_1 - \theta_M) = -x_1$, where q_1 is the link-side position and θ_M is the motor-side position as shown in Figure 6. In case of zero motor torque $\tau_{mot} = 0$, namely in the equilibrium position, the deformations are zero.

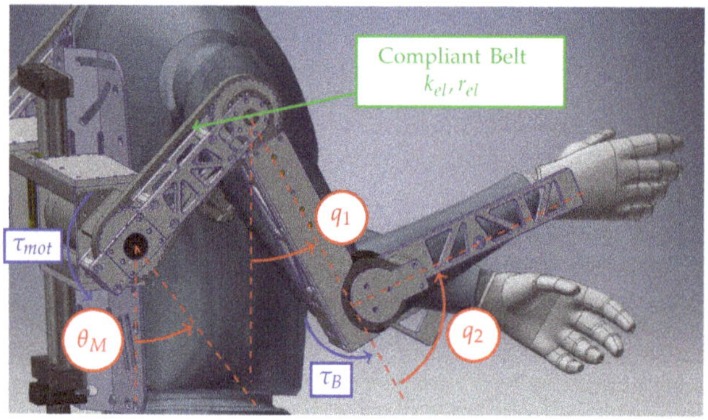

Figure 6. Elastic belt and joint torques are highlighted on the proposed exoskeleton model.

Overall torque transmission is given by the effect of pulling forces F_i on the two springs. Being l_{p_i} the pre-loaded lengths (equal for both springs), the forces are computed as $F_i = k_{el}(l_p + x_i) + r_{el}\dot{x}_i$. Therefore, the relation between motor torques and elastic belt deformations can be derived:

$$\tau_{mot} = -2R_{sea}^2 k_{el}(q_1 - \theta_M) - 2R_{sea}^2 r_{el}(\dot{q}_1 - \dot{\theta}_M). \tag{1}$$

3.2. Exoskeleton Dynamic Model

The overall mechanical system can be represented as a 3-DoFs system that moves in the sagittal plane. The device DoFs are represented by θ_M (the shoulder motor joint position), q_1 (the shoulder joint position) and q_2 (the exoskeleton elbow joint position). θ_M and q_1 are connected by the elastic belt, transmitting the torque by the relation (1). According to the *Euler-Lagrangian* formulation, the dynamics of the exoskeleton connected to the human arm can be expressed as follows:

$$\boldsymbol{M}(\boldsymbol{q})\ddot{\boldsymbol{q}} + \boldsymbol{C}(\boldsymbol{q},\dot{\boldsymbol{q}}) + \boldsymbol{G}(\boldsymbol{q}) + \boldsymbol{f}(\dot{\boldsymbol{q}}) + \boldsymbol{K}(\boldsymbol{q}) + \boldsymbol{D}(\dot{\boldsymbol{q}}) = \boldsymbol{\tau} - \boldsymbol{J}_e^T \boldsymbol{F}_e, \tag{2}$$

where:

- $q \in \boldsymbol{R}^{3x1}$ is the vector of the DoFs $[\theta_M, q_1, q_2]^T$;

- $M(q) \in R^{3x3}$ is the system inertia matrix;
- $C(q, \dot{q}) \in R^{3x1}$ is the Coriolis and centrifugal vector;
- $G(q) \in R^{3x1}$ is the gravitational vector;
- $f(q, \dot{q}) \in R^{3x1}$ is the vector of friction forces;
- $K(q) \in R^{3x1}$ is the system elasticity vector;
- $D(\dot{q}) \in R^{3x1}$ is the system damping vector;
- $\tau \in R^{3x1}$ is the vector of applied torques at the actuated joints;
- J_e^T is the transposed extended Jacobian matrix;
- F_e is the vector of external forces applied by the human and/or external load.

The formulation in (2) represents three coupled second-order differential equations that relate the joint positions, velocities and accelerations to the joint torques $\tau = [\tau_{mot},\ 0,\ \tau_B]^T$. In particular, τ_{mot} is the torque applied by the shoulder motor and τ_B is the torque applied by the elbow motor.

Combining the exoskeleton dynamics of the 2-DoFs manipulator with the compliant belt actuator dynamics in Section 3.1, it is possible to obtain the final overall 3-DoFs configuration. The resulting matrices comprehend terms coming both from the rigid mechanical system and the elastic coupling with the shoulder actuator. Angular velocities and positions of the shoulder joint q_1 and motor joint θ_M are coupled through the first two elements of the damping vector $D(\dot{q})$ and the stiffness vector $K(q)$. The torque transmission to the rigid system is explained in Section 3.1.

Figure 6 represents the CAD model of the exoskeleton and human arm, highlighting the joints (red label), the actuation torques (blue label) and the compliant belt (green label).

3.3. Compliant Shoulder Joint Design

In order to select the elastic belt stiffness k_{el}, a preliminary analysis have been performed considering the following stiffness values (peculiar for *off-the-shelves* elastic belts): $k_{el} = [25{,}000, 50{,}000, 75{,}000]$ N/m. Pre-load has also been varied from zero to the maximum pre-load force $F_{p_{max}} = k_{el}\frac{l_{max}}{2}$ (considering 5 values in such range). By increasing k_{el}, both the maximum transmittable torque and the equivalent shoulder joint stiffness increase (where the equivalent shoulder joint stiffness can be calculated as $K_{eq} = \frac{\partial \tau_{mot}}{\partial(q_1 - \theta_M)} = 2k_{el}R_{sea}^2$).

To select the elastic belt, a shoulder equivalent stiffness $K_{eq} \in [150, 200]$ Nm/rad has been imposed. Such equivalent stiffness is a medium-level value capable to give a degree of compliance to the exoskeleton while avoiding too high deformations. To satisfy the above requirement, k_{el} has been selected equal to 50,000 N/m. Figure 7a shows the stiffness characteristic curve of the selected *ELATECH® SIT Spa* belt and Figure 7b its equivalent shoulder joint stiffness. The stiffness characteristic variation is due to the pre-load applied to the belt.

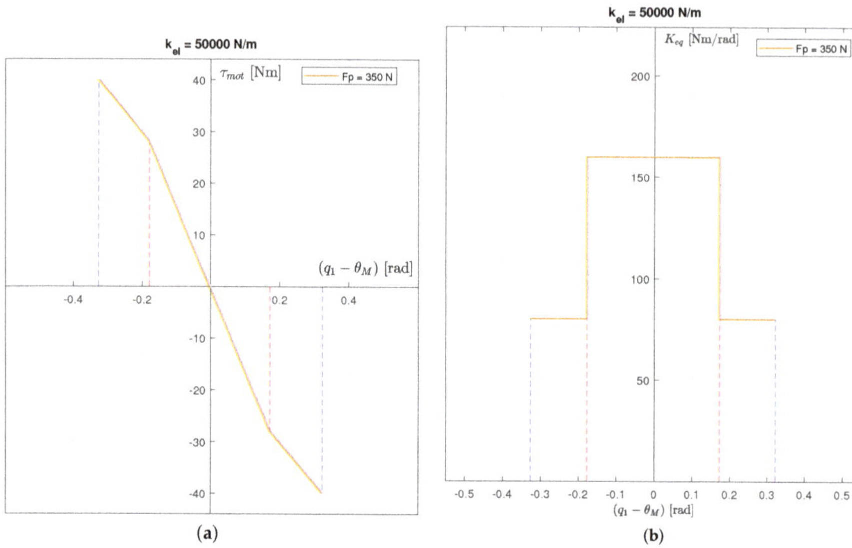

Figure 7. (**a**) Stiffness characteristic of the chosen belt. The stiffness characteristic variation is due to the pre-load applied to the belt. (**b**) Equivalent stiffness at the link-side shoulder joint.

3.4. Design Solution

The 2 DoFs kinematics has been selected, with the *Denavit-Hartenberg parameters* in Table 1.

In the presented section, the selected hardware components are detailed, satisfying the above define design specifications.

3.4.1. Brushless Motor Maxon EC60 FLAT/MILE/PM72

The Maxon motor *EC60 FLAT/MILE/PM72* (24 V) has been selected as the shoulder motor, including:

- speed reducer IMS PM 72 C Ø72 mm, 4 stage, ratio 305:1.
- EC60 flat, brushless, with Hall sensors.
- encoder Maxon MILE, 1-1024 pulses, 3 differential channels.
- nominal torque 56.30 Nm.

The proposed motor satisfy the required static torque within its nominal torque and allows to apply higher pick torques for control purposes. The weight of the motor is 1.5 kg.

3.4.2. Transmission

The elastic belt *ELATECH® SIT Spa* has been selected. The belt has a bielicoidal teeth with a progressive and continuous meshing in order to reduce vibrations and noises. 1 to 1 ratio (*EGLE*) pulleys have been selected.

3.4.3. Elbow Actuator

robolink®D High End robotic joint, size 20, symmetrical, provided by *Igus* has been selected as the elbow actuator. This motor is composed by a stepper motor (NEMA17/23/23XL) which drives a revolute joint (code: RL-D-20-101-38-01033). The weight of the motor is 0.9 kg.

3.4.4. Exoskeleton CAD

Figure 8 shows the proposed exoskeleton design. The thickness of the arm link is 70 mm and the elbow motor is placed inside the link. The shoulder motor is fixed to the back-plate through an aluminium support. The position of the support can be adjusted to increase the device wearability. The shoulder motor torque is transmitted to the human's shoulder through the compliant actuation. The link-side shoulder joint mounts an encoder in order to measure q_1 and, therefore, the angular deformation $q_1 - \theta_M$. On the basis of such measurement and on the basis of the elastic belt stiffness, the external torque applied by the human/external load can be estimated. Such estimate is used in the control loop for the active assistance definition. The shoulder pulley has to be aligned with the shoulder axis parallel to the frontal plane. In such a way, the flexion/extension degree of freedom is implemented. The elbow motor must be also aligned with the operator's elbow joint to guarantee the elbow rotation. The links length can be regulated in order to adapt the exoskeleton to the user. A range of ± 3 cm has been considered for the adaptability of the links length.

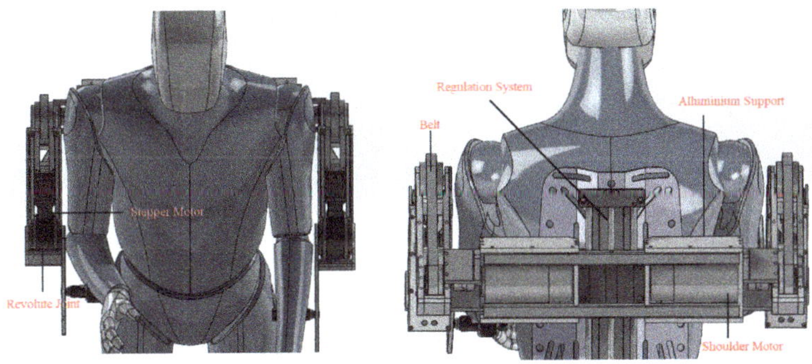

Figure 8. Exoskeleton prototype CAD model.

The total estimated weight of the designed exoskeleton is about 12 kg, not including batteries. The shoulder motors are the heaviest component of the device. On one hand, the weight of the device can be reduced selecting a higher-performance motor for the shoulder joint. However, this will result in higher costs. On the other hand, the exoskeleton payload can be reduced in order to require less performance from motors. Considering the target task detailed in Section 2.1 and considering the budget constraints, the proposed motor is the optimal solution that authors found available on the market in the design phase. It has to be remarked that the payload ratio of the proposed device is >0.8.

Remark 1. *The above mentioned motors have been selected with a nominal torque satisfying the specifications described in Section 2.2.2. The peak torque is even higher in order to compensate for modeling errors or human subject higher arm weight.*

Remark 2. *On the basis of the selected motors, the friction model in (2) implements a two parameters friction model for the shoulder and elbow joints* $f_{q_i} = \alpha_{1,i}\dot{q}_i + \alpha_{2,i}sign(\dot{q}_i)$ *(where i indicates the exoskeleton joint) with parameters* $\alpha_{1,i}$ *and* $\alpha_{2,i}$ *as in Reference [46] (similar motor as the one mounted on the UR10 manipulator). Such values have been used in simulation. Real friction values have to be identified on the real exoskeleton.*

4. Industrial Exoskeleton Control

4.1. Problem Formulation

The goal of the proposed controller is to assist the human in the lifting and transportation of heavy parts. To achieve such goal, the proposed control logic merges together the perks of the *optimal control* (to obtain stability and robustness) with the adaptability proposed by the *fuzzy logic*. A hierarchic controller has been designed, composed by an inner optimal control loop (to track a reference task trajectory) and by an outer fuzzy logic control loop (responsible for updating the commanded trajectory according to the detected intentions of motion of the human). The gain scheduling control allows to have online modification of the control gains of the inner optimal controller according to the commanded trajectory. The interaction torque τ_{int} between the human and the exoskeleton is calculated exploiting the encoders measurements at both the sides of the elastic belt actuation. τ_{int} is exploited by the *fuzzy* controller in order to identify the intention of motion of the human.

4.2. Optimal Control Design

The inner *optimal control* guarantees the tracking and stabilization of the system around the task reference trajectory. Let the state-space dynamic equations of a linear (or linearized) time-varying system be:

$$\begin{aligned} \dot{\boldsymbol{x}}(t) &= \boldsymbol{A}(t)\boldsymbol{x}(t) + \boldsymbol{B}(t)\boldsymbol{u}(t), \\ \boldsymbol{y}(t) &= \boldsymbol{C}(t)\boldsymbol{x}(t), \end{aligned} \tag{3}$$

where $\boldsymbol{x} \in \mathbb{R}^n$ is the state vector, $\boldsymbol{A}$ and $\boldsymbol{B}$ are the state and input matrix of the linearized system respectively, $\boldsymbol{u} \in \mathbb{R}^m$ is the control action vector and $\boldsymbol{y} \in \mathbb{R}^p$ is the output vector. According to the optimal control theory on *Linear Quadratic Regulator* ([47]), it is possible to define a quadratic cost function J as:

$$J = \frac{1}{2}\Delta\boldsymbol{x}^T(t_f)\boldsymbol{P}\Delta\boldsymbol{x}(t_f) + \frac{1}{2}\int_{t_0}^{t_f} \left(\Delta\boldsymbol{x}^T(t)\boldsymbol{Q}\Delta\boldsymbol{x}(t) + \Delta\boldsymbol{u}^T(t)\boldsymbol{R}\Delta\boldsymbol{u}(t)\right)\, dt, \tag{4}$$

where $\Delta x = \boldsymbol{q}_{ref} - \boldsymbol{q}$, $\boldsymbol{P}$ and $\boldsymbol{Q}$ are symmetric and positive semi-definite weight matrices and $\boldsymbol{R}$ is a symmetric and positive definite weight matrix. An optimal LQR controller can be designed in order to minimize this cost function, being the weight matrices previously defined, as shown in Figure 9.

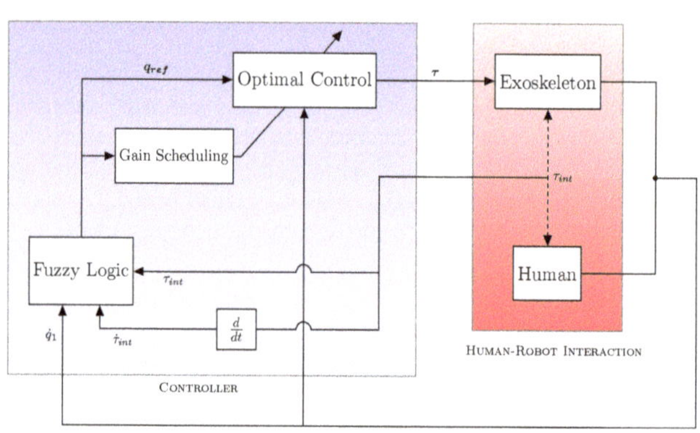

Figure 9. Overall control scheme showing the inner optimal controller, the outer fuzzy logic controller and the gain scheduler, highlighting the feedbacks to the control loops.

4.3. Gain Scheduling Control Design

A *gain scheduling LQR* approach has been used in order to have a performing and stabilizing controller for all the possible configurations of the system. This offline controller, in fact, modifies the control gain matrix on the basis of the desired pre-defined trajectory.

Basically, the nonlinear system outlined in Section 3 is quasi-linearized into several linear time-invariant (LTI) subsystems with respect to different operating points. Let the nonlinear state equation of the system be represented as:

$$\dot{x}(t) = \eta(x(t), u(t), t) = \left[\ddot{\theta}_M, \dot{\theta}_M, \ddot{q}_1, \dot{q}_1, \ddot{q}_2, \dot{q}_2\right]^T. \quad (5)$$

The Taylor series expansion using the appropriate operating points (x^*, u^*) can be obtained as $\dot{x} = Ax + Bu + E_{rr}(x^*, u^*, t)$, where:

$$A = \left.\frac{\partial \eta(x,u)}{\partial x}\right|_{\substack{x=x^* \\ u=u^*}} \quad \text{and} \quad B = \left.\frac{\partial \eta(x,u)}{\partial u}\right|_{\substack{x=x^* \\ u=u^*}}. \quad (6)$$

As demonstrated by Reference [48], the system approximation error term $E_{rr}(x^*, u^*, t)$ can be neglected, having the nonlinear system simply linearized as $\dot{x} = Ax + Bu$.

Due to the fact that the linearization around the operating points holds, the plant and the weight matrices are assumed to be time-invariant and, therefore, t_f in the time interval can be assumed to be equal to infinite.

If $P(t)$ does converge, $\dot{P} = 0$ can be assumed for $t \ll t_f$ and the resulting equation for the LQR problem is the so-called *algebraic Riccati equation* (ARE):

$$0 = -Q - PA - A^T P + PBR^{-1}B^T P. \quad (7)$$

Therefore , if P_∞ exists, the corresponding steady-state feedback gain matrix is given by $K_\infty = R^{-1}B^T P_\infty$. The resulting optimal control law is then obtained as: $\Delta u^*(t) = -K\Delta x(t)$.

The adopted strategy aims to apply the quasi-linearization approach to the several system configurations related to the typical arm lifting trajectory as detailed in Section 2.1. The feedback control gains are computed for each configuration of the exoskeleton. Such control gains are stored and, based on the reference set-point q_1^{ref}, online applied by the gain scheduler. The proposed strategy is schematized in Figure 10.

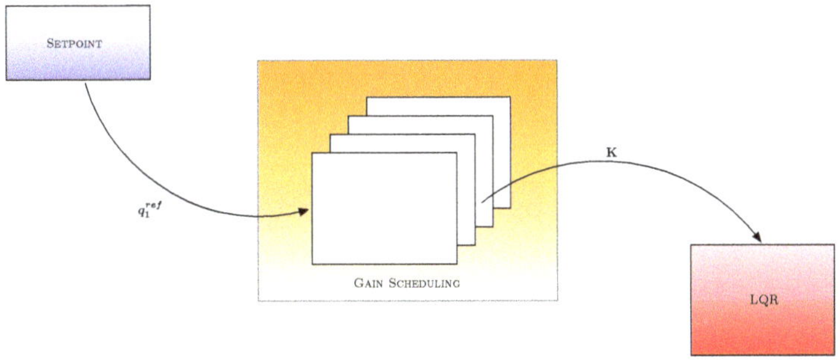

Figure 10. Gain scheduler online updating the control matrix to the inner optimal controller.

4.4. Empowering Fuzzy Controller Design

The main motivation to adopt a *fuzzy logic* for the outer human's intentions-based control is to deal with complex, ill-defined, uncertain and dynamic processes, which are intrinsically difficult to being modelled mathematically. In order to account for human's behaviour inside the control architecture while establishing a human-oriented input-output relations, a fuzzy table is built. The empowering fuzzy controller, therefore, is the highest level controller and it is responsible for the online modification of the reference trajectory, based on the human's intentions of motion.

The inputs of the developed fuzzy controller are the *interaction torque* τ_{int}, the *interaction torque derivative* $\dot{\tau}_{int}$ and *shoulder joint angular velocity* $\dot{q}_1$. The output is the *assistance level* A_L that is yielded to the controller in order to modify the reference set-point of the optimal control.

The strategy behind the choice of the *inputs* membership functions is to obtain an adequate map of the general motion intentions from the human operator, classifying whether the applied torque is intentional or not, thus recognizing if the shoulder joint is willing to move from the current configuration or not. At the same time, an appropriate definition of safety between human and exoskeleton is achieved. More in details, the speed of the shoulder and the interaction torque are monitored to avoid the excess of the specified threshold. Moreover, the embedded strategy allows to recognize whether the human's intention is willing to lift (*lifting assistance*) or lower (*lowering assistance*) the arm, accordingly defining the deformation of the optimal control trajectory.

4.4.1. Membership Functions

The states of the membership function characterizing the shoulder joint velocity are: *stop*, *slow*, *move* and *fast*, whose aim is to decompose the velocity range into different states, to know whether the shoulder angular variation is too fast, too slow, in the desired motion range or to understand if the human wants to stop the motion. The states of the membership function characterizing the interaction torque are: *N*, *S* and *NS*, which stand respectively for *no torque* (below a specified threshold the control is not activated), *safe* and *not safe*. Finally, the states of the membership function characterizing the interaction torque derivative are: *variation* (*V*) or *no variation* (*NV*), meaning if the operator wants to move the shoulder or change the motion by increasing or reducing the applied torque or not.

The output membership functions for the assistance level A_L range over four different fuzzy sets: *none*, *low*, *medium* and *high*.

4.4.2. Rule Base

The rule base for the current control strategy can be summed up with the following nine rules for the lifting assistance and eight rules for the lowering assistance (rule #8 is omitted for the lowering phase). The rule base adjusts the level of assistance depending on how much the operator is willing to move from the current arm position, respectively yielding null to high assistance to the shoulder motion:

$$
\begin{cases}
\#1 \quad \text{IF } \dot{q}_1 \text{ is } \mathit{stop} \text{ AND } \tau_{int} \text{ is } S \text{ AND } \dot{\tau}_{int} \text{ is } NV \text{ THEN } A_L \text{ is } \mathit{none} \\
\#2 \quad \text{IF } \dot{q}_1 \text{ is } \mathit{stop} \text{ AND } \tau_{int} \text{ is } S \text{ AND } \dot{\tau}_{int} \text{ is } V \text{ THEN } A_L \text{ is } \mathit{low} \\
\#3 \quad \text{IF } \dot{q}_1 \text{ is } \mathit{slow} \text{ AND } \tau_{int} \text{ is } S \text{ AND } \dot{\tau}_{int} \text{ is } NV \text{ THEN } A_L \text{ is } \mathit{low} \\
\#4 \quad \text{IF } \dot{q}_1 \text{ is } \mathit{slow} \text{ AND } \tau_{int} \text{ is } S \text{ AND } \dot{\tau}_{int} \text{ is } V \text{ THEN } A_L \text{ is } \mathit{medium} \\
\#5 \quad \text{IF } \dot{q}_1 \text{ is } \mathit{move} \text{ AND } \tau_{int} \text{ is } S \text{ AND } \dot{\tau}_{int} \text{ is } NV \text{ THEN } A_L \text{ is } \mathit{medium} \\
\#6 \quad \text{IF } \dot{q}_1 \text{ is } \mathit{move} \text{ AND } \tau_{int} \text{ is } S \text{ AND } \dot{\tau}_{int} \text{ is } V \text{ THEN } A_L \text{ is } \mathit{high} \\
\#7 \quad \text{IF } \dot{q}_1 \text{ is } \mathit{fast} \text{ THEN } A_L \text{ is } \mathit{none} \\
\#8 \quad \text{IF } \tau_{int} \text{ is } NS \text{ THEN } A_L \text{ is } \mathit{none} \\
\#9 \quad \text{IF } \tau_{int} \text{ is } N \text{ THEN } A_L \text{ is } \mathit{none}.
\end{cases}
$$

Figure 11 shows the process of fuzzy rule base for the lifting assistance developed in MATLAB®.

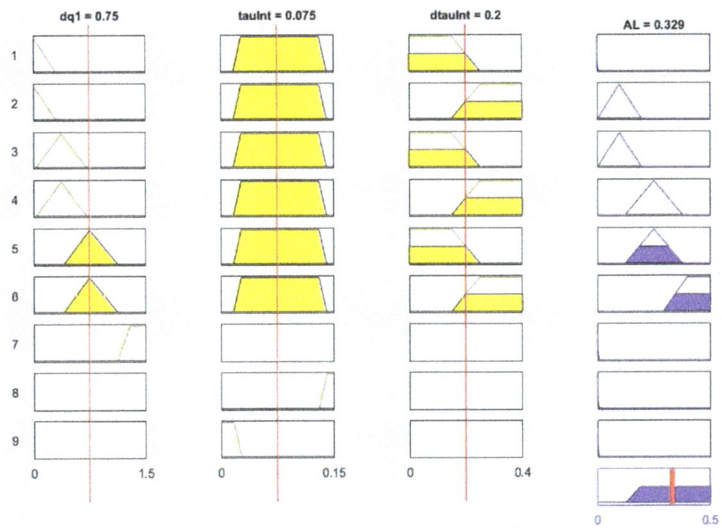

Figure 11. The nine fuzzy rules for the lifting assistance represented in MATLAB®.

The online trajectory generation to be provided to the inner optimal controller and gain scheduler is, therefore, deformed by the required assistance by the human:

$$q_1^{ref} = q_1 - A_L \text{sign}(\tau_{int}). \tag{8}$$

In particular, the shoulder reference angular position is computed from the stored value of q_1 that is updated with a frequency of 50 Hz, only if the value of assistance level is different from zero. The value $A_L\text{sign}(\tau_{int})$ allows to establish whether the reference set point needs to be decreased or increased with respect to the previous value, by considering the sign of the interaction torque.

Remark 3. *It has to be underlined that the only use of the force measurements cannot allow the fuzzy logic to identify an intention of motion of the human (the weight of the arm/payload affects this estimation). Including the velocity and the force derivative in the fuzzy controller can instead allow to understand if the human is intended to move the exoskeleton. In this case, in fact, it is possible to monitor the full interaction state between the human and the robot.*

Remark 4. *The proposed controller extends the work in Reference [49] including improved fuzzy membership functions and rules, together with the proposed lower level gain scheduling optimal controller.*

5. Simulation Validation

The effectiveness of the proposed control architecture has been validated in simulation. All the analyses have been carried out considering the nominal parameters of the system (Section 2.2), affected by uncertainties, in order to take into account modeling errors (e.g., unknown upper limb anthropometric characteristics, involuntary tremor, etc.).

The simulations are performed using MATLAB®/Simulink.

5.1. Empowering Human in Lifting Task

The proposed controller has been tested in simulation assisting the operator in a lifting task of a 10 kg part (i.e., 5 kg for each exoskeleton arm as defined in Section 2.1). In the proposed simulation the payload is grasped at time equal to 1 s. At time equal to 4.5 s a torque is applied by the human shoulder to interact with the exoskeleton. Three interaction torque levels are simulated in Figure 12: 8 Nm, 12 Nm and 16 Nm. Such three torque levels have been implemented in simulation in order to show the different activation of the fuzzy rules. The simulation considers an initial joint velocity equal to zero (i.e., the corresponding state of the velocity membership function is *stop*). All the applied torques make the derivative of the interaction torque resulting in the *variation* state V. Considering the applied torques, the first interaction torque level is not able to activate the fuzzy controller to assist the human (i.e., the corresponding state of the interaction torque membership function is *no torque* N). The second and third interaction torque levels, instead, activate the assistance (i.e., the the corresponding state of the interaction torque membership function is *safe* S), deforming the set-point to the optimal controller until the torque decreases to zero. The proposed controller is, therefore, capable to distinguish from required assistance, empowering the human operator.

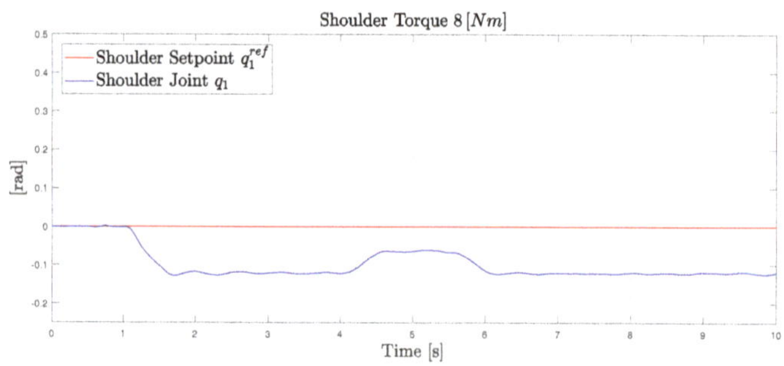

(a)

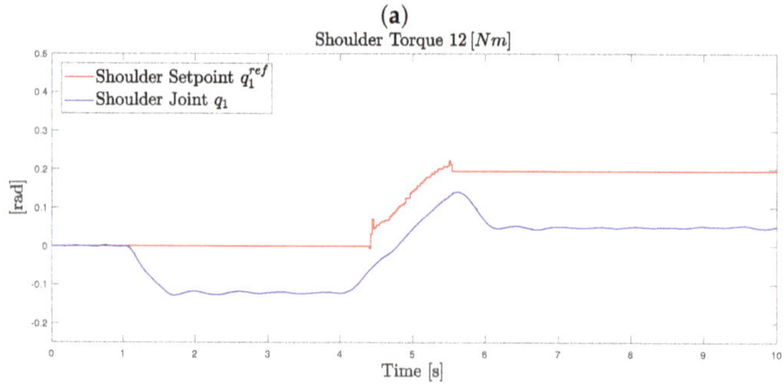

(b)

Figure 12. *Cont.*

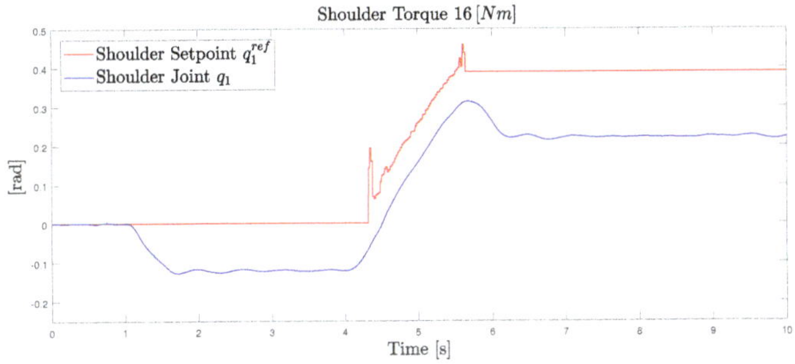

(c)

Figure 12. (**a**) The interaction torque level 8 Nm is applied while manipulating the 10 kg payload added (i.e., 5 kg for each exoskeleton arm as defined in Section 2.1). The applied interaction torque is not able to activate the fuzzy controller to deform the set-point to the optimal control to assist the operator. (**b**) and (**c**) are related respectively to the 12 Nm and 16 Nm interaction torque levels. The fuzzy controller is activated, empowering the operator. In (**c**) the assistance is higher then in (**b**) due to the higher interaction torque level.

5.2. (Partially) Unknown Part Manipulation Task

Concerning the real industrial scenario, where the operator lifts and carries an external weight (like a car's bumper, Section 2.1), the proposed exoskeleton control logic has to guarantee an adequate support to the arm even manipulating (partially) unknown weight parts. The here presented simulation evaluates the performance of the model-based controller for an unknown load-handling task scenario.

The simulation in Figure 13 shows the shoulder joint position time history when an unknown external weight of 4 kg is applied on the exoskeleton (at time $t = 1$ s). The proposed controller has been compared with a PID controller that compute the control action τ_{PID} as follows:

$$\tau_{PID}(t) = K_p \left(q_1^{ref}(t) - q_1(t)\right) + K_d \left(\dot{q}_1^{ref}(t) - \dot{q}_1(t)\right) + K_i \int \left(q_1^{ref}(t) - q_1(t)\right) dt, \tag{9}$$

where $K_p = 50$ Nm/rad is the proportional gain, $K_d = 15$ Nms/rad is the derivative gain and $K_i = 4$ Nm/rad/s is the integral gain. The gains have been experimentally determined to achieve the maximum simulation performance while maintaining the system stability.

The fuzzy logic controller identifies that such load application is not resulting from a human intention of motion, therefore, not updating the reference trajectory to the inner optimal controller and gain scheduler. Comparing the results with the PID controller, the proposed controller is capable to suppress the vibrations while resulting in less deformation of the joint position.

Figure 13. Comparison between the proposed controller and PID controller compensating a 4 kg payload unknown weight. The proposed controller is capable to suppress vibrations, resulting in less deformation.

6. Discussion

The proposed methodology have considered both mechanical design and control design of a low-cost industrial upper limbs exoskeleton. On the basis of the provided analysis, it has been possible to define the mechanical design specifications, taking into account the reference task. Off-the-shelves components (such as motors, elastic belt, etc.) have been identified to satisfy the design requirements. In particular, the belt has been selected to satisfy the compliance actuation requirements, in order to ensure safety at the hardware level. It has to be underlined that a trade-off between design requirements and available solutions is required. In particular, the economic requirement and the torque requirements have been prioritized, with respect to weight and size requirements. The resulting solution, in fact, has a weight of 12 kg without batteries. The device weight can be reduced of about 3 kg by changing the shoulder motors (e.g., purchasing the ANYdrive actuators [50]). However, different motors will increase considerably the hardware costs (costs can vary from 1500 Euro to 5000 Euro per motor). The empowering controller has been designed with emphasis on safety. Safety rules embedded in the outer empowering fuzzy logic have been design on the basis of the interaction state. The designed safety rules allow to achieve a trade-off between empowering performance and safety requirements, avoiding to high dynamics motions, while guaranteeing a safe human-robot interaction. The inner optimal gain scheduling controller allows for the tracking of the reference provided by the outer control loop.

7. Conclusions

The here presented paper describes the mechanical and control design solutions for (i) a low-cost hardware industrial exoskeleton (ii) with high payload ratio to be adopted in lifting and transportation of heavy parts. Mechanical design specifications have been derived from the task, allowing to design (iii) an intrinsic compliant 2 DoFs exoskeleton exploiting e SEA actuation at the shoulder joint to intrinsically increase human-robot interaction safety. The proposed control architecture has been described, defining (iv) a safety-based control framework. The inner gain scheduling optimal controller allows for task trajectory tracking. The outer safety-based fuzzy logic controller allows for human empowering. Simulation results show promising performance in the assistance of human operators (damping vibrations and empowering workers) and in the manipulation of unknown payloads.

The prototype of the proposed solution is under realization and it will be experimentally tested in the proposed task to evaluate the proposed approach. In particular, 20 subjects will be considered in the experimental tests. Both cognitive evaluation (based on questionnaires) and quantitative evaluation (based on EMG measurements) will be performed as in Reference [49]. In addition, authors will apply to the second call of the *EUROBENCH project* to test the proposed exoskeleton in the EUROBENCH exoskeletons benchmark facility, where the leading author Roveda is also leading the *STEPbySTEP project* [51].

8. Current and Future Work

Considering the mechanical design of the device, current work is focusing on a 3 DoFs shoulder joint concept, implementing 2 additional passive DoFs. In this new concept, the shoulder motor is considered aligned with the shoulder joint. The resulting exoskeleton implements, therefore 4, DoFs (Figure 14). The main advantage of the proposed new design is related to the increased mobility of the shoulder. However, such joint requires a different design of the compliant actuation. Therefore, the here mentioned solution is still under evaluation. Additional work is devoted to design a passive ergonomic back support for the exoskeleton to increase the ergonomics of the device. Considering the control design of the device, machine learning techniques are investigated to optimize the outer controller parameters.

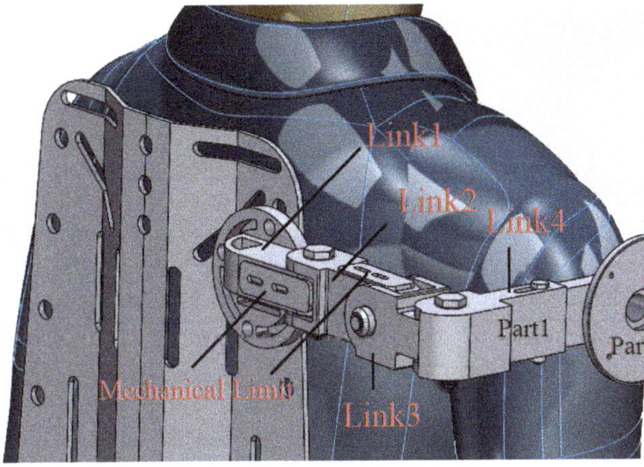

Figure 14. Three DoFs shoulder joint: new concept to increase exoskeleton mobility and task DoFs.

Author Contributions: Conceptualization, L.R., J.L., G.F., D.F., M.M. and A.M.; methodology, L.R.; software, A.M., J.L. and L.R.; validation, A.M., J.L. and L.R.; formal analysis, A.M., J.L. and L.R.; investigation, J.L. and L.R.; resources, L.R.; data curation, L.R.; writing–original draft preparation, J.L., A.M. and L.R.; writing–review and editing, L.R.; visualization, L.R.; supervision, M.M., F.B. and G.L.; project administration, L.R.; funding acquisition, L.R.

Funding: The work has been developed within the EFFORTLESS project, funded from CNR-STIIMA. This project has received funding from the European Union's Horizon 2020 research and innovation programme, via an Open Call issued and executed under Project EUROBENCH (grant agreement No 779963).

Acknowledgments: Authors would like to thank Tito Dinon (CNR-STIIMA) for his expertise and support in the project.

Conflicts of Interest: The authors declare no conflict of interest.

References

1. Pons, J.; Rocon, E.; Ruiz, A.; Moreno, J. Upper-limb robotic rehabilitation exoskeleton: Tremor suppression. In *Rehabilitation Robotics*; Itech Education and Publishing: Vienna, Austria, 2007; pp. 453–470.
2. Lo, H.S.; Xie, S.Q. Exoskeleton robots for upper-limb rehabilitation: State of the art and future prospects. *Med. Eng. Phys.* **2012**, *34*, 261–268. [CrossRef] [PubMed]
3. Jha, P.; Savla, K.; Shah, D. Exoskeleton Arm. In Proceedings of the 2018 International Conference on Smart City and Emerging Technology (ICSCET), Mumbai, India, 5 January 2018; pp. 1–6.
4. Bogue, R. Exoskeletons—A review of industrial applications. *Ind. Robot Int. J.* **2018**, *45*, 585–590. [CrossRef]
5. De Looze, M.P.; Bosch, T.; Krause, F.; Stadler, K.S.; O'Sullivan, L.W. Exoskeletons for industrial application and their potential effects on physical work load. *Ergonomics* **2016**, *59*, 671–681. [CrossRef] [PubMed]
6. Stadler, K.S.; Scherly, D. Exoskeletons in industry: Designs and their potential. In Proceedings of the AUTSYM 2017-8th International Symposium on AUTOMATICCONTROL, Wismar, Germany, 21–22 September 2017; ZHAW Zürcher Hochschule für Angewandte Wissenschaften: Winterthur, Switzerland, 2017.
7. Bosch, T.; van Eck, J.; Knitel, K.; de Looze, M. The effects of a passive exoskeleton on muscle activity, discomfort and endurance time in forward bending work. *Appl. Ergon.* **2016**, *54*, 212–217. [CrossRef] [PubMed]
8. Spada, S.; Ghibaudo, L.; Gilotta, S.; Gastaldi, L.; Cavatorta, M.P. Investigation into the applicability of a passive upper-limb exoskeleton in automotive industry. *Procedia Manuf.* **2017**, *11*, 1255–1262. [CrossRef]
9. Noonee. Available online: https://www.noonee.com/ (accessed on 8 July 2019).
10. ESOSCHELETRO MATE. Available online: https://www.comau.com/it/mate (accessed on 8 July 2019).
11. Work Safer with EksoVest. Available online: https://eksobionics.com/eksoworks/ (accessed on 8 July 2019).
12. Levitate. Available online: https://www.levitatetech.com/ (accessed on 8 July 2019).

13. Imagine If the Human Body Was Not Limited by Its Physical Abilities? Available online: https://www.skelex.com/(accessed on 8 July 2019).
14. Gopura, R.; Bandara, D.; Kiguchi, K.; Mann, G.K. Developments in hardware systems of active upper-limb exoskeleton robots: A review. *Robot. Auton. Syst.* **2016**, *75*, 203–220. [CrossRef]
15. Sylla, N.; Bonnet, V.; Colledani, F.; Fraisse, P. Ergonomic contribution of ABLE exoskeleton in automotive industry. *Int. J. Ind. Ergon.* **2014**, *44*, 475–481. [CrossRef]
16. Stadler, K.S.; Altenburger, R.; Schmidhauser, E.; Scherly, D.; Ortiz, J.; Toxiri, S.; Mateos, L.; Masood, J. Robo-mate an exoskeleton for industrial use—Concept and mechanical design. In *Advances in Cooperative Robotics*; World Scientific: Singapore, 2017; pp. 806–813.
17. Ebrahimi, A. Stuttgart Exo-Jacket: An exoskeleton for industrial upper body applications. In Proceedings of the 2017 10th International Conference on Human System Interactions (HSI), Ulsan, Korea, 17–19 July 2017; pp. 258–263.
18. Li, R.Y.M.; Ng, D.P.L. Wearable Robotics, Industrial Robots and Construction Worker's Safety and Health. In *International Conference on Applied Human Factors and Ergonomics*; Springer: Cham, Switzerland, 2017; pp. 31–36.
19. Huysamen, K.; de Looze, M.; Bosch, T.; Ortiz, J.; Toxiri, S.; O'Sullivan, L.W. Assessment of an active industrial exoskeleton to aid dynamic lifting and lowering manual handling tasks. *Appl. Ergon.* **2018**, *68*, 125–131. [CrossRef]
20. Atoun. Available online: http://atoun.co.jp/products (accessed on 8 July 2019).
21. Innophys. Available online: https://innophys.jp/? (accessed on 8 July 2019).
22. Sankai, Y. HAL: Hybrid assistive limb based on cybernics. In *Robotics Research*; Springer: Berlin/Heidelberg, Germany, 2010; pp. 25–34.
23. Sarcos. Available online: https://www.sarcos.com/products/guardian-xo/ (accessed on 8 July 2019).
24. Vanderborght, B.; Albu-Schäffer, A.; Bicchi, A.; Burdet, E.; Caldwell, D.G.; Carloni, R.; Catalano, M.; Eiberger, O.; Friedl, W.; Ganesh, G.; et al. Variable impedance actuators: A review. *Robot. Auton. Syst.* **2013**, *61*, 1601–1614. [CrossRef]
25. Vitiello, N.; Cempini, M.; Crea, S.; Giovacchini, F.; Cortese, M.; Moisè, M.; Posteraro, F.; Carrozza, M.C. Functional design of a powered elbow orthosis toward its clinical employment. *IEEE/ASME Trans. Mechatron.* **2016**, *21*, 1880–1891. [CrossRef]
26. Bianchi, M.; Cempini, M.; Conti, R.; Meli, E.; Ridolfi, A.; Vitiello, N.; Allotta, B. Design of a series elastic transmission for hand exoskeletons. *Mechatronics* **2018**, *51*, 8–18. [CrossRef]
27. Yu, H.; Huang, S.; Chen, G.; Pan, Y.; Guo, Z. Human–robot interaction control of rehabilitation robots with series elastic actuators. *IEEE Trans. Robot.* **2015**, *31*, 1089–1100. [CrossRef]
28. Li, X.; Pan, Y.; Chen, G.; Yu, H. Adaptive human–robot interaction control for robots driven by series elastic actuators. *IEEE Trans. Robot.* **2016**, *33*, 169–182. [CrossRef]
29. Fernández, J.; Sprengel, H.; Mallwitz, M.; Zipper, M.; Yu, B.; Bargsten, V. Designing modular series-elastic actuators for safe human-robot collaboration in industrial settings. In *Advances in Cooperative Robotics, Proceedings of the 19th International Conference on Clawar 2016, London, UK, 12–14 September 2016*; World Scientific: Singapore, 2016; p. 135.
30. Lenzi, T.; Vitiello, N.; De Rossi, S.M.M.; Roccella, S.; Vecchi, F.; Carrozza, M.C. NEUROExos: A variable impedance powered elbow exoskeleton. In Proceedings of the 2011 IEEE International Conference on Robotics and Automation, Shanghai, China, 9–13 May 2011; pp. 1419–1426.
31. Kim, B.; Deshpande, A.D. An upper-body rehabilitation exoskeleton Harmony with an anatomical shoulder mechanism: Design, modeling, control, and performance evaluation. *Int. J. Robot. Res.* **2017**, *36*, 414–435. [CrossRef]
32. Abdallah, I.B.; Bouteraa, Y.; Rekik, C. Design and development of 3d printed myoelectric robotic exoskeleton for hand rehabilitation. *Int. J. Smart Sens. Intell. Syst.* **2017**, *10*, 341–366. [CrossRef]
33. Bowers, M.; Goldfarb, N.; Jagetia, A.; Khajuriwala, R.; Kumar, A.; Lam, B.; Shah, N. Design of a Low Cost Robotic System to Aid in the Rehabilitation of Stroke Patients. 2017. Available online: https://adjagetia.github.io/Documents/design-low-cost.pdf (accessed on 8 July 2019).
34. Anam, K.; Al-Jumaily, A.A. Active exoskeleton control systems: State of the art. *Procedia Eng.* **2012**, *41*, 988–994. [CrossRef]

35. Noda, T.; Sugimoto, N.; Furukawa, J.; Sato, M.A.; Hyon, S.H.; Morimoto, J. Brain-controlled exoskeleton robot for BMI rehabilitation. In Proceedings of the 2012 12th IEEE-RAS International Conference on Humanoid Robots (Humanoids 2012), Osaka, Japan, 29 November–1 December 2012; pp. 21–27.
36. Li, Z.; Wang, B.; Sun, F.; Yang, C.; Xie, Q.; Zhang, W. sEMG-based joint force control for an upper-limb power-assist exoskeleton robot. *IEEE J. Biomed. Health Inform.* **2014**, *18*, 1043–1050. [PubMed]
37. Secciani, N.; Bianchi, M.; Meli, E.; Volpe, Y.; Ridolfi, A. A novel application of a surface ElectroMyoGraphy-based control strategy for a hand exoskeleton system: A single-case study. *Int. J. Adv. Robot. Syst.* **2019**, *16*, 1729881419828197. [CrossRef]
38. Li, Z.; Huang, Z.; He, W.; Su, C.Y. Adaptive impedance control for an upper limb robotic exoskeleton using biological signals. *IEEE Trans. Ind. Electron.* **2017**, *64*, 1664–1674. [CrossRef]
39. Yu, W.; Rosen, J.; Li, X. PID admittance control for an upper limb exoskeleton. In Proceedings of the 2011 American Control Conference, San Francisco, CA, USA, 29 June–1 July 2011; pp. 1124–1129.
40. Ryser, F.; Bützer, T.; Held, J.P.; Lambercy, O.; Gassert, R. Fully embedded myoelectric control for a wearable robotic hand orthosis. In Proceedings of the 2017 International Conference on Rehabilitation Robotics (ICORR), London, UK, 17–20 July 2017; pp. 615–621.
41. Ding, Y.; Galiana, I.; Siviy, C.; Panizzolo, F.A.; Walsh, C. IMU-based iterative control for hip extension assistance with a soft exosuit. In Proceedings of the 2016 IEEE International Conference on Robotics and Automation (ICRA), Stockholm, Sweeden, 16–21 May 2016; pp. 3501–3508.
42. Langtree, I. Height Chart of Men and Woman in Different Countries. In *Canada: Disabled*; 2018. Available online: https://www.disabled-world.com/calculators-charts/height-chart.php (accessed on 1 December 2017).
43. Walpole, S.C.; Prieto-Merino, D.; Edwards, P.; Cleland, J.; Stevens, G.; Roberts, I. The weight of nations: An estimation of adult human biomass. *BMC Public Health* **2012**, *12*, 439. [CrossRef] [PubMed]
44. Legnani, G.; Palmieri, G. *Fondamenti di Meccanica e Biomeccanica del Movimento*; CittàStudi: Milano, Italy, 2016.
45. Lens, T.; Kirchhoff, J.; von Stryk, O. Dynamic modeling of elastic tendon actuators with tendon slackening. In Proceedings of the 2012 12th IEEE-RAS International Conference on Humanoid Robots (Humanoids 2012), Osaka, Japan, 29 November–1 December 2012; pp. 779–784.
46. Roveda, L.; Pallucca, G.; Pedrocchi, N.; Braghin, F.; Tosatti, L.M. Iterative learning procedure with reinforcement for high-accuracy force tracking in robotized tasks. *IEEE Trans. Ind. Inform.* **2017**, *14*, 1753–1763. [CrossRef]
47. Lewis, F.L.; Vrabie, D.; Syrmos, V.L. *Optimal Control*; John Wiley & Sons: Hoboken, NJ, USA, 2012.
48. Tao, C.W.; Taur, J.S.; Chen, Y. Design of a parallel distributed fuzzy LQR controller for the twin rotor multi-input multi-output system. *Fuzzy Sets Syst.* **2010**, *161*, 2081–2103. [CrossRef]
49. Roveda, L.; Haghshenas, S.; Prini, A.; Dinon, T.; Pedrocchi, N.; Braghin, F.; Tosatti, L.M. Fuzzy impedance control for enhancing capabilities of humans in onerous tasks execution. In Proceedings of the 2018 15th International Conference on Ubiquitous Robots (UR), Honolulu, HI, USA, 26–30 June 2018; pp. 406–411.
50. Drive for Real-World Robots. Available online: https://www.anybotics.com/anydrive-robotic-actuator/ (accessed on 30 July 2019).
51. EUROBENCH. Available online: http://eurobench2020.eu/ (accessed on 30 July 2019).

robotics

MDPI

Article

Optimization of a Kitting Line: A Case Study

Nicola Comand *, Riccardo Minto, Giovanni Boschetti, Maurizio Faccio and Giulio Rosati

Department of Management and Engineering, University of Padova, Via Venezia 1, 35121 Padova, Italy
* Correspondence: nicola.comand@phd.unipd.it; Tel.: +39-049-827-6810

Received: 1 May 2019; Accepted: 5 August 2019; Published: 7 August 2019

Abstract: Assembly kitting lines typically deal with components that present different quantities in the final kit. Re-configurable feeders are not advisable for low quantity components since this solution leads to high inefficiency, due to the time required for retooling and for the feeding rate unbalance between the components. On the other hand a fully flexible feeder can increase the flexibility of the system at the cost of reducing the throughput. An assembly kitting problem was therefore addressed in different ways for low quantity components and high quantity components, studying the two different subsystems that compose a hybrid flexible assembly system. To optimize the re-configurable feeders, which supply the high quantity components, the opening sequence of a system composed by several hoppers was analyzed. We propose a solution which replaces the weighing device with a vision inspection system, showing its impact on the productivity of the line. A model coded into a Matlab script was developed to perform the optimization of the system and understand its behavior. Furthermore a fully flexible assembly system was developed in the laboratory in order to test the kitting of the low quantity components with the proposed subsystem. Finally the Overall Equipment Effectiveness of the line was calculated to evaluate the possible improvements obtained by the proposed solution.

Keywords: hopper; optimization; sequencing; kitting; H FAS

---

1. Introduction

In order to compete in a global market, nowadays companies need to offer a wide range of different products. The production technology most sensitive to this evolution is assembly, because of its position at the end of production processes, where the whole product variety is present. As a consequence, companies try to use adaptive flexible assembly systems [1]. As reported by Barbazza et al. [2], the basic requirements for the winning design and management of a flexible assembly system concern the optimization of:

- unit direct production cost (€/part), that is, the ratio of the hourly costs of the workcell and the average throughput;
- mix flexibility, that is, the ability to handle a wide variety of part types (components), and manage a wide mix of components and products;
- volume flexibility, that is, the ability to change the productivity of the system without reducing the efficiency of the system.

The basic assembly production strategies are traditionally classified as manual assembly systems, flexible (automated) assembly systems, and dedicated (automated) assembly systems [3]. According to Heilala and Voho [4], the transition from a dedicated to a flexible automated assembly system and then to a manual assembly system is due to the increase in the requirements of flexibility and number of variants. On the other hand, pure manual assembly has different drawbacks. The accuracy of the tasks performed and the activity repeatability need improvements. Ergonomic problems, as well as occupational injuries, could occur [5]. At last absenteeism can create production lack and low efficiency, especially in the case of balanced assembly lines [6]; for these reasons a pure manual assembly system cannot be very competitive, especially in countries with high labor cost levels. As affirmed by Edmondson and Redford [7], the driving factor behind the definition of a suitable assembly system configuration is, first of all, economy. Automation is usually a way to lower the costs of labor in Western countries and the key issue is to achieve and optimize flexibility and a suitable degree of automation in a fast changing market situation [8]. Therefore, in the last decades, a subset of flexible automated assembly systems have been developed with the aim of increasing flexibility and decreasing production costs if compared with the traditional manual and automated assembly systems.

The traditional flexible assembly system (FAS) is typically composed of one programmable manipulator, which picks the parts from predefined positions, places them on the assembly station and perform the assembly task and one or more re-configurable feeding devices, typically vibratory bowl feeders, for each component needed to complete the production process. An evolution of the concept of the flexible assembly systems is presented in Reference [9], fully FAS (F-FAS). The F-FAS is a single-station robotized assembly system, able to perform mixed model assembly, where the unique feeding device is a fully flexible feeder, ensuring a higher level of flexibility than the traditional automated FAS. The feeder system, since it is composed by a bulk, a vibratory plane and a camera, can feed several different components without the need for reconfiguration. As far as this solution presents high levels of flexibility, it has a lower throughput than a traditional FAS, since image processing for shape recognition is time consuming and the feeding process is stochastic.

In order to bridge the gap between the presented solutions, the hybrid FAS (H-FAS) has been developed [10]. The H-FAS consists of one programmable manipulator, one flexible feeder and one or more re-configurable feeder devices like the vibratory bowl feeder. Lowering the number of parts deployed by the fully flexible feeder leads to a reduction in the complexity of the system, thus increasing the throughput. It should be noted that the H-FAS is able to reach performances, in terms of unit direct production costs, mix flexibility and volume flexibility, higher than those of the other systems presented. The advantages of using this type of solution can be noticed when the feeding process is optimized for the different components, as a function of the characteristics of each one.

The proposed work arises from a case study, which aims to improve the efficiency of an assembly kitting line. The assembly kitting process is widely used for different purposes, for example, sales kits or production kit. A typical application of the studied assembly kitting line is the production of sales kits of connectors, such as screws and bolts, ready to use in the product assembly process.

Such a process is performed by means of a traditional FAS. The production line taken as a reference is indeed composed by 8 bowl feeders, which through a conveyor belt and three hoppers delivers the components onto the inferior belt, which moves towards the packing machine. The line is depicted in Figure 1.

A literature review analysis has led to a limited number of studies on increasing the efficiency of an assembly kitting line. Bevilacqua et al., in Reference [11], present a case study in the pharmaceutical industry, where they applied new procedures based on the lean production approach; moreover, they evaluated the increase in efficiency based on the Overall Efficiency Equipment (OEE, [12]) value, an approach suitable also for our study. In Reference [13] the authors described the design process for an automated assembly line; it should be noted that the authors have carried out a redesign of the deployed components, a solution which could not be applied in our case study.

In order to increase the efficiency in our case study, we observed that the deployment of low quantity components by means of the bowl feeders leads to high inefficiency, due to the time required for retooling and for the feeding rate unbalance between the components. For such reasons this paper investigates the possibility of applying an H-FAS model on the considered process, since we believe that for its characteristics it is more suitable than traditional ones. The aim of this work is therefore to propose an alternative solution to the kitting line layout taken as reference from the case study. The study will be divided into two problems: the first will consider the feeding of the low quantity components and will be solved with the development of a flexible robotic workcell, the second will deal with the feeding of the high quantity components and will lead to the improvement of the original feeding system.

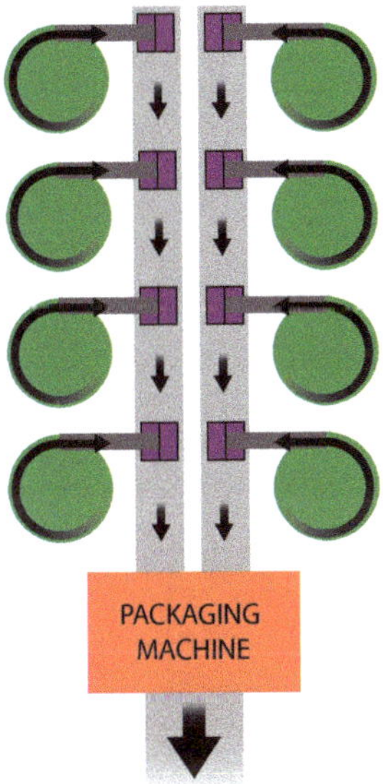

Figure 1. Line layout.

With regard to the feeding system, a focus on the hoppers cascade system was considered advantageous in the proposed case study. No one, to the authors' knowledge, has studied the impact of a specific configuration of a hoppers cascade system and its parameters (e.g., opening/closing time, output feed-rate, etc.) on its performance. Similar studies [14] have focused on continuous processing with the need for control flow rates of materials and whose models were not suitable for the presented case study.

The aim of the study will therefore be:

- reconfiguration of a kitting line with H-FAS systems, applied to kitting without special assembly requirements;
- modeling of the hoppers of the bowl feeders of the H-FAS system;
- demonstration of the improvement in line performance by removing some of the traditional feeders;

The novelty of the presented work, besides suggesting two different improvements for an assembly kitting line, is the study of the behavior of a hoppers cascade system based on its configuration. In this way we can estimate the throughput of a feeding system and evaluate more convenient configurations. Moreover the proposed solution was applied in a case study, validating the proposed results. The presented work is organized as follows: Section 2 presents the systems used to improve the current assembly kitting line, divided in Section 2.1 for the robotics F-FAS and Section 2.2 for the hoppers cascade system. Section 4 shows the impact of the proposed solutions and Section 3 presents results provided by the application of the model to the case study. Lastly, Section 5 concludes the work.

2. Materials and Methods

This section describes the systems used to address the low and the high quantity component feeding problem. To solve the former a flexible robotic workcell was developed and experimental tests were performed to find the mean time for a pick and place operation. This is described in detail in Section 2.1. To solve the high quantity components, three configurations were modeled and by mean of simulations was evaluated the impact of some parameters on the system performance. The system analyzed is described in Section 2.2 and the sensitivity analysis of the model is presented in Section 2.2.1. The simulation is treated in particular in Section 2.2.2 and the vision inspection system in Section 2.2.3.

2.1. Flexible Robotic Workcell

An experimental setup is needed to evaluate the effectiveness of the usage of the adopted devices, such as flexible feeders and vertical packaging machines. A F-FAS prototype was therefore developed at the Industrial Robotics Laboratories of the University of Padua, where a robot carried out a pick and place job. The tested workcell is composed of the following devices, as represented in Figure 2:

- the *manipulator*, an Adept Viper 650 anthropomorphic arm with a 650 mm reach; the device is equipped with a Schunk MPG 25 pneumatic gripper with two parallel jaws.
- The *controller* Adept CX controller supported by PC-based applications for image processing (Halcon) and work cycle scheduling (Matlab); the two subsystems communicate through a TCP-IP connection.
- The *feeding system*, a Flex Factory Anyfeed SX240; this commercial flexible feeder presents a back-lighted vibratory plane where the parts from the vibratory bulk are distributed through controlled shakes.
- The *vision system*, consisting of ab AVT Pike F-505 monochrome 2452 $\times$ 2054 FireWire 5 Megapixel camera; the focal axis of the camera is orthogonal to the working plane and with a 35 mm focal length lens.

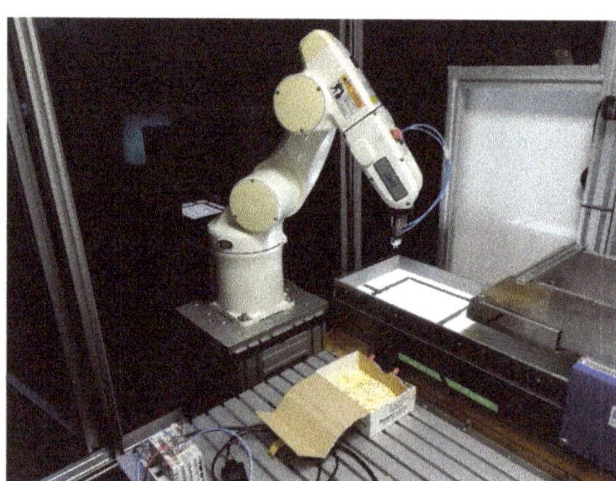

Figure 2. Prototype workcell used during the tests.

The robot picks parts of four different shapes, as represented in Figure 3, from the flexible feeder; the picking area, represented by the 170 $\times$ 220 mm area enclosed within a black boundary in Figure 2, was designed in order to be the largest area where the robot could reach any point. The center of symmetry of the area is therefore placed with X, Y, Z coordinates from the robot base equal to respectively 400 mm, 365 mm, 18 mm.

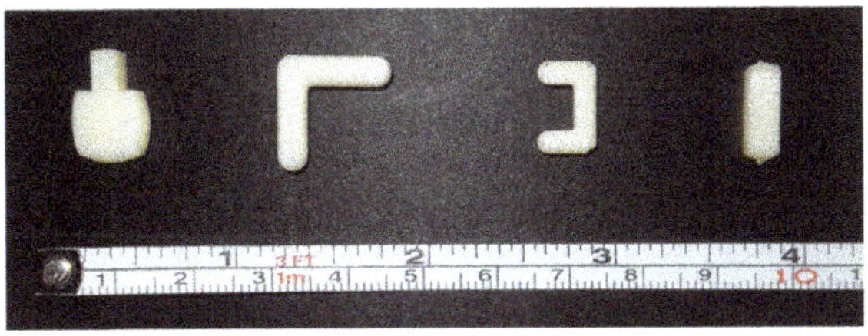

Figure 3. Parts picked by the robot in the experimental workcell.

The parts on the vibratory plane are identified by the shape matching method provided by the Halcon software [15]. The method allows the identification of an object from the image by recognizing the gray-scale level of the reference photo, with the score parameter, which is set equal to 0.7, representing how faithful is the image to the recognized part. The software is also provided of a function that identifies the pickup location on the part which differs for each shape and it is defined in order to avoid to lose grip.

After the image processing was carried out, the system initiates the pick and place job, which consists of the following tasks:

- the robot moves from a starting position to the one of the requested part;
- after picking the part, the manipulator places it in a box, put at the starting point *home*;
- when all the parts corresponding to the first model are picked, the robot picks the parts corresponding to the following one;
- if the robot cannot reach any part on the vibratory plane, the feeder shakes it and, if there is not a sufficient number of distributed parts, it supplies more.

The placing position is defined as *home*, since it corresponds to the first position undertaken by the robot in order to not obstruct the sight of the camera. This position is set at the same height of *viapoint*, defined at 100 mm from the edge of the feeder: this was defined not only to avoid potential collision between the gripper and the feeder, but also to lower the cycle time by optimizing the robot trajectory. The other parameters that characterize the robot movement were defined in the Matlab script as shown in Table 1.

Table 1. Value for the motion parameter used in the test.

Parameter	Value (Unit)
Speed	100 (%)
$Speed_{appro/depart}$	50 (%)
$H_{appro/depart}$	15 (mm)
offset from feeder	1.5 (mm)
Viapoint (x, y, z)	420, 214, 140 mm
Viapoint (α, β, γ)	0° 180° 0°
home (x, y, z)	460, −102, 140 mm
home (α, β, γ)	0° 180° 0°

It should be noted that these parameters could be set in the controller of the robot in an easier way, however in order to increase the flexibility of the system without the code becoming more complex, it was preferable to send them via the TCP-IP connection.

Since the joint 6 was sweeping a wide angle in a short linear movement, thus increasing the cycle time, its rotation at the viapoint was set equal to 60% of the rotation at the picking point. Moreover, since the parts could be reached with two different rotations due to the symmetry of the gripper, the value of the joint 6 at the picking point has been chosen as the nearest to zero; these solutions resulted in a decrease of the cycle time of about 10.6%. The experimental picking time has been evaluated by means of a 103 test, as shown in Figure 4; each time interval was measured by Matlab since the instant the coordinates are sent to the controller to the one when the controller has notified Matlab that the robot has returned to the *home* position. The medium value results equal to 1.41 s; an analysis of the distribution of the measured data, Figure 5, showed an interval of estimation equal to ±0.12 s (95%).

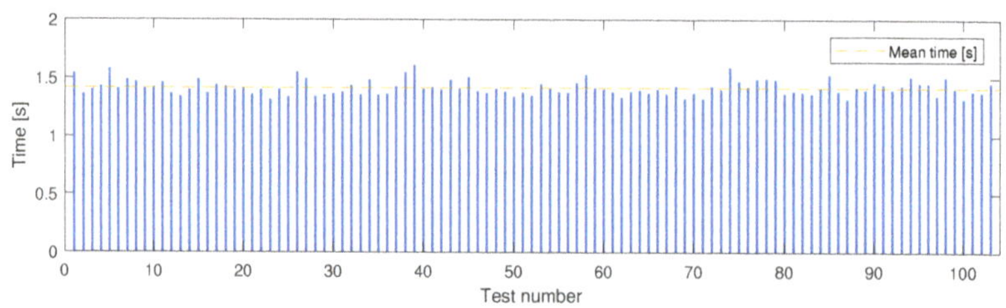

Figure 4. Measured time for each picking during the experimental test.

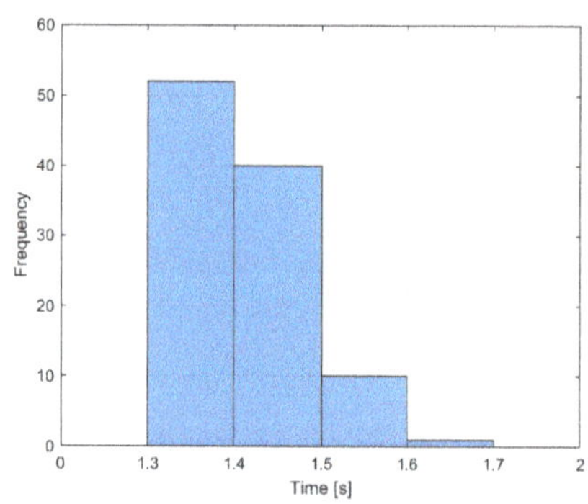

Figure 5. Frequencies histogram for the measured time.

2.2. Hoppers Sequence Model

The system considered is generally composed of an input belt, a suitable sequence of hoppers and an output belt. In this paper, we just focus on these devices, assuming that the system is fed by a bowl feeder with limited capability that delivers the kits to a packing machine. The maximum throughput of the feeder is set to 199.8 pieces/min, measured from the case study line. The motion of the pieces between a device and the subsequent is assumed to be a free falling motion, and is assumed that the discharging feed rate is infinite, that is, the pieces move together as a bulk from one hopper to its successor. The falling time can be so calculated as:

$$T_c = \sqrt{\frac{2 \cdot h}{g}} \tag{1}$$

where h is the distance between two devices, here set to 200 mm and g the gravity acceleration, assumed 9.81 m/s². We assumed that the pieces are delivered evenly spaced of a distance d by the superior belt with speed v_{belt} , so that the time delay between two subsequent pieces falls is:

$$T_d = \frac{d}{v_{belt}} \tag{2}$$

The time between the opening command and the following idle state of a hopper is calculated assuming that the hopper can close after a percentage p of the space between the hoppers has been covered by the falling components. The percentage here is set equal to 20%, for a value of $d = 200$ mm, means a distance of 40 mm. Substituting this value into Equation (1), the time between the open command and the successive idle state is evaluated as:

$$T = \cdot T_a + \sqrt{p} \cdot T_c \quad (3)$$

In order to find the filling time of the first hopper we use the values calculated with Equations (1) and (2):

$$T_f = n \cdot T_d + T_c \quad (4)$$

where n is the number of components per kit delivered by the subline.

The values which could not be evaluated were obtained from the state of the art. Table 2 shows the values for the parameters used in the tests, where T_w is the weighing time and T_{mov} is the time required to the inferior belt to shift of one position to another.

Table 2. Variable values taken into account for the simulation.

Parameter	Value (s)
T_a	0.15–0.25–0.32
T_c	0.202
T_{mov}	0.3–0.6
T_w	0.8
Configuration	*3–2–2x2*

Three configurations reported in Figure 6, were considered to perform the study of the hoppers opening sequence:

- The first configuration is composed of three hoppers. The first is the pool hopper, the second is a weighing hopper to perform the quality inspection and the third is a boost/memory hopper. This layout will be referred as the 3 configuration. It is depicted in Figure 6a.
- The second configuration is composed of two hoppers. A pool hopper and a boost/memory one. The quality inspection is performed by a camera set upon the superior belt. This layout will be referred as the 2 configuration. It is depicted in Figure 6b.
- The third configuration is composed of two columns with two hoppers each and a selector between the superior belt and the hoppers. The quality inspection is performed by a camera set upon the superior belt. This layout will be referred as the *2x2* configuration and is depicted in Figure 6c.

The proposed case study inspired the 3 configuration, which is assumed to be the starting point of the study. In this configuration, the weighing hopper is needed to perform the quality inspection, verifying if the correct number of pieces has been delivered by the superior belt, in order to avoid defective kits in output. Such weighing device is equipped with a load cell used to estimate the weight, which provides a raw electrical signal. The main constraint in achieving high accuracy is a superimposed noise, which is eliminated thanks to a filter. However a trade-off between the measurement time and a high precision estimation has to be accepted. Thus the weighing time needed to obtain a stable measurement tends to become the bottleneck of the system, preventing it from reaching high efficiency and saturating the input devices. The 2 configuration addresses the problem eliminating the weighing device and replacing it with a vision system, capable of performing in its place the quality inspection. This solution is depicted in Figure 6b, and since it does not contemplate the use of a weighing device, it is not affected by the aforementioned trade-off, allowing to perform a 100% quality control ignoring the weighing time and to discard single components instead of a full batch recognized as defective by the weighing device at the end of the sequence.

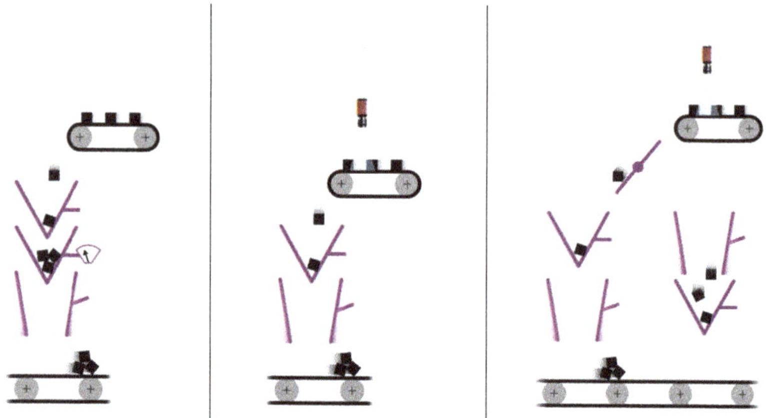

(**a**) Layout *3* (**b**) Layout *2* (**c**) Layout *2x2*

Figure 6. Layouts considered.

The throughput increase in the simulation was relevant, but the input feeder could not be saturated. Therefore we developed the *2x2* configuration depicted in Figure 6c, which is capable of saturating the input feeder, thanks to a higher throughput.

In order to find the throughput of the system, the bottleneck of each configuration was analyzed and its takt time described with equations. In the 2 configuration the bottleneck is the first hopper, whose takt time is:

$$TaktTime = T_a + (1 + \sqrt{p}) \cdot T_c + n \cdot T_d \tag{5}$$

In the 3 configuration the bottleneck is the weighing hopper and a superior limit is set by the takt time of the 2 configuration:

$$TaktTime = \max \begin{cases} T_a + (1 + \sqrt{p}) \cdot T_c + n \cdot T_d \\ T_a + T_w + (1 + \sqrt{p}) \cdot T_c \end{cases} \tag{6}$$

In the *2x2* configuration the bottleneck is the first hopper, which is filled by the belt:

$$TaktTime = n \cdot T_d + T_c \tag{7}$$

A superior limit for all the configurations is set by the inferior belt whose minimum takt time is:

$$TaktTime = T_c + T_{mov} \tag{8}$$

From these equation is possible to calculate the throughput of the system for each configurations with different parameters with the following equation:

$$Q = \frac{1}{TaktTime} \tag{9}$$

Figure 7a shows the throughput for different number of parts and for each studied configuration using parameters in Figure 7. As expected, increasing the number of parts for each kit decreases the throughput, since $n \cdot T_d$ increases. It should be noted that the *2x2* configuration results in the maximum throughput compared to the others.

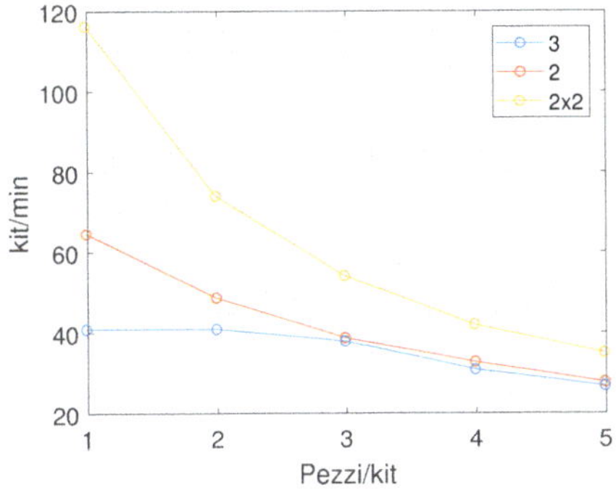

(a) Plot of the modelled throughput of the configurations

Parameter	Value [s]
T_a	0.32
T_c	0.2
T_d	0.3
T_w	0.8
T_{mov}	0.3

(b) Parameters used for the modeled throughput in the presented figure

Figure 7. Modeled throughput of the configurations.

2.2.1. Sensitivity Analysis

A sensitivity analysis was carried out in order to observe the sensitivity of the presented model to different values for the considered parameters. Since T_a, T_c, T_d and T_w were obtained from the state of the art, we believe that they represent a source of uncertainty; therefore, we tested the system behavior for different values of the aforementioned parameters. Following Equations (5)–(7), a test was run by changing the values as represented in Table 3, following an One-at-a-time approach.

Table 3. Values considered for the parameter of the sensitivity analysis.

Parameter	Value (s)
T_a	0.10–0.30–0.40
T_c	0.14–0.20–0.24
T_d	0.20–0.30–0.40
T_w	0.50–0.80–1.00

The plots in Figure 8 are obtained by evaluating the takt time with the aforementioned model for the different values.

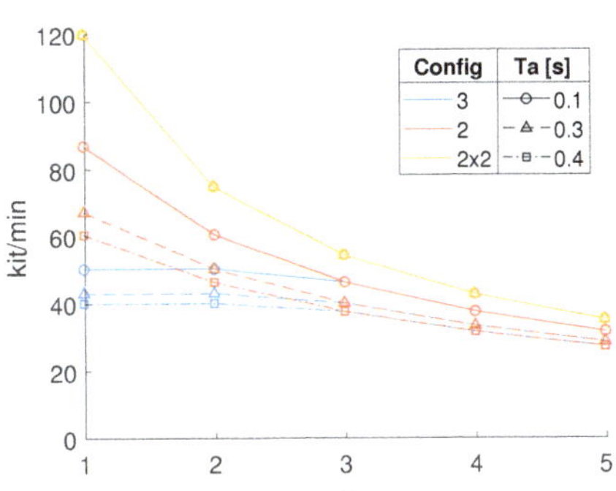

(a) Sensitivity analysis for T_a

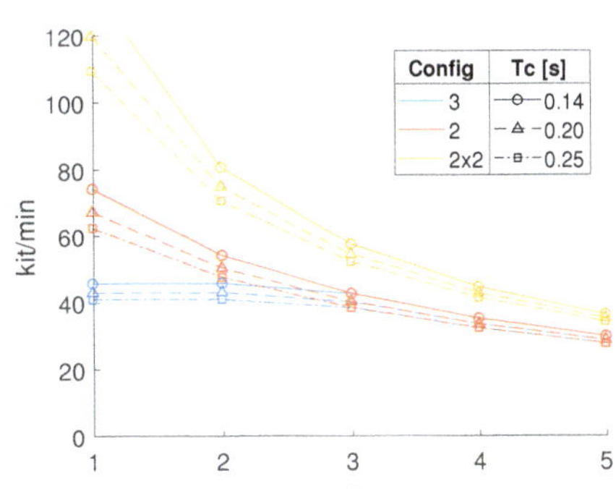

(b) Sensitivity analysis for T_c

Figure 8. *Cont.*

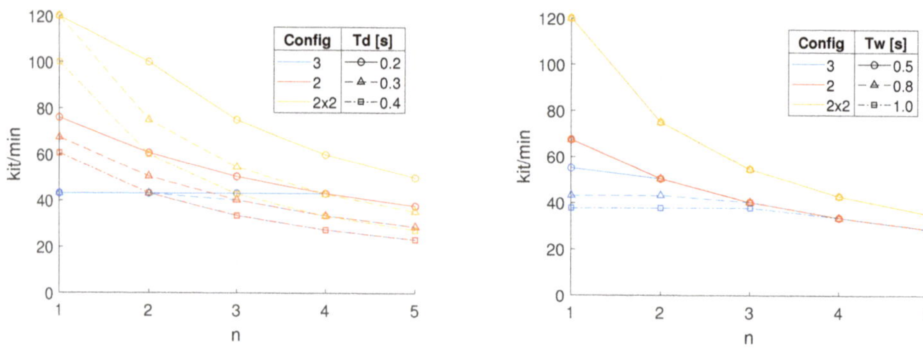

(**c**) Sensitivity analysis for T_d (**d**) Sensitivity analysis for T_w

Figure 8. Results of the sensitivity analysis for the parameters T_a, T_c, T_d, T_w.

2.2.2. Simulation

A simulation of the configurations was performed with a finite state machine, which uploads the states of the different devices over time. Since the simulation is affected by transient and/or random effects, considered as random variations on the parameters, it could be used as a reference for the proposed model; moreover, it provides for further data, such as the task scheduling and the availability of the feeders. Each device is linked to its predecessor and its successor with the relationships illustrated in Figure 9, which describes the transition between states. The superior belt is the input device of the system and switches just between two states, *ready* and *filling*, with a logic described in Figure 9a. It interacts only with the first hopper, so its state depends only on the state of this device. The generic hopper switches between many states, *filling, idle, weighing, ready, closing* and *opening*. The non weighing hopper presents the same graph without the state *weighing*. The generic hopper can be related to one other hopper or a belt, referred as *predecessor* if it is in a previous position on the feeding chain or *successor* if it is in a next position on the feeding chain. Its graph is depicted in Figure 9c. The inferior belt, that is, the output device of the system, can switch between the states *filling, moving* and *idle*. The *moving* state is due to the need of simulate the time required from the output belt of the case study line to switch between two subsequent position. The states are described in detail in Table 4. Each transition is due to the change of state of an adjacent device or for the expiration of the time needed for a specific action (e.g., the opening and closing movement of a hopper).

The transition logic just described was implemented into a Matlab script in order to perform time varying simulations of the system. To reflect the effect of the uncertainties on the process in a real environment we introduced some random variation of the parameters of the simulation. A random variation in the interval [−0.1 s, +0.1 s] was added to T_w to simulate the dynamic effect of the components falling on the weighing hopper and a Gaussian distribution ($\mu = 100$ mm; $\sigma = 15$ mm) of the components spacing on the input belt was assumed to simulate some variability in the bowl feeder feeding rate.

Table 4. The states implemented for the devices in the simulation.

States	Description
Ready	The device is ready, but its successor not
Idle	The device is ready, but its predecessor not
Opening	The hopper is opening
Closing	The hopper is closing
Filling	The hopper or the inferior belt is being filled, or the superior belt is filling
Weighing	The weighing hopper is performing the weight measure
Moving	The inferior belt is shifting to the next position

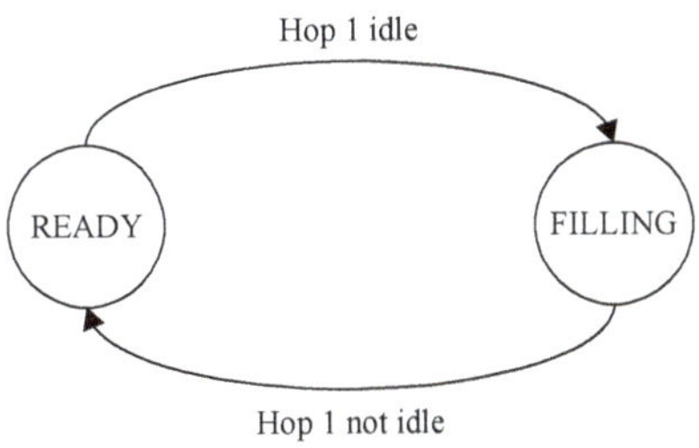

(**a**) Superior belt graph

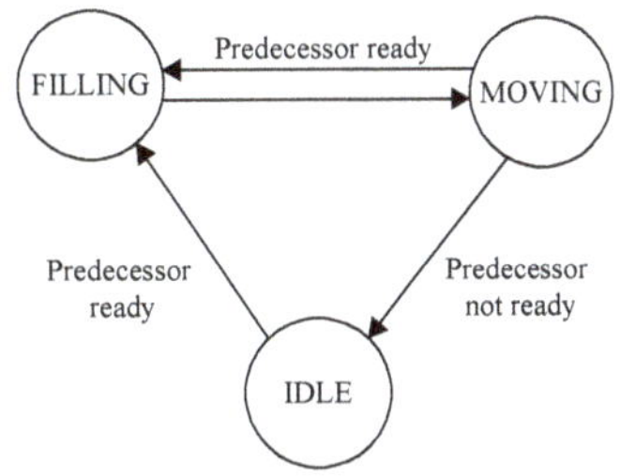

(**b**) Inferior belt graph

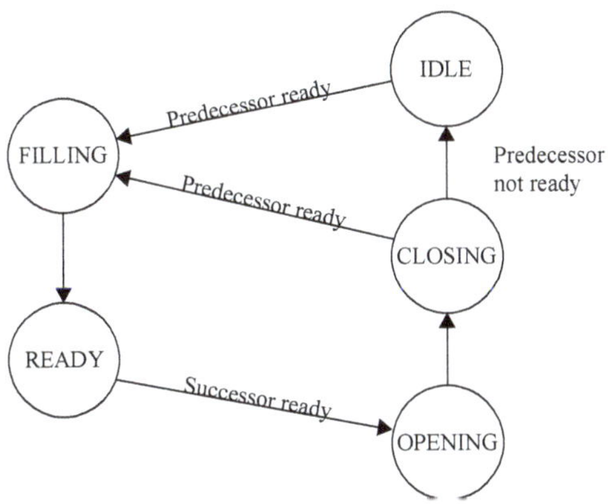

(**c**) Hopper graph

Figure 9. Finite state graphs of the devices.

A further simulation with the same parameters in Figure 7 was performed and the throughput for the three configurations is shown in Figure 10. The relative error between the model prediction and the simulated throughput was calculated and presented in Table 5. The values state a good agreement between the model and the simulation, except for the tendency of overestimate the throughput, which can be seen from the negative mean error.

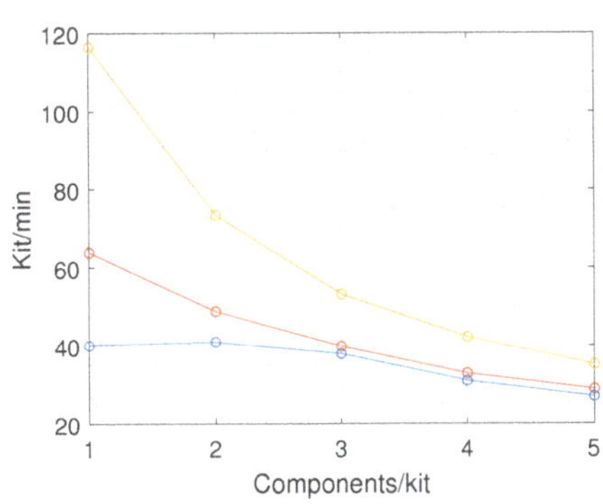

Figure 10. Plot of the throughput found with the simulations.

Table 5. Relative error (%) between the model prediction and the simulated throughput.

Configuration	n					Mean
	1	2	3	4	5	
3	−6.04	−3.69	−4.40	−6.51	−5.08	−5.14
2	−3.31	−1.61	0.24	−0.98	1.44	−0.84
2x2	−2.96	−2.28	−2.55	−1.51	−0.48	−1.96

One output of the simulation is the state of any device in every moment, which could be used to plot the state chart of the devices, like the one reported in Figure 11, plotted for the values $T_a = 0.15$ s, $T_{mov} = 0.30$ s and $n = 5$.

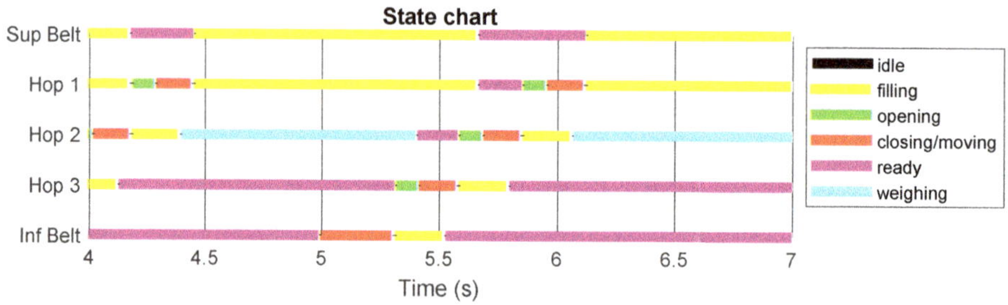

Figure 11. An example of state chart of the devices obtained with the simulation.

The *availability* is the ratio of run time to planned production time. With the data used to plot such state charts was also possible to evaluate the availability for the devices as:

$$A = \frac{RT}{PPT}$$

where A is the availability, RT is the run time, i.e., the time the device did not spend in ready or idle state, and PPT is the planned production time, i.e., the total time the device is used.

In Figure 12 we can see the results of the simulations performed varying the opening and closing time of the hoppers to understand their influence on the system. To the left the data calculated with a $T_a = 0.15$ s, while to the right the data related to $T_a = 0.32$ s. The resulting plot shows that the opening time of the hoppers slightly affects the performance of the system, which decreases with low kit size, due to the increasing frequency of opening/closing operations. The worsening interests mainly the *3* and *2* configurations, but not change substantially the comparison between the configurations, therefore from now on we will refer indistinctly to the plot to the left or to the right column in Figure 12. The plot Figure 12b shows that the A of the superior belt for the *2x2* configuration (yellow solid line) is 1 for any kit size considered. It should be noted that only the *2x2* configuration is able to reach this value, which mean the saturation of the superior belt. For a kit size equal to 1 the inferior belt is close to saturation. From the plot in Figure 12d we can see that the kit throughput of the *2x2* configuration rises with the decrease of the kit size. Figure 12f is the product of the kit throughput, that is, the plot in Figure 12d, and the kit size.

The A of the superior belt for the *2* configuration, as can be seen in plot Figure 12b, starts from values near to 0.82 and drops for decreasing kit size. From the plot Figure 12d is clear that its performance is worse than the *2x2* configuration but outperforms the *3* configuration for a low kit size (1 and 2). Finally the A of the superior belt for the *3* configuration, as can be seen from the plot Figure 12b, is similar to the one of the *2* configuration, but for decreasing kit size it shows a heavier drop. From the plot Figure 12d is clear that for a kit size lower or equal than 2 the throughput of the system is stable, due to the saturation of the weighing feeder. This means a reduction of component throughput, as can be seen from plot Figure 12f. It is to be noticed that all the plots in Figure 12 are of minor importance for small kit sizes, due to the assumption made in the introduction that the feeding of low quantity components is addressed with the development of a flexible robotic workcell.

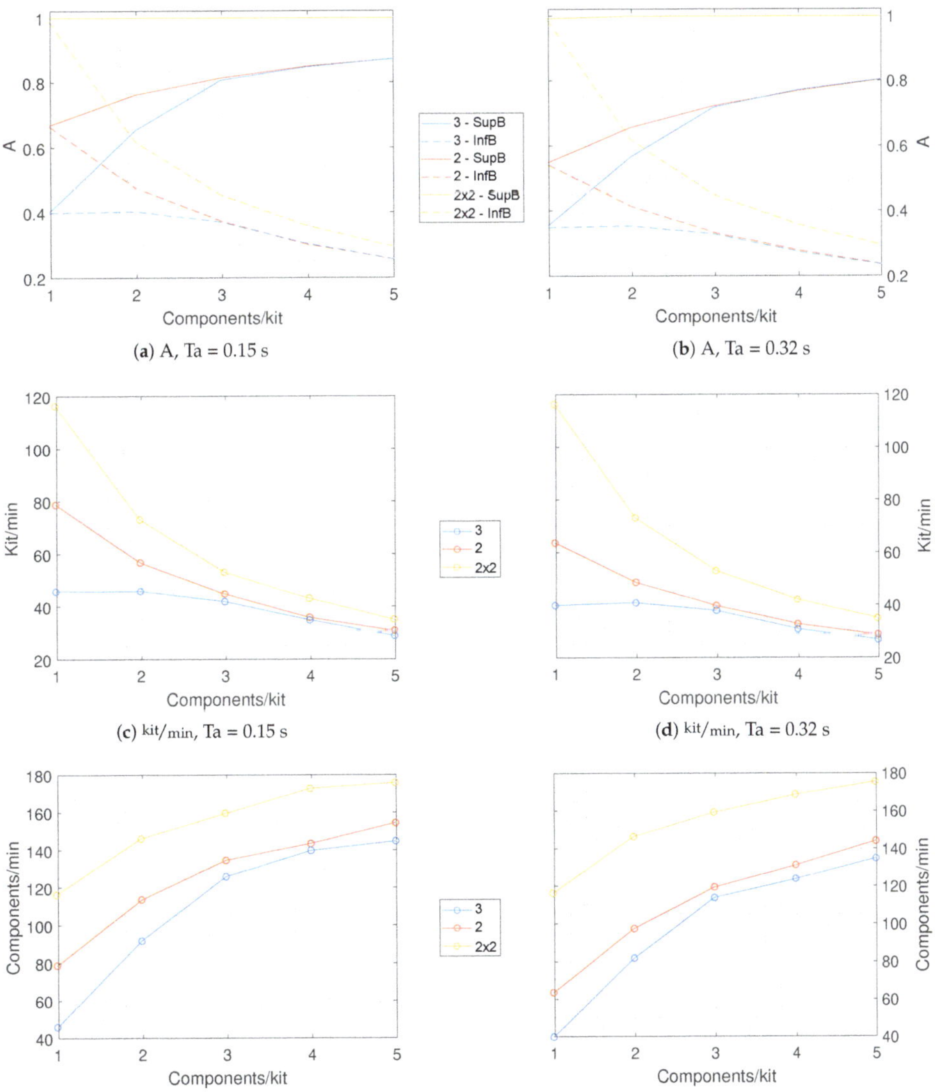

(**a**) A, Ta = 0.15 s (**b**) A, Ta = 0.32 s

(**c**) kit/min, Ta = 0.15 s (**d**) kit/min, Ta = 0.32 s

(**e**) *pieces/min*, Ta = 0.15 s (**f**) *pieces/min*, Ta = 0.32 s

Figure 12. Throughput and availability of the system varying the opening and closing time of the hoppers.

Figure 13 depicts the behavior of the system for different T_{mov}. The left column of plots is defined for $T_{mov} = 0.30$ s, while the right column is defined for $T_{mov} = 0.60$ s, the last taken from the case study line. From Figure 13b it is clear that the *2x2* configuration is strongly affected by a slower belt and its *A* drops for a kit size smaller than one. The dashed yellow line shows that for that kit size the inferior belt starts to be saturated and cannot accept higher throughput, causing a decrease in the *A* of the previous devices. It is noticeable that the *3* and the *2* configuration in the two cases are almost not affected by the time required by the inferior belt to perform a shift, as we can see from the comparison between Figure 13c,d.

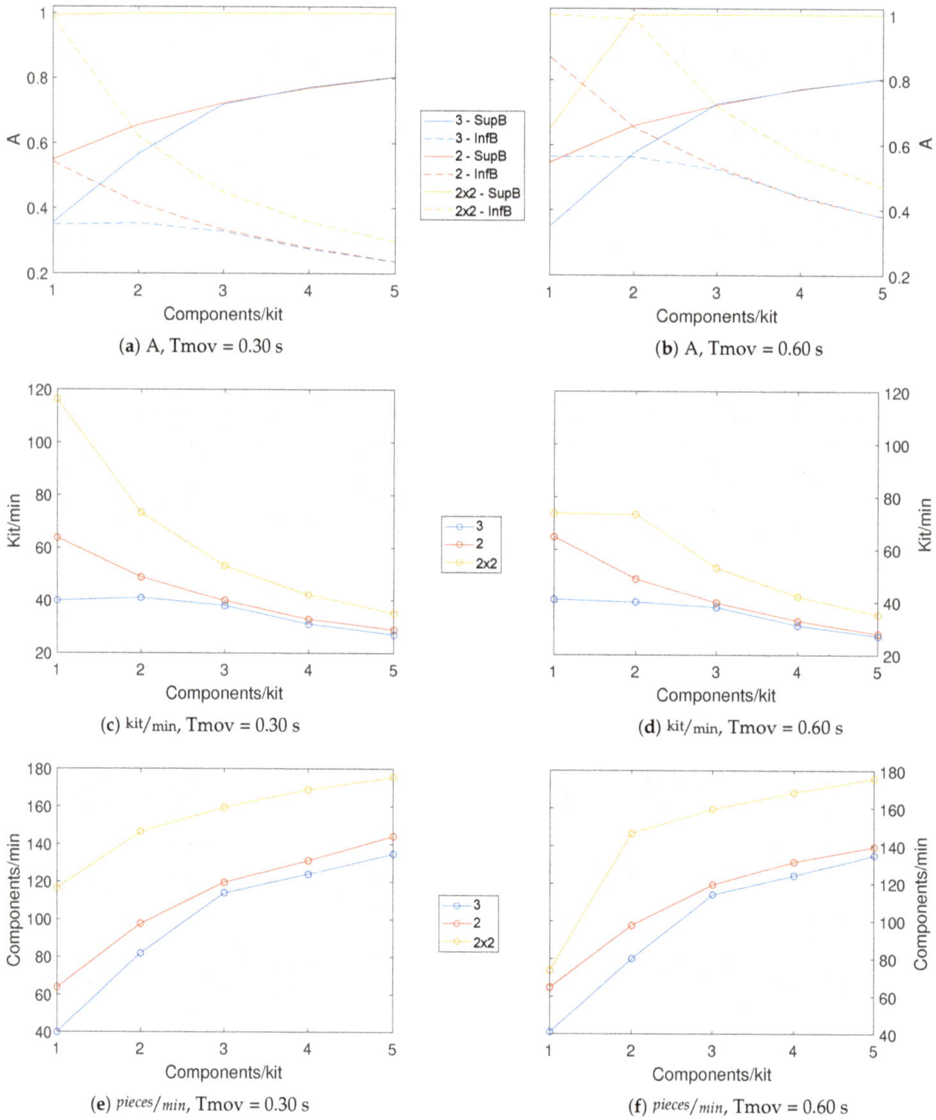

(**a**) A, Tmov = 0.30 s (**b**) A, Tmov = 0.60 s (**c**) kit/min, Tmov = 0.30 s (**d**) kit/min, Tmov = 0.60 s (**e**) *pieces/min*, Tmov = 0.30 s (**f**) *pieces/min*, Tmov = 0.60 s

Figure 13. Throughput and availability of the system varying the time required to the inferior belt to shift to one position to another.

It is important to notice that the *2x2* configuration would be capable of feed two inferior belts as depicted in Figure 1, so that the T_{mov} would be approximately half compared to the other two configurations. The comparison between the configurations would make the *2x2* configurations even more valuable.

2.2.3. Vision Inspection System

In the Industrial Robotics Laboratories at the University of Padua a prototype of the vision quality inspection system placed onto the superior belt was developed. The system is composed by a AVT *Pike F032-b* camera, triggered by a Sick *DRS61* incremental encoder. The belt's width is 400 mm, its speed is set equal to 200 mm/s. The pieces are put on the conveyor belt and they run on the backlighted area, while the camera acquires images, as illustrated in Figure 14c.

The matrix Pike camera is used as a quasi-linear camera, acquiring 6 lines at a time. It is synchronized with the belt thanks to the encoder. The acquisition is performed by a C++ algorithm which stores one hundred acquisition into a buffer and then send all the content of the buffer to Matlab, where is performed the image elaboration and then controlled the air blow to discard defective pieces.

The image processing consists in a sum of the pixel intensity value performed by row, obtaining the profile illustrated in Figure 14b. A check on the length of the piece was used to detect defects, but further works could lead to the development of more sophisticated methods, thanks to the clear profile of the objects obtained with this setup.

It is really important to notice that with the introduction of a vision inspection system that is able to discard the wrong or defective pieces, the retooling also becomes easier. In fact, bowl feeders require an amount of pieces higher than the actual production lot to be handled, in order to guarantee a uniform output flow [16]. This leads to high retooling time, due to the need to empty the feeder of the previous lots, before filling it with the new components. The vision inspection system could lead, together with the adoption of a flexible feeder, to a dynamic lot change, removing the need to empty the feeder and letting the vision inspection system discard the previous components.

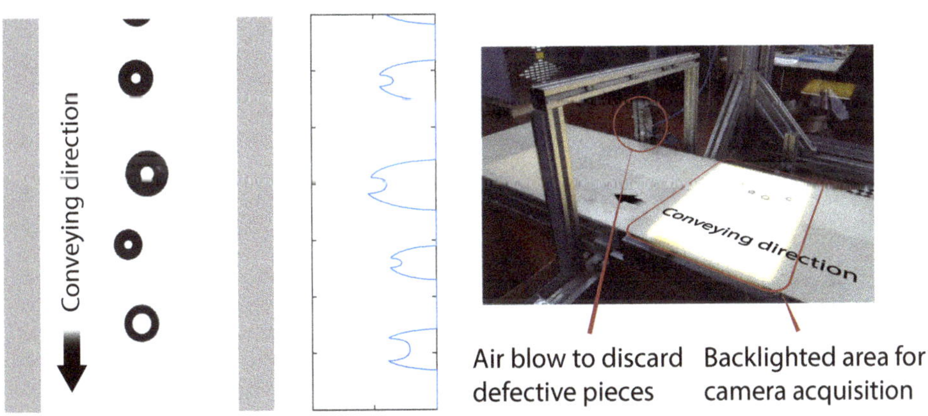

(**a**) Snaphshot of the camera. (**b**) Profile obtained from image processing. (**c**) Experimental setup.

Figure 14. Vision system experiment.

3. Case Study

To demonstrate the effectiveness of the proposed solutions, these were applied to the considered case study, with a specific focus on the hoppers cascade system model. The actual system, as described in Section 2.2 as system *3* and represented in Figure 15, is composed of three hoppers of which the middle one performs the weighing. The predictions of the proposed model were compared to the current solution, validating the expected results with the real data. Moreover the possible throughput improvements adopting the *2* and *2x2* configurations were evaluated. Table 6 shows the parameters used in the model.

The feeders of the considered assembly kitting line feed the same number of components and n will be equal to 2 or 4. The comparison will then involve two values for each considered configurations, reported in Table 7.

The throughput obtained for the 3 configuration from the model, corresponding to the first row of Table 7, can be compared with that measured from the case study, which is equal to 22.05 kit/min and to 32.56 kit/min, respectively for n equal to 2 and 4. The resulting error for the 3 configuration is therefore equal to 6.07% and 1.62%. We believe therefore that it is possible to affirm that the model is validated by the case study; moreover it is clear how the *2x2* solution is the most convenient one in terms of throughput, showing an increment in productivity of about 49.6%, which potentially could lead to the removal of two bowl feeders.

Figure 15. Case study hoppers cascade system. In this application the system is composed of three hoppers and a weighing device in the middle one.

Table 6. Values of the parameters for the case study.

Parameter	Value (s)
T_a	0.36
T_c	0.28
T_d	0.45
T_w	1.0
T_{mov}	0.6
n	2–4

Table 7. Throughput obtained from the model for the parameters taken from the case study.

Configuration	Throughput (kit/min)	
	n = 2	n=4
3	23.39	33.99
2	23.39	36.03
2x2	28.85	50.85

4. Results Discussion

Since the model has been verified, the impact on the efficiency of the assembly kitting line can be evaluated. The introduction of the flexible feeder device allows to remove two bowl feeders in the case study considered, the ones designated to supply the parts with low quantity for kit. Moreover, the model of the hoppers of the rigid feeders, as stated by the analysis of the case study, allowed to remove two more, therefore the resulting H-FAS is composed by:

- four rigid feeders with hoppers devised to supply the high volume components as described before;
- one fully flexible feeder to supply low volume components;
- one manipulator in charge of picking the part from the flexible feeder.

In order to evaluate the improvements of the proposed solution, the OEE was chosen as a proper parameter. This is defined as the product of three different values: the *availability*, that takes into account any events that stop the planned production, the *quality*, i.e., the number of parts that does not match the requirements, and the *performance*, i.e., the devices speed.

The first parameter considered, the *availability* D, was evaluated as:

$$D = \frac{T_{batch}}{n \cdot T_{setup,i} \cdot T_{batch}} - 0.05 \tag{10}$$

where T_{batch} (h) is the production time for the single batch, $T_{setup,i}$ (min) is the time needed for the setup of each of the n feeders, such as $T_{setup,i}$ times n equals to the total T_{setup}. We considered n equal to 8 in the original case and 4 in the proposed development, since the flexible feeder does not need any setup. Lastly, the 0.05 value is a corrective constant defined in order to consider potential unexpected stops. Figure 16 represents the percentage difference of D between the original system and the proposed one in case of different values of T_{batch} and T_{setup}.

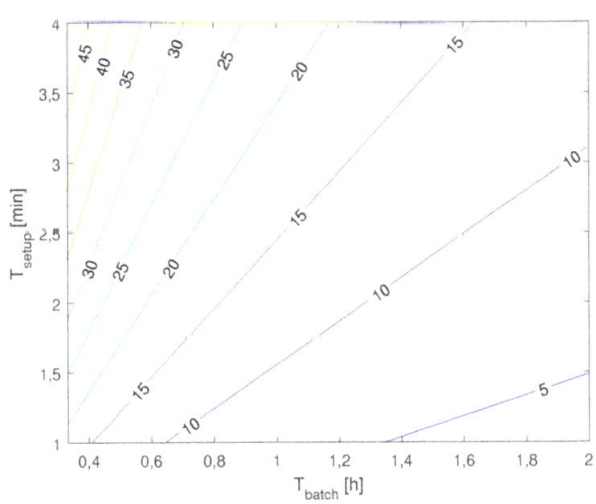

Figure 16. Availability increase (%) by changing the number of re-configurable feeders.

For the quality and performance parameter, a different equation was needed. Defined P_d the probability of defects and P_b the probability of slow cycles for each feeder considered, we could define the total quality parameter Q and the total performance parameter P as

$$Q = 1 - P_D \tag{11}$$

$$P = 1 - P_B \tag{12}$$

where P_D and P_B are evaluated as stated by the law of total probability [17] applied for n equal to 8 and 4 feeders:

$$P_D = P_d \cdot n + P_d^n - (\sum_{k=2}^{n-1} (-1)^k \binom{n}{k} \cdot P_d^k) \tag{13}$$

$$P_B = P_b \cdot n + P_b^n - (\sum_{k=2}^{n-1} (-1)^k \binom{n}{k} \cdot P_b^k) \tag{14}$$

The range value used for P_d is restrained between 0.05% and 0.5%, while the values of P_b are evaluated as 0.7 times P_d, based on the data from the case study. Figure 17 represents the difference of the product of Q and P evaluated for n equal to 8 and 4 in the case of different values of P_d and P_b defined from the examined case study in the considered range.

It is possible to notice a significant increment in the OEE and therefore in the throughput, especially in the *availability* parameter, which could reach an increase of over 40%. This proves the starting hypothesis that the proposed reconfiguration of the feeding system in the case study kitting assembly process could improve the throughput.

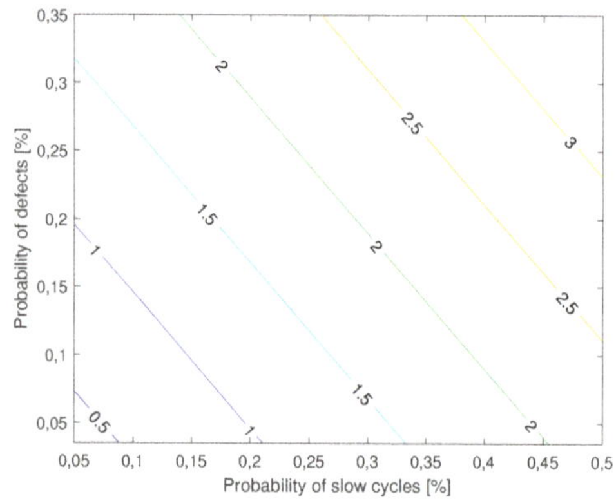

Figure 17. Quality and performance increase (%) by changing the number of re-configurable feeders.

5. Conclusions

We presented a solution in order to increase the efficiency of an assembly kitting line based on an H-FAS system. Firstly, we noticed that a low quantity of components could be supplied by a fully flexible feeder, thus reducing further the number of vibratory feeders. An F-FAS subsystem was studied in the laboratory in order to test the performance of the system in a kitting assembly line.

With regards to the high quantity components, the current bowl feeders could be preserved, re-configuring only the hoppers cascade system. Several studies have been proposed in this field of research; however, they are focused on the material characteristics and not on the the system itself; thus, our work suggested an innovative solution based on several reconfigurations of the feeding system. A mathematical model has been developed and its robustness was verified by means of a sensitivity analysis; moreover, we developed a simulation to evaluate the accuracy of the proposed model, since the simulation is a finite state machine which performs by uploading the states of the different devices.

We noticed that, for a certain range of opening time, the throughput is not so much influenced; we believe that rising the speed of the lower belt could present further improvements for the alternative configurations developed from 3. The requested weighing device can be replaced with the adoption of a vision system, similar to the one suggested in this work. This was tested in the laboratory and showed an initial but effective solution for quality inspection.

Lastly, we evaluated the increased throughput of the system by calculating the percentage increase in the OEE for the original and the proposed configuration. While the increase in quality and performance could be considered not so interesting (at maximum an increase of about 3%), the availability could reach an increase of over 40%.

Future research aims to develop the vision system in order to be more effective by studying different methods for image processing and data processing in order to recognize further defects. Moreover, the hoppers could present other configurations that were not considered in the current study. Finally, in future works the presented model will be tested in further case studies.

Author Contributions: Conceptualization, N.C. and G.R.; Formal analysis, M.F. and R.M.; Investigation, N.C.; Methodology, N.C. and R.M.; Project administration, G.B. and G.R.; Resources, G.B., M.F. and G.R.; Software, N.C.; Supervision, G.B., M.F. and G.R.; Writing—Original Draft, N.C., M.F. and R.M.; Writing—Review & Editing, G.B., N.C., M.F., R.M. and G.R.

Funding: This research received no external funding.

Conflicts of Interest: The authors declare no conflict of interest.

References

1. Azzi, A.; Battini, D.; Faccio, M.; Persona, A. Sequencing procedure for balancing the workloads variations in case of mixed model assembly system with multiple secondary feeder lines. *Int. J. Product. Res.* **2012**, *50*, 6081–6098. [CrossRef]
2. Barbazza, L.; Faccio, M.; Oscari, F.; Rosati, G. Agility in assembly systems: A comparison model. *Assembl. Autom.* **2017**, *37*, 411–421. [CrossRef]
3. Rampersad, H.K. *Integrated and Simultaneous Design for Robotic Assembly: Product Development, Planning...*; John Wiley & Sons, Inc.: Hoboken, NJ, USA, 1994.
4. Heilala, J.; Voho, P. Modular reconfigurable flexible final assembly systems. *Assembl. Autom.* **2001**, *21*, 20–30. [CrossRef]
5. Battini, D.; Faccio, M.; Persona, A.; Sgarbossa, F. New methodological framework to improve productivity and ergonomics in assembly system design. *Int. J. Ind. Ergon.* **2011**, *41*, 30–42. [CrossRef]
6. Cohen, Y.; Faccio, M. The combined effect of multiple models sequencing and absenteeism/turnover on the throughput of assembly lines. In Proceedings of the 17th Industrial Engineering and Management Conference, Tel Aviv, Israel, 27–28 March 2012; pp. 27–28.
7. Edmondson, N.; Redford, A. Generic flexible assembly system design. *Assembl. Autom.* **2002**, *22*, 139–152. [CrossRef]
8. Heilala, J.; Montonen, J.; Väätäinen, O. Life cycle and unit-cost analysis for modular reconfigurable flexible light assembly systems. *Proc. Inst. Mech. Eng. Part B J. Eng. Manuf.* **2008**, *222*, 1289–1299. [CrossRef]
9. Rosati, G.; Faccio, M.; Carli, A.; Rossi, A. Fully flexible assembly systems (F-FAS): A new concept in flexible automation. *Assembl. Autom.* **2013**, *33*, 8–21. [CrossRef]
10. Rosati, G.; Faccio, M.; Barbazza, L.; Rossi, A. Hybrid fexible assembly systems (H-FAS): Bridging the gap between traditional and fully flexible assembly systems. *Int. J. Adv. Manuf. Technol.* **2015**, *81*, 1289–1301. [CrossRef]
11. Bevilacqua, M.; Ciarapica, F.; De Sanctis, I.; Mazzuto, G.; Paciarotti, C. A Changeover Time Reduction through an integration of lean practices: a case study from pharmaceutical sector. *Assembl. Autom.* **2015**, *35*, 22–34. [CrossRef]
12. "Origin of OEE", OEE Foundation. Available online: http://www.oeefoundation.org/origin-of-oee/ (accessed on 15 July 2015).
13. Cox, D.J.; Lewandowski, W. Design for automation and automated manufacturing of drug testing kits. In Proceedings of the 2005 IEEE International Conference on Systems, Man and Cybernetics, Waikoloa, HI, USA, 12 October 2005; Volume 2, pp. 1194–1199.
14. Engisch, W.E.; Muzzio, F.J. Feedrate deviations caused by hopper refill of loss-in-weight feeders. *Powder Technol.* **2015**, *283*, 389–400. [CrossRef]
15. HALCON Solution Guide II-B, Shape-Based Matching. Available online: http://download.mvtec.com/halcon-9.0-solution-guide-ii-/b-shape-based-matching.pdf (accessed on 4 August 2019).
16. Boothroyd, G. *Assembl. Autom. and Product Design, Second Edition (Manufacturing Engineering and Materials Processing)*; CRC Press, Inc.: Boca Raton, FL, USA, 2005.
17. Zwillinger, D.; Kokoska, S. *CRC Standard Probability and Statistics Tables and Formulae*; CRC Press, Inc.: Boca Raton, FL, USA, 1999.

robotics

MDPI

Article

Analysis of the Compliance Properties of an Industrial Robot with the Mozzi Axis Approach

Alberto Doria [1], Silvio Cocuzza [1,*], Nicola Comand [2], Matteo Bottin [2] and Aldo Rossi [1]

1. Department of Industrial Engineering, University of Padova, 35131 Padova, Italy; alberto.doria@unipd.it (A.D.); aldo.rossi@unipd.it (A.R.)
2. Department of Management and Engineering, University of Padova, 36100 Vicenza, Italy; nicola.comand@phd.unipd.it (N.C.); matteo.bottin.1@phd.unipd.it (M.B.)

* Correspondence: silvio.cocuzza@unipd.it; Tel.: +39-049-827-6793

Received: 2 August 2019; Accepted: 9 September 2019; Published: 11 September 2019

Abstract: In robotic processes, the compliance of the robot arm plays a very important role. In some conditions, for example, in robotic assembly, robot arm compliance can compensate for small position and orientation errors of the end-effector. In other processes, like machining, robot compliance may generate chatter vibrations with an impairment in the quality of the machined surface. In industrial robots, the compliance of the end-effector is chiefly due to joint compliances. In this paper, joint compliances of a serial six-joint industrial robot are identified with a novel modal method making use of specific modes of vibration dominated by the compliance of only one joint. Then, in order to represent the effect of the identified compliances on robot performance in an intuitive and geometric way, a novel kinematic method based on the concept of "Mozzi axis" of the end-effector is presented and discussed.

Keywords: robot; compliance; machining; modal testing; Mozzi axis

1. Introduction

The compliance properties of industrial robots are very important for many industrial applications, such as automatic robotic assembly and material removal processes (e.g., machining and deburring). On the one hand, in robotic assembly, joint compliance can be useful for compensating dimensional errors in the parts to be assembled; on the other hand, in material removal processes, a low Cartesian compliance (high stiffness) of the end-effector is required. Indeed, still very few robots have been applied in this economically important application area [1] mainly due to their low stiffness. Moreover, the compliance properties of robots appear very important in emerging fields such as flexible assembly systems [2,3] and collaborative robotics [4].

From a static point of view, low stiffness causes imprecise products, due to the robot deflections during the robotic task. From a dynamic point of view, low frequency chatter vibrations [5–7] can be induced when low-stiffness robots are used, with an impairment in the quality of the machined surface. Moreover, induced vibrations cause a reduction of tool life and can damage robot joint transmission.

In particular, the low joint stiffness of the robot is one of the main issues in using robotic machining instead of CNC machining. Indeed, it is well established in the literature (see, for example, [8]) that the dominant contribution factor for a large displacement of the robot end-effector is joint compliance (mainly due to gear transmission elasticity), while link flexibility can be neglected. Moreover, the stiffness of an industrial robot is usually on the level of 10^6 N/m (with a base natural frequency of robot around 10 Hz), while a standard CNC machine has stiffness on the level of 10^8 N/m (with a base natural frequency of several hundred Hz or even more than 1000 Hz) [5]. Sometimes, combinations of serial and parallel kinematic chains have been proposed to increase robot stiffness [9].

The static force/torque-deformation experiments and identification are widely used to obtain the joint stiffness of industrial robots with good accuracy [10]. During the static experiments, a set of forces/torques is applied to the robot end-effector (or to a single joint) in different robot configurations, while the displacement sensors (e.g., vision system, CMM, or laser sensors) measure the steady-state deformation of the end-effector. Therefore, the Cartesian stiffness of the robot can be calculated. Then, through the analytical relation between the joint and Cartesian stiffness based on the kinematic model of robot, many types of identification methods (e.g., least squares or genetic algorithms) can be employed to obtain the joint stiffness. Most of the works in the literature (e.g., see [8,11–14]) neglect link flexibility and identify the joints' rotational stiffness (modeled with linear rotational springs) using this method. In [15], the robotic arm is modeled considering rigid links and three lumped rotational springs for each joint to take into account joint compliance, bearings compliance, and link deflections, which are all identified with the above method. In [16], an analysis approach in which both joint and link stiffness are considered is presented.

It is worth mentioning that in [13], the identified joint stiffnesses are then used in an analytical model, in which natural frequencies and mode shapes are validated through modal testing experiments. In particular, a good match is demonstrated between the first two experimental natural frequencies and related mode shapes and the corresponding ones predicted by the analytical model.

In [17], experimental modal analysis is used to identify the joint and base stiffness of an industrial robot represented with a four-degrees-of-freedom (four-DOF) planar model.

In robotic processes, the directions along which the robot arm is compliant are very important. The stiffness matrix in the joint space does not directly give this information; moreover, for a serial robot, the stiffness matrix in the Cartesian space is not diagonal and it is configuration-dependent. This means that the force and deformation in the Cartesian space are coupled (e.g., a force applied in one direction will cause a deformation in all possible directions) and this can generate some counter-intuitive results. Moreover, due to the difference in compliance in different directions, machining accuracy can be different in different directions.

Since the end-effector of the robot, which is grasping a tool or an object, can be considered a rigid body mounted on the compliant robot, the basic ideas of the screw theory [18] can be adopted to describe its motion.

In this paper, first, a novel method for the identification of robot stiffness is presented (Section 2). It is based on the selective excitation of modes of vibration that chiefly involve only one joint. Since in this specific case modal stiffness coincides with joint stiffness, the latter can be identified from the measured natural frequency and the calculated value of moment of inertia. A serial six-DOF robot is considered (Omron Adept Viper s650), since it is representative of industrial robots used for a wide range of operations. Then, a new method for representing linear and rotational compliance using the concept of screw or Mozzi axis is presented (Section 3). The proposed method is a useful tool to understand how the robot deforms with respect to the working surface, and to highlight the joints that give the largest contribution to robot compliance in specific configurations. Furthermore, robot configurations that place the Mozzi axis in the most favorable position with respect to the working surface can be found. Finally, the compliance properties of the tested robot are analyzed with the proposed method in different robot configurations (Section 4), and conclusions are drawn.

2. Identification

2.1. Testing Equipment and Method

Stiffness (or compliance) properties of robot joints are usually measured by means of static tests [13] carrying out measurements in the Cartesian space. Then, a mathematical model is used to correlate force vector $\{F\}$ and deformation vector $\{\Delta x\}$ in Cartesian space with the stiffness matrix $\left[K_q\right]$ in the joint space. $\left[K_q\right]$ is diagonal and the following equation holds [8]:

$$\{\Delta x\} = [J(q)]\left[K_q\right]^{-1}[J(q)]^T\{F\} \tag{1}$$

where $[J(q)]$ is the Jacobian matrix that depends on robot configuration. Matrix $\left[K_q\right]$ can be obtained from Equation (1) by least-square method considering some different configurations of the robot [8,13].

In this paper, a different approach based on modal testing is adopted. The basic idea is to excite modes of vibration dominated by the compliance of one joint and to identify stiffness (or compliance) from the measured modal frequency and the moment of inertia about the joint axis. The moment of inertia is calculated by means of the CAD (Computer Aided Design) model of the robot.

The equipment needed to carry out this kind of test is rather simple and cheap; it includes a hammer for modal testing, an accelerometer, and a data acquisition board. In the framework of this research, a PCB 086C01 hammer (with load cell sensitivity 0.2549 mV/N), a PCB 356A17 triaxial accelerometer (sensitivity 50.5 mV/(m/s^2)) built by PCB Piezotronics, Inc., Depew, NY, USA, and a NI9234 data acquisition board were used built by National Instruments, Austin, TX, USA. After some preliminary tests, a sampling frequency of 1024 Hz and 2048 samples were selected. Measured signals were analyzed by means of ModalVIEW R2, a specific software for modal analysis.

Since motions of the base of the robot may have a negative effect on the quality of measurements, the robot was rigidly fastened to a large steel base (see Figure 1) and an accelerometer was used to monitor residual base motions.

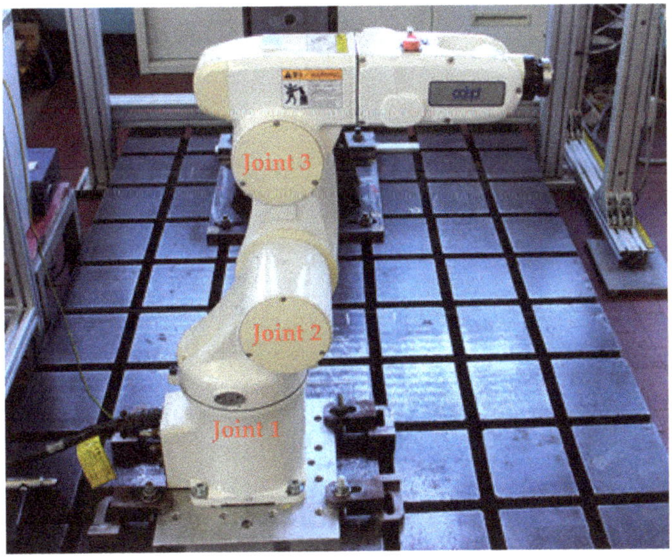

Figure 1. The robot in the testing rig.

When a serial six-joint robot is loaded by a force applied on the end-effector, on the one hand, this force produces the largest moments about the first joints, and on the other hand, the rotations of the first joints due to compliance have the largest effect on end-effector positioning errors, due to the large lever arms [19]. For these reasons, the present research is limited to the identification of the compliance properties of the first three joints of the robot. The last three joints are considered ideal revolute joints that contribute to define the pose of the end-effector, but are infinitely stiff. In the future, the method will be used to identify the compliance of all the joints, if the compliance of the wrist joints will appear relevant in some configurations.

Since in actual working conditions the joints of the robot are held in the desired configuration by the servomotors, the tests were carried out with the servomotors switched on. The identified stiffnesses are the result of the mechanical stiffness of the transmission plus the stiffness of the servo.

In order to identify the stiffnesses of the first three joints of the robot, three specific test conditions were defined (see Table 1). The first condition, which is represented in Figure 2, aimed to highlight the stiffness of the first joint. For this reason, the end-effector flange was excited by a lateral force (y direction) and joint rotations were set in order to achieve the maximum arm extension. In this configuration, the first mode of vibration should be dominated by the compliance of the first joint.

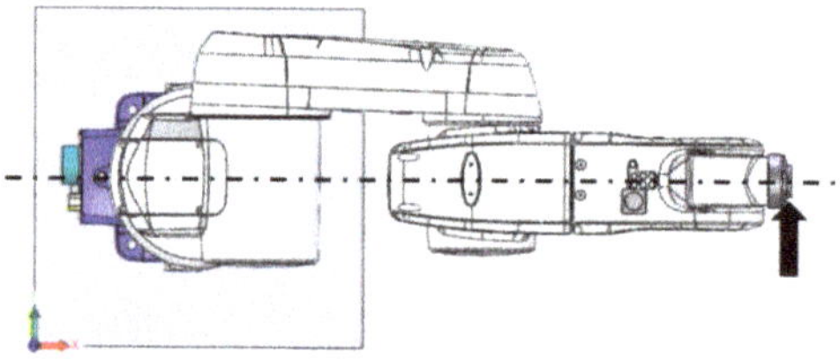

Figure 2. Test configuration 1 for the identification of the stiffness of joint 1.

Table 1. Robot configurations.

Configuration	Joint 1 [°]	Joint 2 [°]	Joint 3 [°]	Joint 4 [°]	Joint 5 [°]	Joint 6 [°]
Test 1	0	0	90	0	0	0
Test 2	0	0	16.96	0	−16.96	0
Test 3	0	−37.54	176.73	0	49.20	0
Validation	0	30	30	0	0	0

The second test condition (Figure 3) aimed to highlight the stiffness of joint 2. For this reason, the robot arm was configured with the approach axis of the end-effector pointing upwards and intersecting the axis of joint 3, and a vertical force was exerted by the hammer. Since in this test configuration the impact force intersected the axis of joint 3 and was parallel to the axis of joint 1, the compliances of these joints should not have influenced the fundamental mode of vibration.

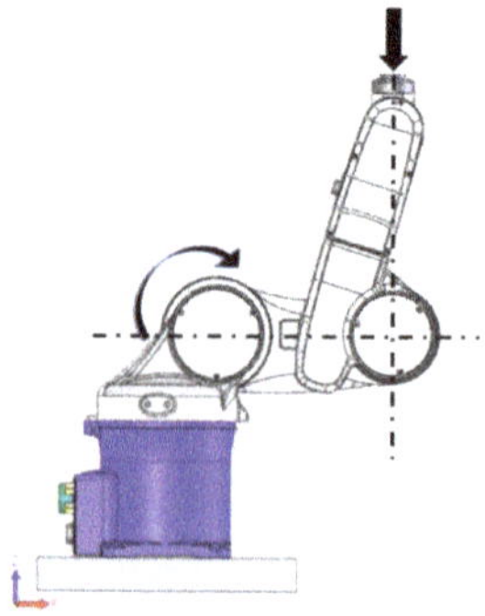

Figure 3. Test configuration 2 for the identification of the stiffness of joint 2.

Figure 4 shows the third test configuration. In this case, the approach axis of the end-effector was aligned to the x axis and intersected both the axis of joint 2 and the axis of joint 1. Therefore, the fundamental mode of vibration excited by a hammer force in x direction should have been dominated by the stiffness of joint 3.

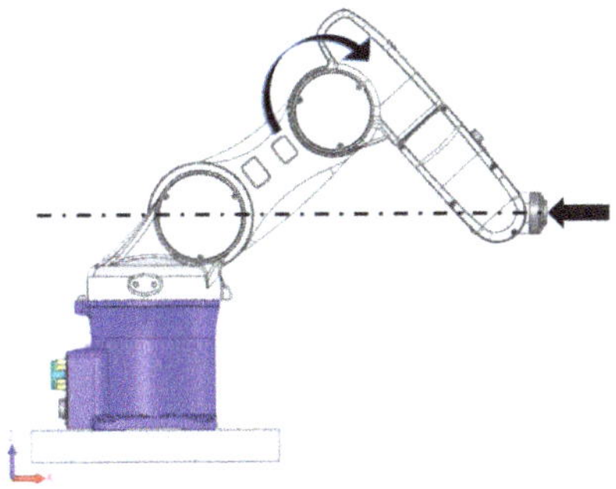

Figure 4. Test configuration 3 for the identification of the stiffness of joint 3.

The tests were carried out with the modal analysis approach [20], which is widely used in the fields of automotive engineering [21,22], aerospace engineering [23], automatic machines [24], and robotics [25]. A grid of measurement points was defined on the robot (see Figure 5); points 10 and 11 were used to monitor base vibrations. In each test configuration, the hammer impact was always exerted on the end-effector flange in the defined direction, whereas the triaxial accelerometer was moved to the various grid points (rowing response approach). In this way, for each configuration, 33 frequency response functions (FRFs) were measured between the 3 acceleration components of the 11 grid points and the hammer impact force. In order to improve the repeatability and quality of measurements, each FRF was calculated averaging the results obtained with three hammer blows. Measured data were then processed with ModalVIEW in order to identify natural frequencies, modal dampings, and modal shapes.

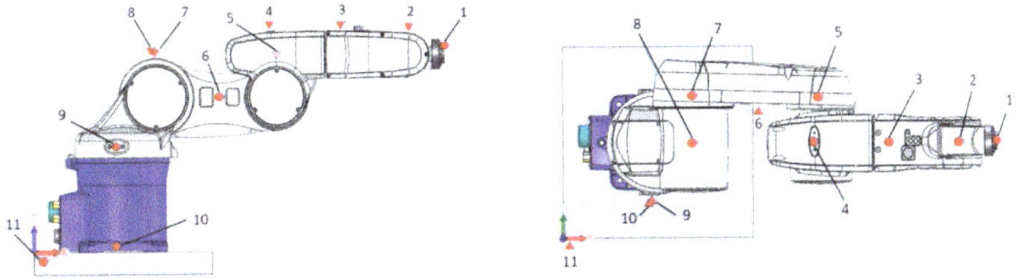

Figure 5. Set of measurement points: (**a**) side view; (**b**) top view.

2.2. Experimental Results

Figure 6 shows the overlay of FRFs measured in configuration 1. The modulus plot highlights the presence of the fundamental mode at about 13 Hz, and the phase plot corroborates this fact, showing large phase changes at the same frequency.

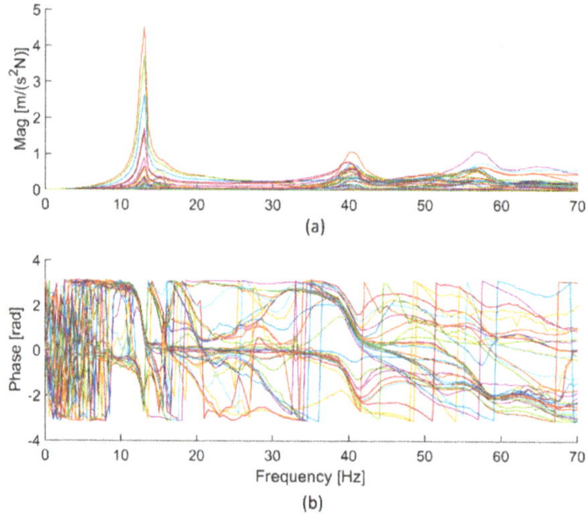

Figure 6. FRFs measured in the first test configuration (modulus (**a**) and phase (**b**)).

The identification, carried out with the algorithm Quick Fit of ModalVIEW, made it possible to find a natural frequency of 13.0 Hz, with a viscous damping ratio [26] of 2.3%. The natural frequency is the main parameter needed to identify the stiffness of the joint, but the shape of the identified mode is important as well, since it makes possible the verification of the hypothesis that the selected joint dominates the mode of vibration. The mode of vibration corresponding to the natural frequency of 13.0 Hz is represented in Figure 7; it is dominated by displacements in the x–y plane caused by the rotation of joint 1, and the contribution of the other joints and of link flexibility to this mode appears negligible. Therefore, the identified frequency is suited to identify the stiffness of joint 1.

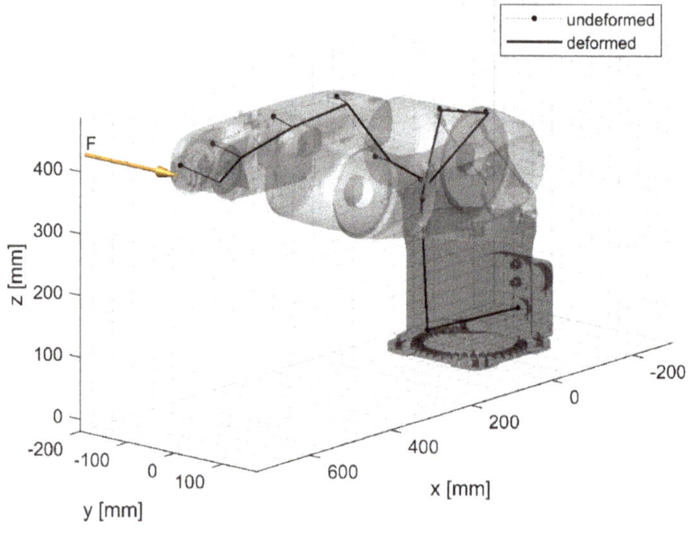

Figure 7. Mode at 13.0 Hz for the identification of joint 1 compliance.

The small resonance peaks at about 40 and 57 Hz in Figure 6 are related to the excitation of higher-order modes that also involve the compliance of the other joints of the robot.

The results of the modal tests carried out in order to identify the compliance of joint 2 are presented in Figures 8 and 9. The overlay of the measured FRFs shows the first peak at about 18 Hz. Modal analysis made it possible to identify a mode of vibration with natural frequency 17.5 Hz and viscous damping ratio of 1.9%. The shape of this mode of vibration (Figure 9) is characterized by displacements dominated by the rotation of joint 2, therefore it is suited to identify the compliance of this joint.

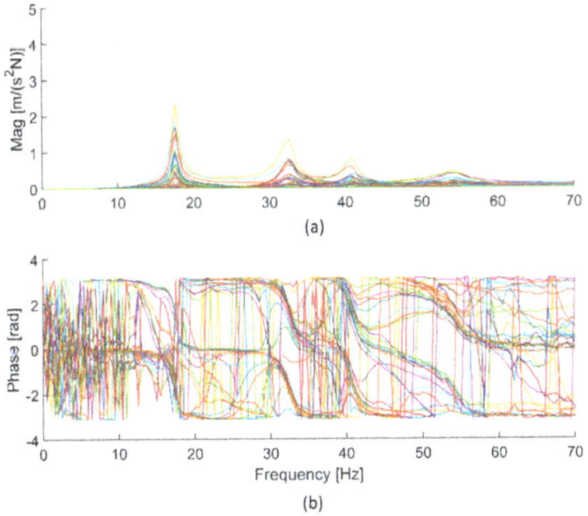

Figure 8. FRFs measured in the second test configuration (modulus (**a**) and phase (**b**)).

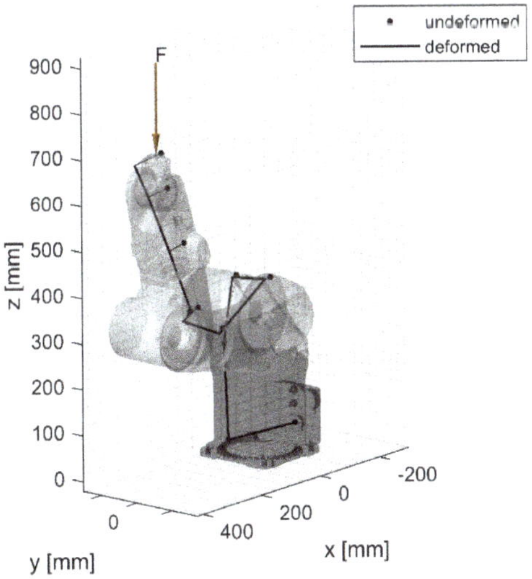

Figure 9. Mode at 17.5 Hz for the identification of joint 2 compliance.

The minor resonance peaks at about 32 and 40 Hz are related to the excitation of higher-order modes that also involve the compliance of the other joints of the robot.

Finally, Figure 10 shows the FRFs measured in the third configuration, in order to identify the compliance of joint 3. Two resonance peaks appear. The first is at about 14 Hz and the second, which is the highest, is at about 33 Hz. Modal analysis results show that the first peak (the minor) corresponds to a mode dominated by the compliance of joint 2; this mode was excited since the hammer force did not intersect exactly the axis of joint 2. The mode at 33.5 Hz (with viscous damping ratio of 2.7%) is dominated by joint 3 compliance and is suited to identify the compliance of this joint (see Figure 11).

It is worth noticing that the damping ratios of the selected modes are similar; this is a good clue that highlights that these modes chiefly involve the joints, which have similar transmissions and servos that are characterized by similar dissipation phenomena.

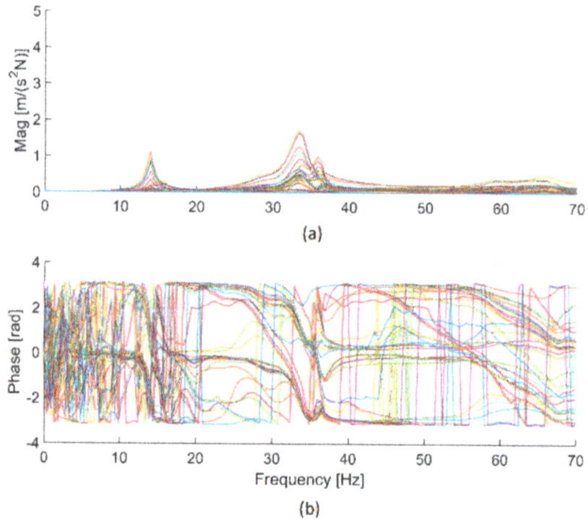

Figure 10. FRFs measured in the third test configuration (modulus (**a**) and phase (**b**)).

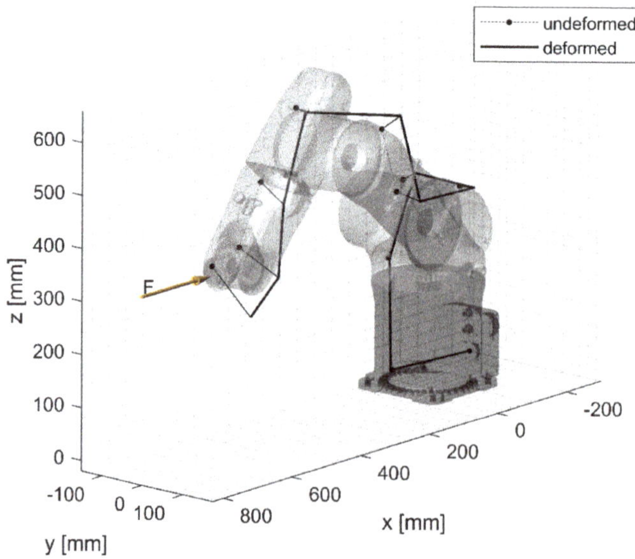

Figure 11. Mode at 33.5 Hz for the identification of joint 3 compliance.

A linear vibrating system with N DOF is described by a system of coupled second-order differential equations in space coordinates. The modal approach transforms the N DOF vibrating system into N independent 1-DOF vibrating systems [20,27] governed by these equations:

$$m_k\ddot{\eta}_k + c_k\dot{\eta}_k + k_k\eta_k = F_k(t) \quad k = 1 \ldots N \tag{2}$$

where m_k, c_k, k_k are the modal mass, damping, and stiffness, respectively. η_k is the kth modal coordinate and $F_k(t)$ is the modal force. The natural frequency is:

$$\omega_k = \sqrt{\frac{k_k}{m_k}} \tag{3}$$

If robot vibrations are considered, in general, modal stiffness does not coincide with joint stiffness, because the modes of vibration involve rotation about several joints. However, if a mode i exists that is dominated by rotation about joint i, the modal stiffness k_i is a good approximation of joint stiffness kq_i. Therefore, joint stiffness can be calculated from Equation (3) with $k = i$. In this case, modal mass m_i coincides with the moment of inertia of the robot about joint i. These moments of inertia were calculated using the CAD model of the robot.

Since the manufacturer gives the total mass of the robot (28 kg) but does not give information about mass distribution, the total volume of the robot and of the various links were carefully calculated and the total mass was distributed to the links proportionally to their volumes. In the future, the calculated values will be checked with other methods.

The stiffnesses identified with the above-mentioned method are summarized in Table 2; they are similar for the three joints and in agreement with literature values [13].

Table 2. Identified parameters.

Joint	Configuration	Natural Frequency [Hz]	Moment of Inertia [kgm^2]	Stiffness [Nm/rad]	Compliance [rad/(Nm)]
1	Test 1	13.0	1.990	12913	7.74×10^{-5}
2	Test 2	17.5	0.809	9730	1.03×10^{-4}
3	Test 3	33.5	0.229	10240	9.76×10^{-5}

2.3. Validation

In order to assess the validity of the identification method, the stiffness values of Table 2 were implemented in a dynamic model of the robot and the first natural frequencies of the robot were calculated in a configuration (validation configuration) different from the configurations used for identification (see Table 1 and Figure 12).

Then, the robot was modally tested in the validation configuration and the first natural frequencies were experimentally identified. Hammer excitation was applied on the end-effector along three orthogonal directions of the end-effector coordinate system. It is worth noticing that the validation configuration was chosen in order to obtain modes of vibration that involved the compliance of various joints simultaneously.

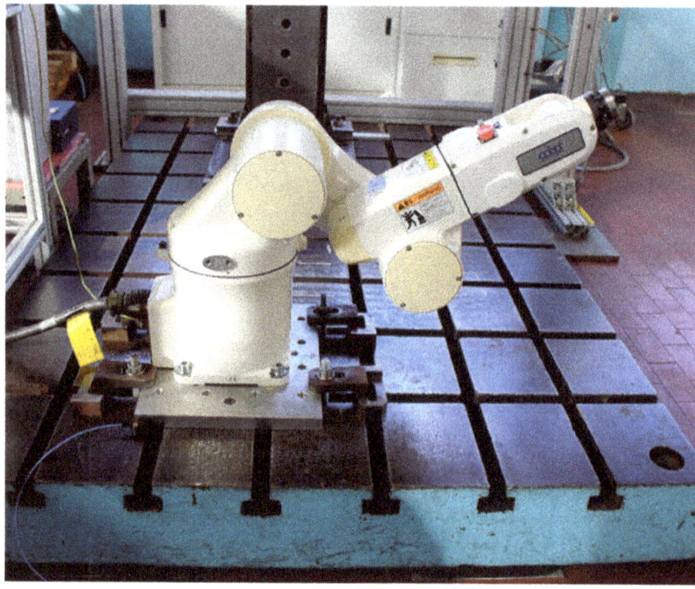

Figure 12. Robot configuration for validation.

The comparison between numerical and experimental values is shown in Table 3. The three testing conditions (with different directions of excitation) essentially lead to the identification of the same natural frequencies. These frequencies are rather close to the numerical ones, with a maximum error of about 3 Hz in the natural frequency of the third mode.

Table 3. Validation tests.

Natural Frequency	Experimental x Excitation [Hz]	Experimental y Excitation [Hz]	Experimental z Excitation [Hz]	Numerical [Hz]
1	16.0	16.0	16.0	15.0
2	-	18.0	17.5	16.5
3	37.0	36.5	35.5	39.0

3. Mathematical Model

When a structure (like a robot) has a polygonal open shape with an end locked to a very stiff structure and an end loaded by a static or dynamic force, the deformation of the loaded end can be studied considering the motion of a rigid body fixed to the loaded end of the structure. The basic concepts of rigid body mechanics can be adopted for describing this motion.

Giulio Giuseppe Mozzi del Garbo was an Italian mathematician who published in 1763 the book "Discorso matematico sopra il rotolamento dei corpi" (Mathematic Discourse on the Roll of Bodies) in which he stated that a generic differential spatial rigid motion can be considered a helical motion around a line, which is called the Mozzi axis [27]. In other words, the rigid motion of a body is represented in every instant by rotation about and translation along the Mozzi axis, which is also known as the instantaneous screw axis. The direction of the Mozzi axis coincides with the instantaneous direction of the angular velocity vector. Afterwards, the ideas of Mozzi were developed by Cauchy in 1827, Poinsot in 1834, Chasles in 1878 [28]. Ball in 1900 developed the modern screw theory introducing the concepts of twist and wrench [29,30]. A twist is the combination of a screw with a parallel linear velocity related to the angular velocity by the pitch of the screw; a wrench is a force acting along the screw combined with a torque related to the force by the pitch of the screw.

In recent years, the interest in the screw theory increased due to the discovery of possible applications in the fields of multibody dynamics [31], vehicle dynamics [32–34], vibration control [35], and robotics. In robotics, the screw theory has been adopted for calculating Jacobians [18], for decomposing the stiffness matrix [36], and for finding its properties [37,38]: the twist-compliant axes and the wrench-compliant axes. A twist about a twist-compliant axis produces only a torque parallel to the rotation axis, whereas a wrench applied on a wrench-compliant axis produces a parallel linear deformation.

In the framework of this research, the concepts of screw theory will be used for a different purpose: the representation of the rotation and translation axis of the compliant motion of the end-effector caused by an external load. This information is very useful when the robot end-effector interacts with an object and the displacement caused by compliance may assist or counteract the performed operation. Sometimes the object is characterized by its own compliance, and in the near future, the same concepts will be adopted to represent the relative rotation and translation between the robot and the object.

In order to accomplish this specific task, the basic concepts of Mozzi or screw axis are sufficient. In Figure 13, a serial six-joint robot is represented. A fixed system of coordinates with origin O_0 and axes $x_0\, y_0\, z_0$ is established. Joint rotations are fixed but compliance is allowed in the first three joints. Point Q is the center of the wrist. Since wrist joints are infinitely stiff, link 3, the wrist, the end-effector, and the tool grasped by the end-effector form a unique rigid body, which is used for Mozzi axis analysis. Point O_t is used for defining the external force on the tool, which causes robot arm deformation. It is worth noticing that if the compliance of some wrist joint were important, the rigid body would begin after the last compliant joint.

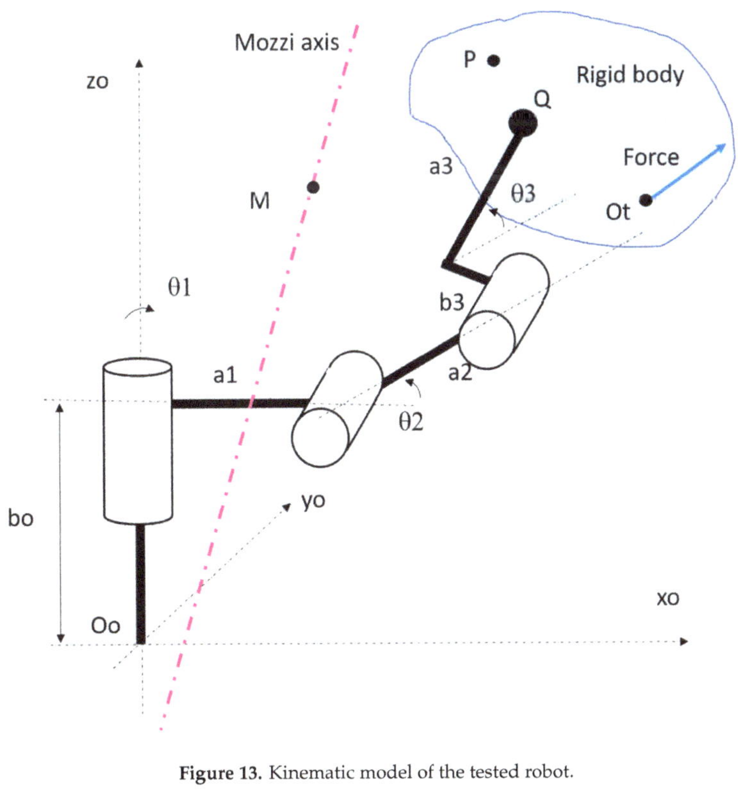

Figure 13. Kinematic model of the tested robot.

The linear velocity $\vec{v}_P$ of a point P belonging to the rigid body can be calculated using the equation of rigid body velocity:

$$\vec{v}_P = \vec{v}_Q + \vec{\omega} \times \overrightarrow{QP} \tag{4}$$

where $\vec{v}_Q$ is the linear velocity of a reference point belonging to the rigid body (wrist center Q), $\vec{\omega}$ is the angular velocity vector, and $\overrightarrow{QP}$ is the vector from Q to P.

The linear velocity of point P can be expressed as a translation along the Mozzi axis plus a rotation about the Mozzi axis according to this equation:

$$\vec{v}_P = k\vec{\omega} + \vec{\omega} \times \overrightarrow{MP} \tag{5}$$

where k is a scalar and M is a point on the Mozzi axis. If the two expressions of $\vec{v}_P$ are equated, the following result holds:

$$\vec{v}_Q + \vec{\omega} \times \overrightarrow{QM} = k\vec{\omega} \tag{6}$$

Velocity $\vec{v}_Q$ in general has a component ($\vec{v}_{Q\|}$) parallel to $\vec{\omega}$ and a component ($\vec{v}_{Q\perp}$) perpendicular to $\vec{\omega}$, whereas the vector product is always perpendicular to $\vec{\omega}$. Hence Equation (6) splits into the two equations:

$$\vec{v}_{Q\perp} + \vec{\omega} \times \overrightarrow{QM} = 0 \tag{7}$$

$$\vec{v}_{Q\|} = k\vec{\omega} \tag{8}$$

The parallel component $\vec{v}_{Q\|}$, which coincides with the linear velocity along the Mozzi axis, can be calculated by means of the inner product:

$$\vec{v}_{Q\|} = \vec{\omega} \cdot \vec{v}_Q \frac{\vec{\omega}}{|\omega|} \tag{9}$$

where $\frac{\vec{\omega}}{|\omega|}$ is the unit vector of $\vec{\omega}$. The perpendicular component ($\vec{v}_{Q\perp}$) can be calculated by difference:

$$\vec{v}_{Q\perp} = \vec{v}_Q - \vec{v}_{Q\|} \tag{10}$$

Moreover, for the perpendicular component, this equation holds:

$$\vec{\omega} \cdot \vec{v}_{Q\perp} = 0 \tag{11}$$

The unknown of the system of Equations (7) and (11) is vector $\overrightarrow{QM}$, which defines the position of a point on the Mozzi axis. Since the angular velocity matrix is skew-symmetric [18], there are infinite solutions for $\overrightarrow{QM}$ that belong to a line; this line is the Mozzi axis. The solution of the system of (7) and (11) can be found using vector algebra as shown in [18]:

$$\overrightarrow{QM} = \frac{\vec{\omega} \times \vec{v}_{Q\perp}}{\omega^2} + \mu\vec{\omega} \tag{12}$$

where μ is an arbitrary scalar parameter. Therefore, the position of a point along the Mozzi axis with respect to origin O_o is given by:

$$\overrightarrow{O_oM} = \overrightarrow{O_oQ} + \overrightarrow{QM} \tag{13}$$

If all the vectors are projected on the fixed coordinate system, these scalar equations define the Mozzi axis:

$$\left\{\begin{array}{c} O_oM_x \\ O_oM_y \\ O_oM_z \end{array}\right\} = \left\{\begin{array}{c} O_oQ_x \\ O_oQ_y \\ O_oQ_z \end{array}\right\} + \left\{\begin{array}{c} (\omega_x\mu)/|\omega| - (\omega_z v_{Qy})/\omega^2 + (\omega_y v_{Qz})/\omega^2 \\ (\omega_y\mu)/|\omega| + (\omega_z v_{Qx})/\omega^2 - (\omega_x v_{Qz})/\omega^2 \\ (\omega_z\mu)/|\omega| - (\omega_y v_{Qx})/\omega^2 + (\omega_x v_{Qy})/\omega^2 \end{array}\right\} \tag{14}$$

When an interval of time dt is considered, Equation (14) can be written in terms of differential displacements (dQ_x, dQ_y, dQ_z) and rotations $(d\theta_x, d\theta_y, d\theta_z)$:

$$\left\{\begin{array}{c} O_oM_x \\ O_oM_y \\ O_oM_z \end{array}\right\} = \left\{\begin{array}{c} O_oQ_x \\ O_oQ_y \\ O_oQ_z \end{array}\right\} + \left\{\begin{array}{c} (d\theta_x\mu)/|d\theta| - (d\theta_z dQ_y)/d\theta^2 + (d\theta_y dQ_z)/d\theta^2 \\ (d\theta_y\mu)/|d\theta| + (d\theta_z dQ_x)/d\theta^2 - (d\theta_x dQ_z)/d\theta^2 \\ (d\theta_z\mu)/|d\theta| - (d\theta_y dQ_x)/d\theta^2 + (d\theta_x dQ_y)/d\theta^2 \end{array}\right\} \tag{15}$$

When differential displacements (dQ_x, dQ_y, dQ_z) and rotations $(d\theta_x, d\theta_y, d\theta_z)$ are calculated by means of Equation (1), Equation (15) can describe the Mozzi axis of a robot with a certain configuration and an assigned force vector $\{F\}$.

The representation of the compliance properties of the robot arm by means of the Mozzi axis has some advantages with respect to the representation based on the Cartesian stiffness matrix. First, the Mozzi axis gives a more intuitive and geometric interpretation of the phenomenon. Second, the Mozzi axis also suggests the origin of the compliance, since its proximity to a joint axis identifies this axis as the most compliant.

4. Mozzi Axis of the Tested Robot

To show the potentialities of the Mozzi axis approach, multiple simulations, in two different scenarios, A and B, were performed. Both A and B configurations were excited by grinding forces along x_0, y_0, and z_0 axes. The two configurations differed in the orientation of the grinding equipment. From a generic point of view, these configurations show how the Mozzi axis changes with a different position of the grinding forces with the same robot joint configuration. Finally, a real case machining scenario [5] was simulated (Figure 14).

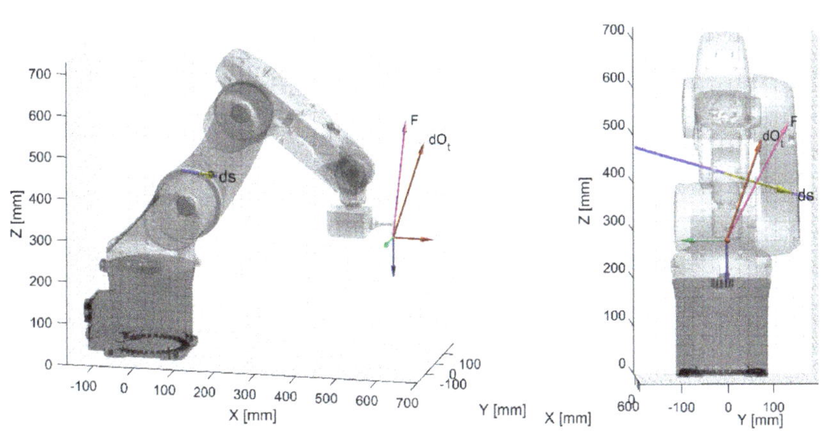

Figure 14. Configuration A excited by a machining force $\vec{F}$ applied on a plane parallel to y_0-z_0 plane, two different points of view.

In Figures 15–17, the robot is excited by a machining force that lies in the vertical plane perpendicular to the axes of joints 2 and 3. The rotation vector of the end-effector is parallel to the axes of joints 2 and 3, therefore the Mozzi axis has the same direction and passes at a distance from the robot base that depends on the compliances of joints 2 and 3. No translation along the Mozzi axis takes place. It is worth noticing that, since in Figure 15 the force has a small lever arm with respect to joint 2, the motion of the end-effector is dominated by compliance of joint 3 and the Mozzi axis passes very close to the center of this joint. In Figures 16 and 17, the vertical force exerts moments about both joint 2 and 3 and the Mozzi axis crosses the robot at an intermediate point.

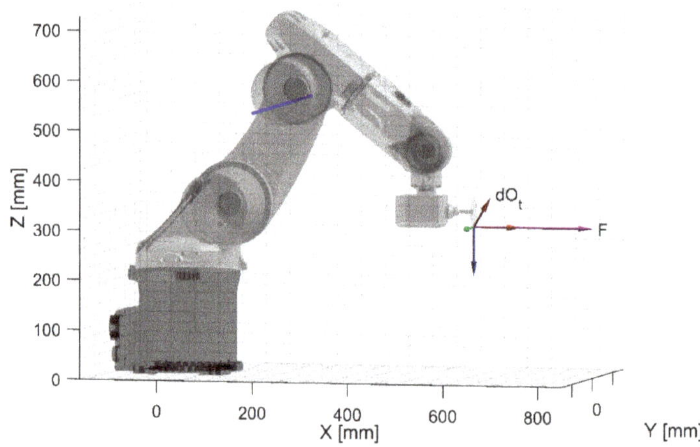

Figure 15. Configuration A with force $\vec{F}$ applied parallel to x_0 axis.

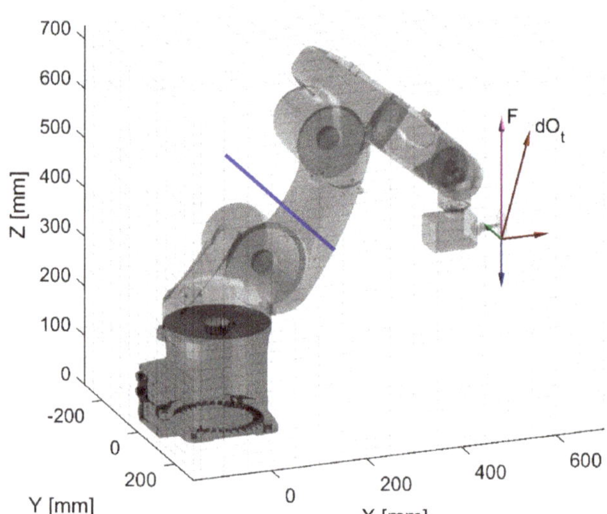

Figure 16. Configuration A with force $\vec{F}$ applied parallel to z_0 axis, bottom view.

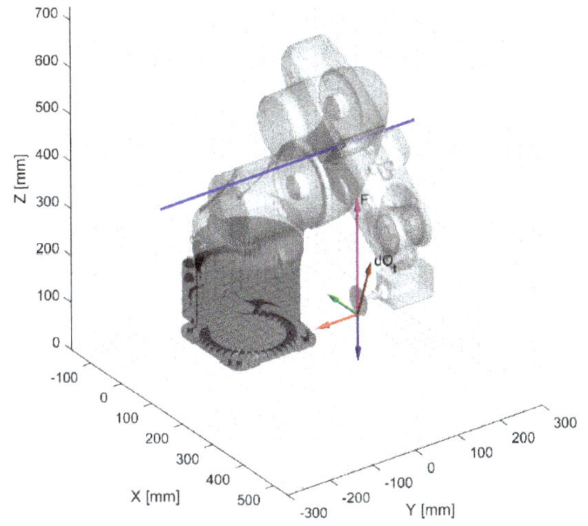

Figure 17. Configuration B with force $\vec{F}$ applied parallel to z_0 axis.

In Figures 18 and 19, the robot is excited by machining forces parallel to the y_0 axis. Since this axis is parallel to the joint 2 and 3 axes, the only joint that can comply to the machining force is joint 1, therefore the Mozzi axis is coincident with the axis of joint 1.

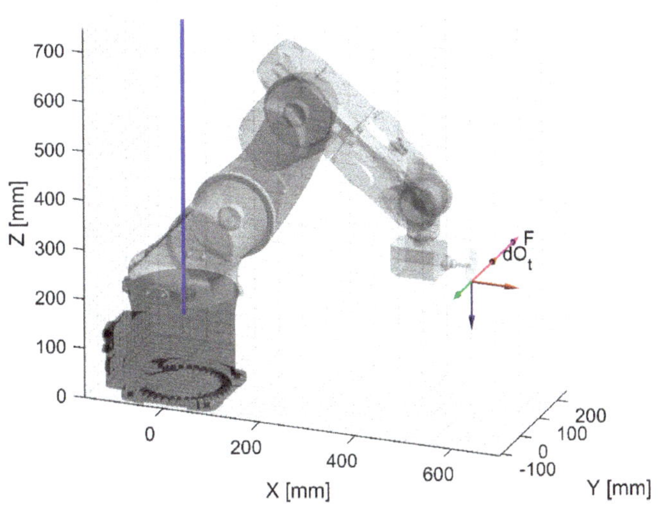

Figure 18. Configuration A with force $\vec{F}$ applied parallel to y_0 axis.

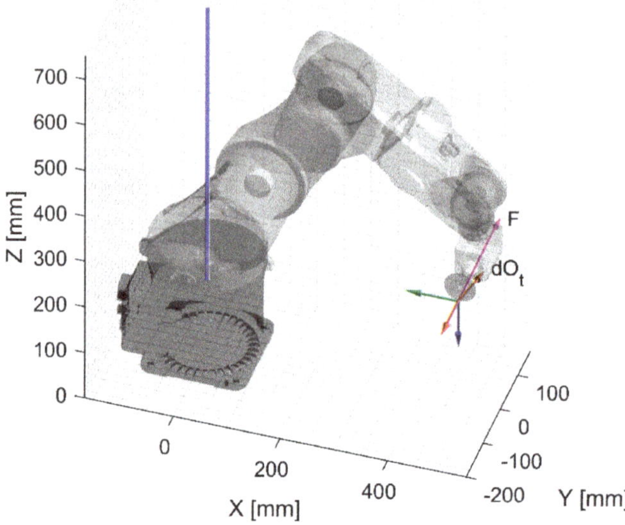

Figure 19. Configuration B with force $\vec{F}$ applied parallel to y_0 axis.

Figure 20 shows how the compliances of all the joints can combine together when excited by an eccentric force. Force $\vec{F}$ is applied parallel to the x_0 axis on a plane perpendicular to the axes of joints 2 and 3, which does not intersect the axis of joint 1. Thus, it excites all the joints and the position of the Mozzi axis results from the combination of all joint compliances. The Mozzi axis is tilted with respect to plane x_0–y_0, and there is a displacement $\vec{ds}$ along the axis as well.

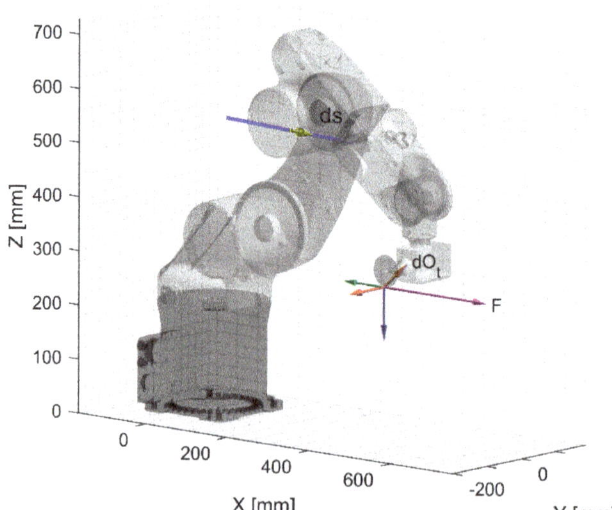

Figure 20. Configuration B with force $\vec{F}$ applied parallel to x_0 axis.

In Figure 14, a real scenario simulation is shown. Force $\vec{F}$ lies on a plane parallel to the y_0–z_0 plane and its direction is defined by the downmilling process described in [5]. In this case, the Mozzi axis has a location similar to the one in Figure 20: the main contribution to the end-effector movement is provided by the compliance of joints 2 and 3, but a small component of force $\vec{F}$ along the y_0 axis results in a non-negligible compliance of joint 1. It is interesting to notice how the high lever arm with respect to joint 1 results in a significant compliance even with a small exciting force.

Another important aspect that can be highlighted by the seven figures is the direction of the Cartesian movement $\vec{dO_t}$ of the point of application of the force. This direction is perpendicular to the Mozzi axis when displacement $\vec{ds}$ along the axis is null and a combination of both rotational and sliding displacements in the general case. Moreover, the direction is not related to the direction of force $\vec{F}$, but it depends on the compliance of the joints. The only relation that can be found between the two vectors is that this movement $\vec{dO_t}$ is placed within the semispace defined by force vector $\vec{F}$.

5. Conclusions

In this paper, two critical issues related to the compliance of industrial robots have been addressed. The first part of the paper has shown that a modal approach based on the selective excitation of some modes of vibration is a valid alternative to static tests for the identification of joint compliances. The second part of the paper has shown that the Mozzi axis of the end-effector of a robot in a certain configuration and with a certain load gives direct and concise information about the location and direction of rotational and linear compliance.

These results have been obtained dealing with a serial six-joint robot with three compliant joints. In the near future, the analysis will be extended to other types of robots with a larger number of joints. The concept of Mozzi axis will be used to explore the possibility of improving the stiffness of a redundant robot in a specific task choosing appropriate configurations of the arm.

Author Contributions: Conceptualization, A.D., S.C., M.B., N.C., and A.R.; methodology, A.D., S.C.; software, N.C., M.B.; validation, N.C.; formal analysis, A.D.; investigation, S.C., N.C., M.B.; writing—original draft preparation, S.C.; writing—review and editing, M.B., A.D.; supervision, A.D., A.R.

Funding: This research was partially funded by Foundation Aldo Gini (call 2018), and by University of Padova DII project BIRD187930.

Conflicts of Interest: The authors declare no conflict of interest.

References

1. Chen, H.; Dong, F. Robot machining: Recent development and future research issues. *Int. J. Adv. Manuf. Technol.* **2013**, *66*, 1489–1497. [CrossRef]
2. Rosati, G.; Faccio, M.; Carli, A.; Rossi, A. Fully flexible assembly systems (F-FAS): A new concept in flexible automation. *Assembly Autom.* **2013**, *33*, 8–21. [CrossRef]
3. Rosati, G.; Faccio, M.; Finetto, C.; Carli, A. Modelling and optimization of fully flexible assembly systems (F-FAS). *Assembly Autom.* **2013**, *33*, 165–174. [CrossRef]
4. Faccio, M.; Bottin, M.; Rosati, G. Collaborative and traditional robotic assembly: A comparison model. *Int. J. Adv. Manuf. Technol.* **2019**, *102*, 1355–1372. [CrossRef]
5. Pan, Z.; Zhang, H.; Zhu, Z.; Wang, J. Analysis of robotic machining process. *J. Mater. Process. Technol.* **2006**, *173*, 301–309. [CrossRef]
6. Gasparetto, A. Eigenvalue analysis of mode-coupling chatter for machine-tool stabilization. *J. Vib. Control.* **2001**, *7*, 181–197. [CrossRef]
7. Gasparetto, A. A system theory approach to mode coupling chatter in machining. *J. Dyn. Syst. Meas. Contr.* **1998**, *120*, 545–547. [CrossRef]
8. Zhang, H.; Wang, J.; Zhang, G.; Gan, Z.; Pan, Z.; Cui, H.; Zhu, Z. Machining with flexible manipulator: Toward improving robotic machining performance. In Proceedings of the IEEE/ASME International Conference on Advanced Intelligent Mechatronics, Monterey, CA, USA, 24–28 July 2005.

9. Carbone, G. Stiffness analysis and experimental validation of robotic systems. *Front. Mech. Eng. Chin.* **2011**, *6*, 182–196.
10. Ni, H.; Zhang, C.; Hu, T.; Wang, T.; Chen, Q.; Chen, C. A dynamic parameter identification method of industrial robots considering joint elasticity. *Int. J. Adv. Rob. Syst.* **2019**, *16*, 1729881418825217. [CrossRef]
11. Dumas, C.; Caro, S.; Garnier, S.; Furet, B. Joint stiffness identification of six-revolute industrial serial robots. *Rob. Comput. Integr. Manuf.* **2011**, *27*, 881–888. [CrossRef]
12. Abele, E.; Weigold, M.; Rothenbücher, S. Modeling and identification of an industrial robot for machining applications. *CIRP Ann.* **2007**, *56*, 387–390. [CrossRef]
13. Rafieian, F.; Liu, Z.; Hazel, B. Dynamic model and modal testing for vibration analysis of robotic grinding process with a 6DOF flexible-joint manipulator. In Proceedings of the International Conference on Mechatronics and Automation, Changchun, China, 9–12 August 2009.
14. Alici, G.; Shirinzadeh, B. Enhanced stiffness modeling, identification and characterization for robot manipulators. *IEEE Trans. Rob.* **2005**, *21*, 554–564. [CrossRef]
15. Abele, E.; Rothenbücher, S.; Weigold, M. Cartesian compliance model for industrial robots using virtual joints. *Prod. Eng.* **2008**, *2*, 339. [CrossRef]
16. Carbone, G.; Ceccarelli, M. A stiffness analysis for a hybrid parallel-serial manipulator. *Robotica* **2004**, *22*, 567–576. [CrossRef]
17. Behi, F.; Tesar, D. Parametric identification for industrial manipulators using experimental modal analysis. *IEEE Trans. Rob. Autom.* **1991**, *7*, 642–652. [CrossRef]
18. Tsai, L.-W. *Robot Analysis: The Mechanics of Serial and Parallel Manipulators*; John Wiley & Sons: New York, NY, USA, 1999.
19. Bompos, N.A.; Artemiadis, P.K.; Oikonomopoulos, A.S.; Kyriakopoulos, K.J. Modeling, full identification and control of the mitsubishi pa-10 robot arm. In Proceedings of the IEEE/ASME International Conference on Advanced Intelligent Mechatronics, Zurich, Switzerland, 4–7 September 2007.
20. Ewins, D.J. *Modal testing: Theory and Practice*; Research studies press: Baldock, Hertfordshir, UK, 1984.
21. Cossalter, V.; Doria, A.; Mitolo, L. Inertial and modal properties of racing motorcycles. *SAE Tech. Pap.* **2002**. [CrossRef]
22. Cossalter, V.; Doria, A.; Basso, R.; Fabris, D. Experimental analysis of out-of-plane structural vibrations of two-wheeled vehicles. *Shock Vib.* **2004**, *11*, 433–443. [CrossRef]
23. Verbeke, J.; Debruyne, S. Vibration analysis of a UAV multirotor frame. In Proceedings of the ISMA 2016 International Conference on Noise and Vibration Engineering, Leuven, Belgium, 19–21 September 2016.
24. Belotti, R.; Caneva, G.; Palomba, I.; Richiedei, D.; Trevisani, A. Model updating in flexible-link multibody systems. *J. Phys. Conf. Ser.* **2016**, *744*, 012073. [CrossRef]
25. Mejri, S.; Gagnol, V.; Le, T.-P.; Sabourin, L.; Ray, P.; Paultre, P. Dynamic characterization of machining robot and stability analysis. *Int. J. Adv. Manuf. Technol.* **2016**, *82*, 351–359. [CrossRef]
26. Inman, D. *Engineering Vibration*, 4th ed.; Prentice Hall: Upper Saddle River, NJ, USA, 2013.
27. Maia, N.M.M.; Silva, J.M.M. *Theoretical and Experimental Modal Analysis*; Research Studies Press: Taunton, Somerset, UK, 1997; Volume 3.
28. Marcolongo, R. Notizie sul Discorso Matematico e sulla vita di Giulio Mozzi. *Boll. Bibliografia Storia delle Sci. Mat.* **1905**, *VIII*, 1–8.
29. Shabana, A.A. *Dynamics of Multibody Systems*; Cambridge university press: Cambridge, UK, 2013.
30. Ball, R.S.A. *Treatise on the Theory of Screws*; Cambridge university press: Cambridge, UK, 1998.
31. Lipkin, H.; Duffy, J. Sir Robert Stawell Ball and methodologies of modern screw theory. *J. Mech. Eng. Sci. C* **2002**, *216*, 1–11. [CrossRef]
32. Wu, X.; Lu, Y.; Duan, X.; Zhang, D.; Deng, W. Design and DOF Analysis of a Novel Compliant Parallel Mechanism for Large Load. *Sensors* **2019**, *19*, 828. [CrossRef] [PubMed]
33. Cossalter, V.; Doria, A. Analysis of motorcycle slalom manoeuvres using the Mozzi axis concept. *Veh. Syst. Dyn.* **2004**, *42*, 175–194. [CrossRef]
34. Cossalter, V.; Doria, A. Instantaneous screw axis of two-wheeled vehicles in typical manoeuvres. *Veh. Syst. Dyn.* **2006**, *44*, 669–678. [CrossRef]
35. Doria, A.; Taraborrelli, L. The twist axis of frames with particular application to motorcycles. *J. Mech. Eng. Sci. C.* **2016**, *230*, 3026–3039. [CrossRef]

36. Blanchet, P.; Lipkin, H. Dual Properties for Vibration Analysis Via Screw Theory. In Proceedings of the DETC'98 1998 ASME Design Engineering Technical Conferences, Atlanta, GA, USA, 13–16 September 1998.
37. Chen, C.; Peng, F.; Yan, R.; Li, Y.; Wei, D.; Fan, Z.; Tang, X.; Zhu, Z. Stiffness performance index based posture and feed orientation optimization in robotic milling process. *Rob. Comput. Integr. Manuf.* **2019**, *55*, 29–40. [CrossRef]
38. Patterson, T.; Lipkin, H. Structure of robot compliance. *J. Mech. Des.* **1993**, *115*, 576–580. [CrossRef]

robotics

MDPI

Article

Estimation and Closed-Loop Control of *COG*/*ZMP* in Biped Devices Blending *CoP* Measures and Kinematic Information

Giuseppe Menga [1,*,†] **and Marco Ghirardi** [2,†]

1 Department of Control and Computer Engineering, Politecnico di Torino, Corso Duca degli Abruzzi 24, 10129 Torino, Italy
2 Department of Management and Production Engineering, Politecnico di Torino, Corso Duca degli Abruzzi 24, 10129 Torino, Italy; marco.ghirardi@polito.it
* Correspondence: giuseppe.menga@formerfaculty.polito.it; Tel.: +39-011-0907261
† These authors contributed equally to this work.

Received: 17 July 2019; Accepted: 16 October 2019; Published: 22 October 2019

Abstract: The zero moment point (*ZMP*) and the linearized inverted pendulum model linking the *ZMP* to the center of gravity (*COG*) have an important role in the control of the postural equilibrium (balance) of biped robots and lower-limb exoskeletons. A solution for balance real time control, closing the loop from the joint actual values of the *COG* and *ZMP*, has been proposed by Choi. However, this approach cannot be practically implemented: While the *ZMP* actual value is available from the center of pressure (*CoP*) measured under the feet soles, the *COG* is not measurable, but it can only be indirectly assessed from the joint-angle measures, the knowledge of the kinematics, and the usually poorly known weight distribution of the links of the chain. Finally, the possible presence of unknown external disturbance forces and the nonlinear, complex nature of the kinematics perturb the simple relationship between the *ZMP* and *COG* in the linearized model. The aim of this paper is to offer, starting from Choi's model, a practical implementation of closed-loop balance control fusing *CoP* and joint-angle measures, eliminating possible inconsistencies. In order to achieve this result, we introduce a model of the linearized inverted pendulum for an extended estimation, not only of *COG* and *ZMP*, but also of external disturbances. This model is then used, instead of Choi's equations, for estimation and balance control, using $\mathcal{H}_\infty$ theory. As the *COG* information is recovered from the joint-angle measures, the identification of a statistically equivalent serial chain (*SESC*) linking the *COG* to the joint angles is also discussed.

Keywords: biped robotics; exoskeletons; postural equilibrium; zero moment point; inverted pendulum; robust control

1. Introduction

The zero moment point (*ZMP*) and linearized inverted pendulum have continued to play a fundamental role in the control of postural equilibrium of biped robots and lower-limb exoskeletons since their introduction by Vukobratovic [1]. Vukobratovic showed that the center of pressure (*CoP*) of reaction forces under the feet soles on a flat horizontal surface coincides with a point he called the zero moment point (*ZMP*) and that postural equilibrium can be guaranteed if the *CoP* (alias *ZMP*) is maintained inside the convex hull of the surface encompassing the supporting foot (or feet in double stance). Moreover, a very simple relationship, based on the linearized inverted pendulum, links the *ZMP* and the center of gravity (*COG*) projection on the ground of the mechanical chain. The goal is to control the *COG* acting on the joint angles as both are algebraically linked by the kinematics of the chain, the target objective being the *ZMP*.

If motion maintaining postural equilibrium is desired (e.g., during a step in gait), to control the ZMP requires a certain degree of anticipation. Hence, in the so-called preview control framework [2], transition of COG, COG velocity, and ZMP were precomputed in advance and applied in open loop as reference to the biped control. However, this approach has not been able to model a closed-loop system and does not face the problems of disturbance rejection and stability.

A solution to track the preview trajectories in closed loop was successively proposed by Choi [3]. The preview $COG - ZMP$ trajectories during rectilinear gait were reviewed, and a closed-loop strategy was devised and proven, using Lyapunov techniques, to guarantee closed-loop stability and a bounded error tracking of the COG and ZMP preview references. The loop was closed from the actual values of COG and ZMP, generating a feedback signal to control the COG velocity. Then, a speed control for the joints from the COG velocity was designed using a COG Jacobian with specified embedded motion. The output measures were not detailed, but presumably, COG was assessed by measuring the joint angles from the weight distribution on the kinematic chain, and ZMP was measured by pressure sensors under the feet of the biped. The strategy, according to theory, guarantees closed-loop stability. However, when the authors tested it in simulations and practical examples, it showed a lack of robustness to disturbances or poor damping of the closed-loop dynamics. Lyapunov theory guarantees stability but does not say how much the resulting closed-loop poles will be damped.

If real-time measures (let us call them COG_m and ZMP_m) of COG and ZMP are independently available, it is reasonable to assume that before closing the loop, a filtering is performed, fusing both data. However, if these are generated, as stated before, the latter by direct measures and the former indirectly from joint measures and a priori information, they are not always consistent with the relationship stated by the linearized inverted pendulum. The main reasons are: uncertainties in the model parameters (especially in the weight distribution when dealing with an exoskeleton interacting with a patient), external forces acting on the biped (crutches or a chair in a sit-to-stand exercise), centrifugal forces in the frontal plane when motion is not rectilinear.

This study was motivated by the intention to improve postural equilibrium in lower-limb exoskeletons for rehabilitation. The same approach is at the basis of all applications needing balance control of biped robots, such as biped walking in rectilinear [4,5] and curved trajectories [6], in haptic lower-limb exoskeletons [7], and in performing sit-to-stand exercises [8], described by the authors in other papers.

1.1. Paper Contributions

The main contribution of the paper is the development of a feedback control more robust than the one offered by Choi. In order to achieve this result, a detailed understanding of the closed-loop dynamics generated by controlling an inverted pendulum is presented, with particular attention devoted to the control design techniques and the engineering problems in closing the loop. Then, in order to make the filtering effective and close the loop from COG_m and ZMP_m, ensuring compatibility, the proposed approach operates at two levels: a nonlinear algebraic function and a a linear dynamic model.

The nonlinear function is a simplified mapping from joint angles to COG, called statistically equivalent serial chain (SESC) [9], to be identified in a priori experiments. As this identification is based on the same force sensors under the feet used for measuring the CoP, it also resolves any calibration mismatch.

The linear model is an extended system based on the inverted pendulum from input u, the reference COG velocity, and output COG_m and ZMP_m, used to estimate, along with the COG, ZMP, and external disturbances affecting the CoP, the model states. Then, using the estimated states, the loop from COG_m and ZMP_m can be closed applying robust control theory. The reasons for estimating the unknown external force disturbances are twofold: (1) to take into account real external disturbances, especially in exoskeletons but also in biped robots (e.g., centrifugal forces in turning while walking); (2) robustness in the $COG - ZMP$ joint estimation, accommodating modeling errors, parameter uncertainties, and the simplifications introduced by the linearized inverted pendulum.

Still recently, the linearized inverted pendulum has continued to be at the basis of the models for balance control ([10,11] and references therein). However, to the authors' knowledge, there are no works introducing, for robustness, an extended system to estimate disturbances, or the need for a SESC identification.

The proposed control is a non-conventional tracking problem, as two separate model states are tracked. Two different control design techniques are proposed and tested to control the extended system: (1) computing a robust estimator and solving the output feedback problem from the estimated states using a numerical approach based on the Levenberg–Marquardt algorithm [12,13]; (2) solving the standard robust regulator, adapted to deal with the preview signal tracking.

In order to test the approaches, three different experiments were performed. First, both observer and state feedbacks were implemented and compared with Choi's original feedback through simulation of the 2D linearized inverted pendulum. In a standing position, a preview shift of the COG from the heels to the tips of the feet was imposed, while in the meantime, an external force disturbance was applied. Then, a real lower-size mechanical mock-up was considered, composed of foot, leg, thigh, and trunk, with three degrees of freedom (DOF) in the sagittal plane to represent the real exoskeleton for implementing the sit-to-stand exercise. The SESC model of this simplified kinematics was identified with a priori experiments and used in the proposed feedback control through the COG Jacobian of the chain. Finally, a non-linear simulation of the full-scale exoskeleton with patient was run on the same exercise executed by the linearized inverted pendulum, emulating the first phase of a stand-to-sit exercise. A complete sit-to-stand exercise with the presence of a chair and switching dynamics exploiting the same control technique can be found in [8].

The paper is organized as follows. Section 2 reviews Choi's results. Section 3 introduces the main contribution of the paper: a $COG - ZMP$ model of the linearized inverted pendulum and an extended system, embedding in the model external disturbances, for applying robust estimation and robust control. This model is also used in the Appendix to show the limitations of Choi's feedback. Sections 3.1 and 3.2 present the robust estimator–estimate state feedback and the standard robust regulator. Sections 3.3 and 4 contain simulated and real control experiments. In particular, Section 4.1 approaches the identification of the SESC model, and Section 4.2 presents the simulation of a stand-to-sit exercise. Section 5 concludes the paper. The appendix discusses the limitations of Choi's original method.

2. Choi's Approach

As in [3], the 3D linearized inverted pendulum is split into two separate, independent 2D models for the sagittal and the frontal planes. However, in this paper only the sagittal plane will be considered, with axes x (horizontal) and z (vertical). The equation linking COG and ZMP, adopting Choi's notation, is

$$p = c - (1/w_n^2)\ddot{c}, \tag{1}$$

$$w_n \triangleq \sqrt{g/c_z}, \tag{2}$$

where p is the coordinate of the ZMP, c is the projection of the COG on the ground, c_z is the height of the COG, and g is the acceleration of gravity. w_n is the only parameter of the model of the simplified biped walking system.

Let p_d, c_d, and $\dot{c}_d$ indicate the desired preview trajectories of the ZMP, of the COG, and of its derivative during a postural exercise, and assume that the pendulum joint servo is controlled in speed by an input signal u according to

$$\dot{c} = u + \epsilon, \tag{3}$$

where ϵ accounts for the speed-tracking error and process disturbances and u is given by the following feedback law:

$$e_c = c_d - c, e_p = p_d - p, u = \dot{c}_d + k_c e_c - k_p e_p. \tag{4}$$

Then, Choi's results prove, with Lyapunov theory ([3], Theorem 1), that if $k_c > w_n$ and $0 < k_p < w_n$, the closed-loop system is bounded disturbance (ϵ) - bounded errors (e_c, e_p) stable.

Anyway, in spite of stability, a feedback implemented using Choi's equations has a poor damping of the closed-loop dynamics in practical operating conditions. A proof and discussion about this topic can be found in Appendix A.

3. An Extended System for COG–ZMP Robust Estimation and Control

Choi does not introduce any input–output dynamic model to prove his results, but only a Lyapunov function directly based on Equations (1)–(4). Here, vice versa, the essence of the feedback control problem involving measures of COG and ZMP with the reference velocity as input is captured by the simple model of the block diagram of Figure 1: a third-order model with states $c, \dot{c}, \ddot{c}$, where the jerk of the COG (in the following with an excess of notation, COG indicates its projection on the ground) is controlled by a reference velocity signal u in an internal partial speed loop with velocity gain k_v and output COG and ZMP. The third-order model is needed to guarantee a realistic strictly proper system for the design of the state estimator and feedback as position and acceleration are both present in the output, and in the meantime, representing the internal speed loop with the servo dynamics. If the gain of the local speed loop is taken relatively high ($k_v > 100$) the COG speed will closely track the reference u, as desired. This model does not take into account external forces acting on the system or internal disturbances, as in the case of a lower-limb exoskeleton with the presence of a patient. External forces are introduced when crutches are used or when the patient is sitting on a chair in a sit-to-stand exercise, or simply to accommodate discrepancies between COG and ZMP measures. Internal disturbances are generated by the involuntary motion of the patient in the small freedom offered by the exoskeleton, independent of the joint motion, and obviously, by modeling errors.

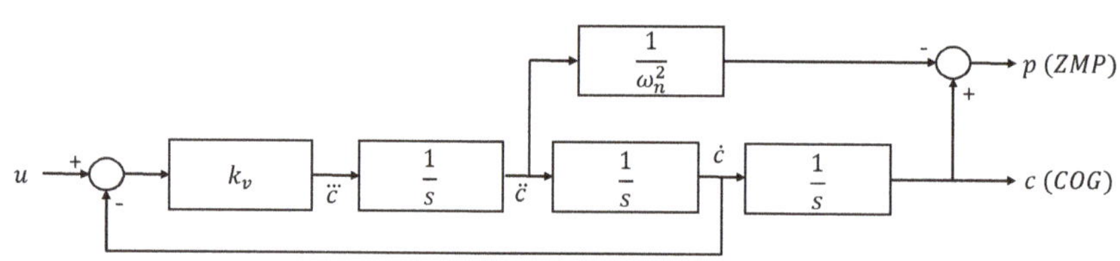

Figure 1. The COG–ZMP model of the linearized inverted pendulum.

Then the model can be completed, generating the extended system of Figure 2, as defined in robust control theory (for the definition of the extended system and its role in robust control and $\mathcal{H}_\infty$ theory, see [14,15]).

F represents low-frequency external forces influencing the CoP. In the model ZMP_{actual} (i.e., CoP), the value measured and ZMP_{ideal}, the one linked to the COG by the linearized inverted pendulum relationship, are defined separately, where δ is the difference between the two, the effect of disturbance F to be estimated. COG and CoP are measured as before, taking into account measurement noise represented by two high pass filters $W_{noise_{COG}}, W_{noise_{ZMP}}$. Output objectives are set on the COG and on the ZMP for sensitivity requirements with respect to process noise ϵ (in a different context, here ϵ has the same interpretation and scope as in Equation (3)) and the effect of the unknown external force F. The weighting functions W_{COG} and W_{ZMP} are chosen to guarantee steady-state gain (i.e., tracking error with respect to disturbances ϵ and F) and the frequency band of the loop transfer function in the designed feedback. In order to have a balanced design, the control activity z_u (with a weighting function W_u) is added as an objective against measurement noises n_{COG} and n_{ZMP}, to set the control activity. This extended system is used to design robust estimators of COG, $\dot{COG}$, ZMP_{ideal}, and δ, as well as robust controls.

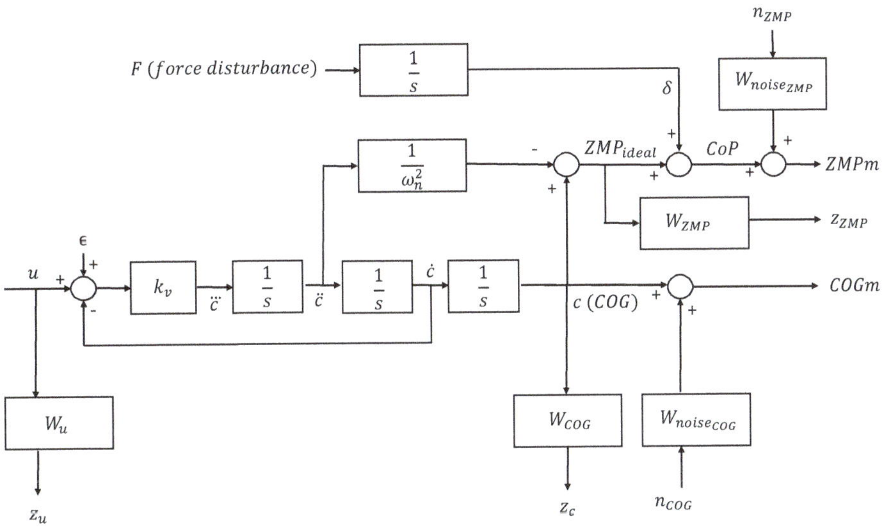

Figure 2. The extended system of COG–ZMP, disturbance model of the linearized inverted pendulum.

Let $\hat{c}, \hat{\dot{c}}, \hat{p}, \hat{\delta}$ be the estimates of the COG projection on the ground, its derivative, the ZMP_{ideal}, and δ. Then the control strategy of Equation (4) is modified as follows:

$$
\begin{aligned}
e_c &= c_d - \hat{c} - \hat{\delta}, \\
e_{cv} &= \dot{c}_d - \hat{\dot{c}}, \\
e_p &= p_d - \hat{p} - \hat{\delta}, \\
u &= \dot{c}_d + k_c e_c - k_p e_p + k_{cv} e_{cv},
\end{aligned}
\tag{5}
$$

with the control scheme represented in Figure 3. This means that c (and in the steady state, p) must track a perturbed reference in order to guarantee that the CoP, and not the ZMP_{ideal} linked to the COG, follows the desired preview signal, despite ZMP_{ideal} and COG converging to the same value in the steady state, independently of the presence of disturbances. A feedback from $\hat{\dot{c}}$ is also introduced as it has a critical influence on the closed loop damping.

In the next subsections, two different approaches, based on robust control theory, to compute the state observer and the state feedback coefficients, are introduced and tested. The estimates and the coefficients in Equation (5) result explicitly from standard $\mathcal{H}_\infty$ techniques by operating a state-space transformation in the extended system of Figure 2, choosing as states $c, \dot{c}, p, \delta$, augmented (for the whole extended system) by the unobservable or uncontrollable states introduced by the dynamics of the weighting functions.

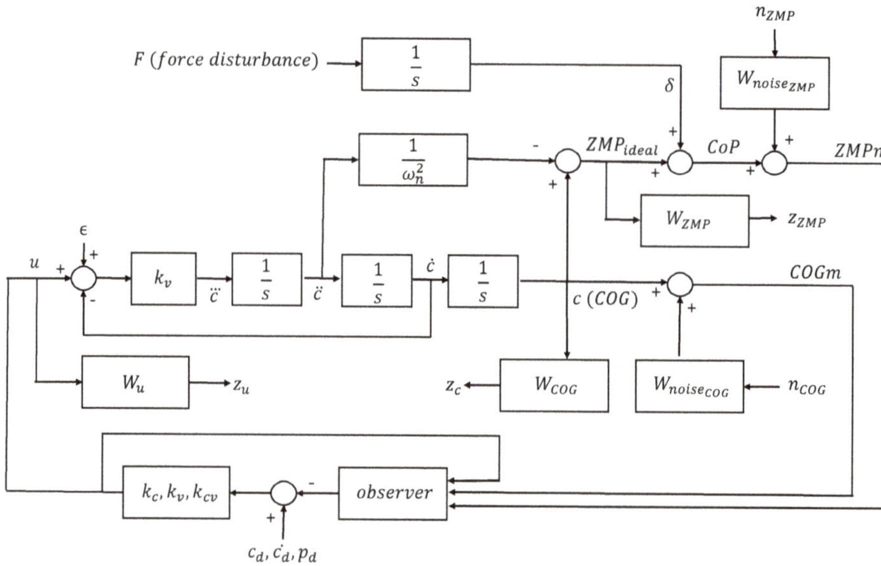

Figure 3. Control of COG–ZMP with observer feedback.

3.1. Separate Estimator and Feedback

In this first approach, the extended system with the new state representation is used to design a state estimator.

Then, a static output feedback problem for constrained pole placement is solved on the cascade of the extended system and observer to derive the gains k_c, k_{cv}, and k_p in (5).

It is known that static output feedback has no analytical solution. Hence, a numerical technique based on the Levenberg–Marquardt algorithms was implemented. No algorithm details are presented here. Just note that by minimizing the sum of the squares of a certain number of penalty functions, the closed-loop poles are brought into a stability region with desired damping, constraining k_c to be greater than a lower bound, and the control activity (measured by the $\mathcal{H}_\infty$ norm of the closed-loop operator from measuring noises to control objectives) to be smaller than an upper bound.

3.2. $\mathcal{H}_\infty$ Robust Control

In the second approach, estimation and feedback are jointly computed solving a $\mathcal{H}_\infty$ robust control regulator. Even if the classical separation of $\mathcal{H}_2$ does not apply in $\mathcal{H}_\infty$ controls [14,15], a weakly coupled state observer and estimated state feedback can still be recognized. Then, maintaining the feedback coefficients of the estimated states of interest ($\hat{c}, \hat{\dot{c}}, \hat{p}, \hat{\delta}$) and setting to zero the remaining ones relative to unobservable or uncontrollable modes (it can be verified that this has very little influence on the closed-loop poles), the strategy of Equation (5) can be implemented. It is interesting to note the similarity of the performances obtained between the two approaches, as shown in the next section.

3.3. Simulation Results

The model of an exoskeleton with a patient used in [5,7] was considered, with parameter $w_n = 3.34$ and choosing a speed gain for the velocity loop of $k_v = 250$. In this experiment, the linear model of the block diagram of Figure 2 was simulated. In an erect posture, with a preview reference computed as suggested by [2], a step transition on the sagittal plane of p of 0.2 m moving the c from the heels to the tips of the feet at time $t = 1.25$ s was imposed. Then at $t = 3.5$ s, an external horizontal force disturbance acting in the sagittal plane on the COG, tries to move the CoP outside the feet support by an additional 0.1 m, causing—if not compensated—a loss of balance.

The experiments are compared using the identical tracking gain k_c of the COG loop and the best (for damping) value of k_p with Choi's control and feedback from a robust estimator obtained from the extended system of Figure 2. The gain parameters adopted in the case of Choi's feedback were $k_c = 60$, $k_p = 5$, with a resulting damping ratio of the dominant poles of 0.05; and in the case of observer feedback, $k_c = 60$, $k_p = 3$, $k_{cv} = 4$, with a resulting damping ratio of the dominant poles of 0.7.

The figures represent reference (dashed) and actual COG (blue), ZMP_{ideal} (red), and CoP (cyan), COG speed estimate (green), and the estimate of the disturbance effect (violet).

Figure 4 shows the results adopting the Choi control. The low damping of the closed-loop poles is clearly visible. Note that, ignoring the disturbances, the CoP does not follow the reference path and exits from the tip of the feet.

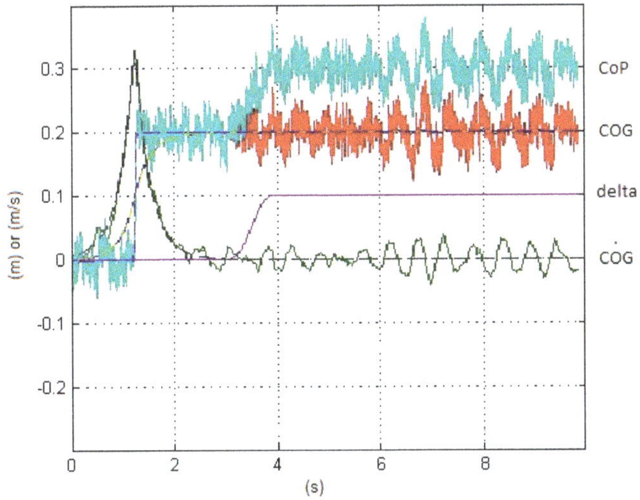

Figure 4. Choi's control without any disturbance compensation.

Vice versa, when disturbances are also estimated, after a short interval of time depending on the filtering bandwidth of the estimator, the ZMP returns to the desired value. Figures 5 and 6 show the feedback from the extended estimator with compensation of disturbances, obtained with the approaches of Sections 3.1 and 3.2, respectively.

The transition of the force disturbance was chosen to be unrealistically steep to evidence that, because of the estimator bandwidth, the compensation of the disturbance can't be perfect, depending on the values assigned to the weighting functions in the extended system.

The delay in the estimation of δ, as a consequence of the disturbance, is shown in Figure 7, where δ is in blue and its estimate is in green.

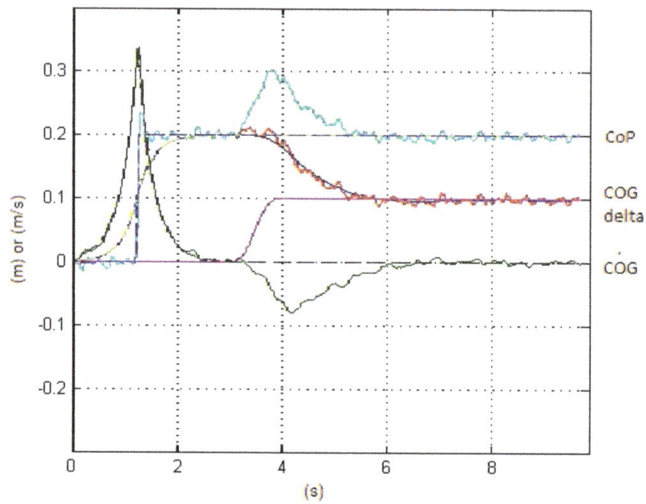

Figure 5. Extended observer feedback with disturbance compensation.

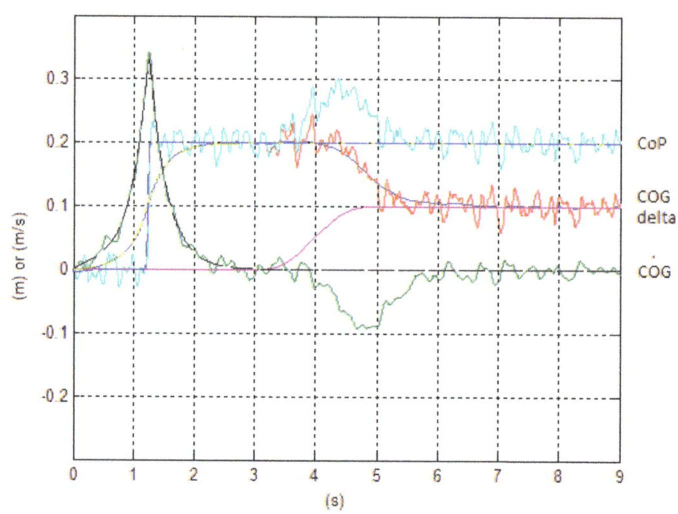

Figure 6. Robust control with disturbance compensation.

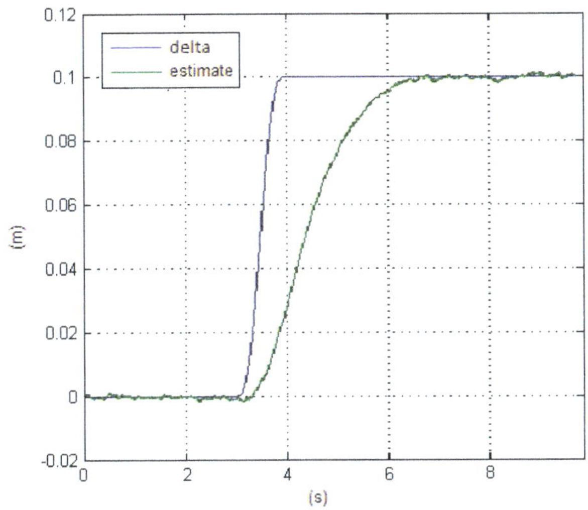

Figure 7. Effect of disturbance on the CoP and its estimation.

4. Control of a 3 DOF Biped

The approach was tested on the real three-DOF small-scale mock-up of an exoskeleton (Figure 8) and on the simulation of the full-scale exoskeleton. The chains, in both cases, are composed of two joint feet, legs, thighs, and one trunk. The 3×3 Jacobian matrix relating COG to joints embedding knee motion and trunk attitude is the following:

$$\begin{bmatrix} \dot{c} \\ \dot{\theta}_2 \\ \dot{\theta}_{trunk} \end{bmatrix} = \begin{bmatrix} & J_{cog} & \\ 0 & 1 & 0 \\ 1 & 1 & 1 \end{bmatrix} \cdot \begin{bmatrix} \dot{\theta}_1 \\ \dot{\theta}_2 \\ \dot{\theta}_3 \end{bmatrix}, \tag{6}$$

where $\theta_1, \theta_2, \theta_3$ are the angles of the ankle, knee, and hip, J_{cog} is the Jacobian of c, and θ_{trunk} is the attitude of the trunk. Joints are controlled by velocity servos, with their references being obtained through the inverse of the Jacobian matrix (6) driven by speed feedback signals. The COG speed feedback is similar to the one used for the linearized inverted pendulum (5), where the measures of the ZMP_m were obtained from the CoP and that of the COM_m from the joint-angle measures $\theta_{1m}, \theta_{2m}, \theta_{3m}$. The remaining two feedbacks, from the knee angle and trunk attitude measures $\theta_{2_m}, \theta_{trunk_m}$, are simply proportional feedbacks, the last measure being obtained from an inertial sensor:

$$\begin{bmatrix} u_{\theta_2} \\ u_{\theta_{trunk}} \end{bmatrix} = \begin{bmatrix} k_{knee} \cdot (\theta_{2_{ref}} - \theta_{2_m}) \\ k_{trunk} \cdot (\theta_{trunk_{ref}} - \theta_{trunk_m}) \end{bmatrix}, \tag{7}$$

where $\theta_{2_{ref}}$ and $\theta_{trunk_{ref}}$ are the references chosen according to the desired postural exercise.

Figure 8. A small-scale 3 degrees of freedom (DOF) leg of the exoskeleton.

4.1. Joint Angles–COG_m Map Fitting and Control

In order to apply the results of the previous section to a biped device, it is necessary to have a reliable evaluation of the COG from the joint angles, consistent with the CoP. Its value depends on the position of the center of masses and weights of each link of the chain. Those data are scarcely known in advance but can be identified with a series of a priori experiments. This approach is called statically equivalent serial chain (SESC) modeling [9] (see also [16,17] for applications to rehabilitation).

Espiau et al., in [9], showed from experiments measuring the projection of the COG on a force table that the physical parameters of a kinematic chain cannot be identified uniquely. What can be identified is only a set of expressions of them, representing classes of equivalent (with respect to the COG) chains. These expressions appear in linear form in the SESC model. Considering a three-joint kinematic model composed of feet, leg, thigh, and HAT (head, arms, trunk) for motions in the sagittal plane, the parameters of the SESC model can be identified using least squares with two equations and two sets of experiments: collecting and recording a series of joint-angle positions with the corresponding measures of the CoP in the steady state and a set of samples of joint angles and CoP trajectories in motion spanning the operating area at random. A slight modification of the model presented in [16] is proposed here, where the first equation, expressing explicitly COG_x, refers to steady-state experiments, while the second equation, expressing COG_z indirectly (Equations (1) and (2) can be rewritten as $COG_x - ZMP_x = \ddot{COG}_x/9.81 \cdot COG_z$), refers to dynamical ones.

The equations are:

$$\begin{bmatrix} COG_x \\ COG_x - ZMP_x \end{bmatrix} = \begin{bmatrix} 1 & 0 \\ 0 & \ddot{COG}_x/9.81 \end{bmatrix} \cdot \begin{bmatrix} 1 & 0 & sin(\theta_1) & sin(\theta_1+\theta_2) & sin(\theta_1+\theta_2+\theta_3) \\ 0 & 1 & cos(\theta_1)\cdot b & cos(\theta_1+\theta_2)\cdot b & cos(\theta_1+\theta_2+\theta_3)\cdot b \end{bmatrix} \cdot \begin{bmatrix} r_{1x} \\ r_{1z} \\ r_2 \\ r_3 \\ r_4 \end{bmatrix}, \quad (8)$$

with

$$
\begin{aligned}
r_{1x} &= (m_0 \cdot x_0 + (m_1 + m_2 + m_3) \cdot x_1)/m_{tot},\\
r_{1z} &= (m_1 + m_2 + m_3) \cdot z_1/(m_1 + m_2 + m_3),\\
r_2 &= (m_1\dot{l}_{10} + (m_2 + m_3) \cdot l_1)/m_{tot},\\
r_3 &= (m_2 \cdot l_{20} + m_3 \cdot l_2)/m_{tot},\\
r_4 &= m_3 \cdot l_{30}/m_{tot},\\
b &= m_{tot}/(m_1 + m_2 + m_3),
\end{aligned}
\tag{9}
$$

where l_1, l_2 are the length of legs and thighs, m_0, m_1, m_2, m_3 are the masses of feet, legs, thighs, and trunk (HAT), ($m_{tot} = m_1 + m_2 + m_3 + m_4$), x_0 is the center of mass of the feet, x_1, z_1 are the coordinates of the ankle, and l_{10}, l_{20}, l_{30} are the distances from the center of mass to the distal joints for the leg, thigh, and proximal joint for HAT. Coefficient b accounts for the difference (the feet do not move during the dynamical experiments) in sensing the COG statically and dynamically. From this model, the six parameters of (9) are identified recursively with a non-linear least squares technique such as Levenberg–Marquardt, where $\ddot{COG}$ is obtained approximately from numerical differentiation of COG. The actual small-scale leg was first identified, with results (statical and dynamical) contained in Figures 9 and 10.

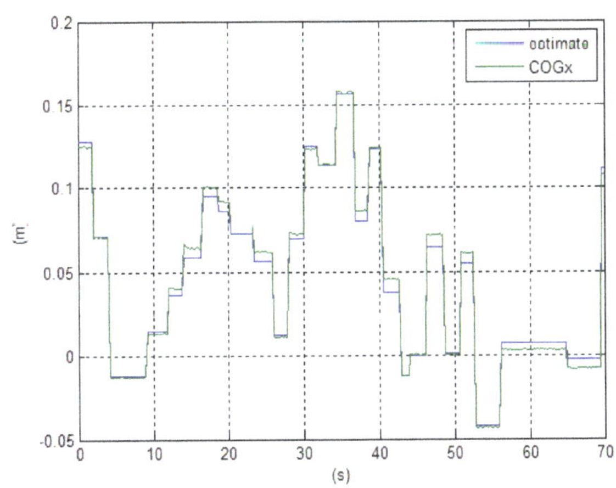

Figure 9. Results of estimating the statically equivalent serial chain (SESC) model in static experiments.

Then a control exercise was carried out, maintaining the CoP position fixed and the posture erect while performing an up-and-down motion (such as sit-to-stand) of the body. The results (CoP_x and COG_x), based on the identified model and the proposed control scheme, when a disturbing force is applied in the sagittal plane are shown in Figure 11. The action of the feedback on the COG to compensate the disturbance is clear.

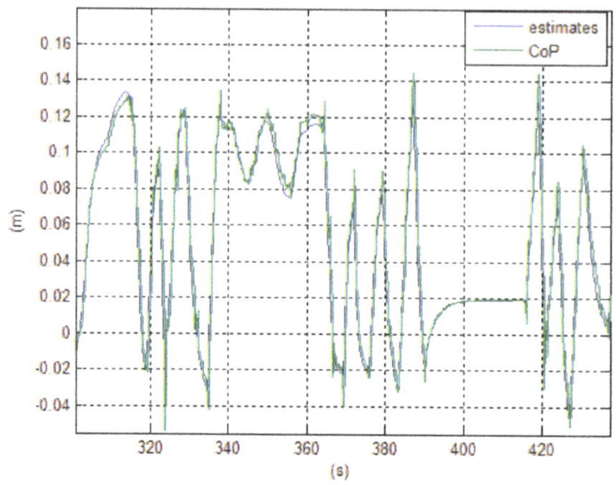

Figure 10. Results of estimating the SESC in dynamic experiments.

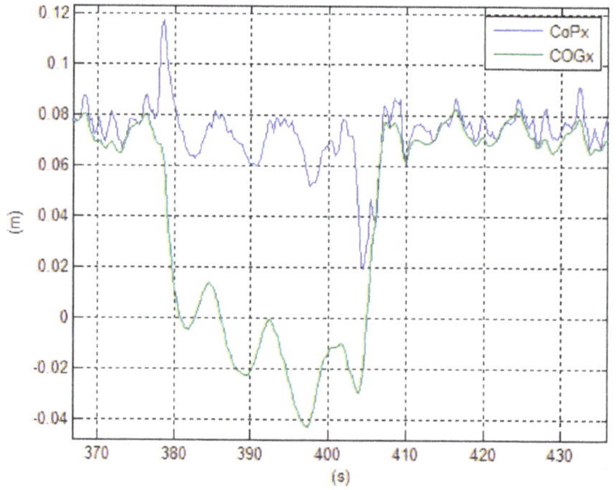

Figure 11. Maintaining the *CoP* position during a stand-to-sit like exercise in the presence of a disturbance force.

4.2. Simulation of a Stand-to-Sit Exercise

In order to validate the results on the linearized inverted pendulum of Section 3.3, the proposed control (with the same estimator and feedback parameters) was applied to the non-linear simulation of a multi-chain with 3 DOF, having as average the same COG_z. It represents a biped in the sagittal plane emulating the first phase of a stand-to-sit exercise to test the balance control of a future full-scale exoskeleton with a patient. While the pelvis is lowered from a standing posture to reach the chair and the trunk attitude assumes a natural bending forward, the COG_x is shifted from heels to tips and a disturbance force is applied, as in the previous experiment of Figures 5 and 6. The animation of the exercise can be seen in Figure 12. The resulting response of the $COG - ZMP$ in Figure 13 is very similar to that of the linearized inverted pendulum. Particularly, in the final phase of the exercise, the reaction of the exoskeleton to preserve equilibrium against the push forward of the disturbing force is particularly visible.

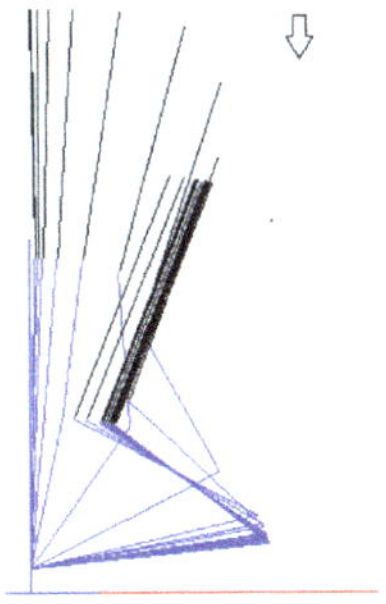

Figure 12. Animation of the 3 DOF kinematics during a stand-to-sit exercise.

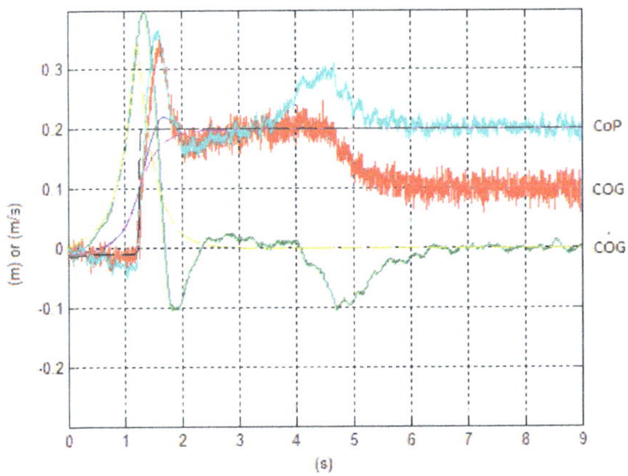

Figure 13. COG–ZMP during stand-to-sit and shift of *CoP* in a nonlinear 3 DOF kinematics-robust control.

Vice versa, Choi's original feedback with identical *COG* gain, applied to the nonlinear simulator, has not been able to guarantee stability.

5. Conclusions

In this paper, Choi's feedback for postural control of a biped robot, based on a linearized inverted pendulum model, has been revised. In practical situations, this feedback generates very undamped closed-loop dynamics. Then, the design problem was reformulated using state estimation and state feedback control. In fact, closing the loop with a state observer of the *COG* and exploiting velocity along with position and acceleration estimates guarantees a greater damping of closed-loop poles, with identical steady-state gain. However, to be effective in fusing *CoP* and *COM* measures, this observer needs to be extended to also estimate external disturbance forces, and the kinematic model of the *COM* needs to be tuned to the actual mass distribution. The former problem was tackled by a robust estimator based on an extended system embedding into the model unknown force disturbances, the latter by identifying a priori the SESC model of the mapping between joint angles and the *COM*. This a priori identification can also be repeated to maintain the mapping up to date in cases of changes in the weight distribution of the biped.

Two approaches to design the feedback were pursued: one is numerical, computing the state feedback for a given observer with a Levenberg–Marquardt algorithm. The second exploits integrally estimator and feedback obtained from a robust control regulator and adapts it to the tracking of the preview signals. The results show similar performances, with good disturbance compensation. It must be emphasized that the adoption of an extended system with its weighting functions offers a formal technique to set the observer and feedback characteristics, guaranteeing the desired loop gain and bandwidth.

Robustness was shown by applying the control designed for a linearized inverted pendulum to two non-linear systems: a three-DOF kinematic chain of an actual mechanical small-scale leg and the simulation of an exoskeleton constraining a patient to perform a joint-legged stand-to-sit exercise in the sagittal plane. The proposed control correctly integrates the COG information, which is poorly reconstructed from joint measures and kinematics of the chain, with actual CoP measures, accommodating uncertainties in the model and unknown external force disturbances. Moreover, an identification procedure of the SESC model was proposed and tested.

COG–ZMP and linearized inverted pendulum models continue to be at the basis of balance control of bipeds. However, extended systems and SESC models have not yet been proposed in order to offer robustness to the approach.

The proposed $COG - ZMP$ control was successfully used by the authors for the balance of turning during walking of biped robots and for a more detailed and complete sit-to-stand exercise described in the companion paper [8]. In particular, future developments will consider haptic exoskeletons, where the action of the patient on some joints, through electromiographical signals, controls the motion of part of the degrees of freedom, while the automatic control discussed here guarantees balance acting on the ankles or on the hips.

Computing robust estimation and robust control, as well as the block diagrams present in the paper, were made with the design environment G++ developed by the authors described in [18] and that can be downloaded from [19]. However, the used technique is fairly standard in the robust control field and can be found in classical textbooks such as [14].

Author Contributions: Methodology, G.M. and M.G.; writing—original draft, G.M.; writing—review & editing, M.G.

Funding: This research has been partially supported by MIUR, the Italian Ministry of Instruction, University and Research through project ESOPO, and the Piedmont Region through project ESROB.

Conflicts of Interest: The authors declare no conflict of interest.

Appendix A. Choi's Feedback Limitations

This appendix is devoted to showing that, in spite of stability, Choi's original feedback has a poor damping of the closed-loop dynamics in practical operating conditions. Applying the control strategy of Equation (4) to the model of Figure 1, we obtain a closed-loop system with three design gains k_v, k_c, k_p and one coefficient w_n, as depicted in Figure A1. In classical linear control theory, it is customary to introduce disturbances in the input and output of the system and to study the closed-loop performance, analyzing the open-loop transfer function (t.f.), and the related closed-loop sensitivity functions linking the output to the input and output noises. The block diagram presents two partial feedback loops, on COG and on ZMP, that can be analyzed separately by opening (indicated with an X in the block diagram) the two feedbacks one at a time and considering the other part of the system.

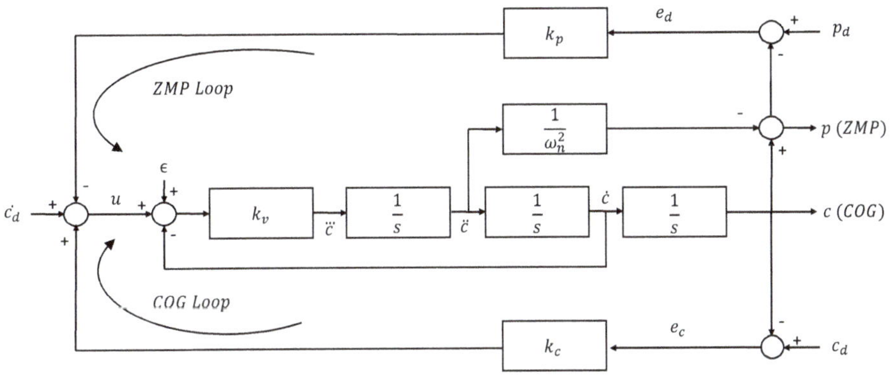

Figure A1. The COG–ZMP model of the linearized inverted pendulum with Choi's feedback control.

The functions related to the ZMP loop are not particularly interesting: the feedback of the ZMP, although important for stability reasons, does not play any role in disturbance rejection and tracking error because its steady-state loop gain is always lower than 1. In fact, it is given by

$$G_{a_{zmp}}(s) = \frac{k_v k_p (1 - s^2/w_n^2)}{s^3 + k_v s + k_v k_c}, \tag{A1}$$

where $k_p < k_c$ always holds.

Vice versa, the following t.f.s of the COG loop are noteworthy: the open-loop transfer function $G_a(s)$, the output tracking error sensitivity $S(s)$, and the output sensitivity to input disturbances $G_{eq_{\epsilon-COG}}(s)$.

$$G_a(s) = \frac{k_v k_c}{s^3 + k_v k_p / w_n^2 s^2 + k_v s - k_v k_p} \tag{A2}$$

$$S(s) = \frac{s^3 + k_v k_p / w_n^2 s^2 + k_v s - k_v k_p}{s^3 + k_v k_p / w_n^2 s^2 + k_v s + k_v (k_c - k_p)} \tag{A3}$$

$$G_{eq_{\epsilon-COG}}(s) = \frac{k_v}{s^3 + k_v k_p / w_n^2 s^2 + k_v s + k_v (k_c - k_p)} \tag{A4}$$

In order to guarantee stability (negative real part of the roots of the third-order, closed-loop, characteristic polynomial appearing as denominator in Equations (A3) and (A4)), the following condition on the parameter k_p must be satisfied:

$$\frac{w_n^2}{w_n^2 + k_v} k_c < k_p < k_c\,, \tag{A5}$$

Note that condition (A5) is slightly different from Choi's result.

In order to have more insights about the closed-loop poles of Equations (A3) and (A4), consider the root locus, function of k_v, of the following open-loop transfer function:

$$G_{a_{k_v}}(s) = \frac{k_v (k_p / w_n^2 s^2 + s + k_c - k_p)}{s^3}. \tag{A6}$$

Note that the numerator of $1 + G_{a_{k_v}}(s)$ is exactly the characteristic polynomial of (A3) and (A4). When $k_v \rightarrow \infty$, one real closed-loop pole $\rightarrow -\infty$, but the dominant closed-loop poles are complex conjugate and approach asymptotically the zeroes of the t.f. (A6), i.e., the root of the polynomial

$$s^2 + w_n^2/k_p s + w_n^2(k_c - k_p)/k_p, \tag{A7}$$

having the damping ratio

$$\zeta = \frac{w_n}{2}\sqrt{\frac{1}{k_p(k_c - k_p)}}. \tag{A8}$$

Moreover, the root locus shows that for any value of $k_v < \infty$, the damping ratio of the pair of dominant complex conjugate poles is always lower than that of these zeroes. From previous results, the following observations can be made:

- with high values of the k_v gain, the dominant closed-loop poles depend on the pair k_c, k_p only, as they are highly insensitive to k_v;
- the steady-state COG loop gain (A2) is proportional to the rate k_c/k_p (independent from k_v);
- the steady-state gain of both sensitivities related to COG are inversely proportional to $k_c - k_p$;
- however, if the gain k_c, or more precisely, the difference $k_c - k_p$, increases, then the damping ratio of the dominant closed-loop poles decreases.

In conclusion, if a sufficiently high loop gain in the $COG - ZMP$ control system is imposed, with the feedback proposed by Choi, even if the closed loop remains stable, its behavior becomes highly undamped. However, a high loop gain, and hence a high value of k_c is needed when a robust control has to be used in exoskeletons to improve postural equilibrium for ill or elderly people, in order to cope with uncertainties.

References

1. Vukobratovic, M.; Borovac, B.; Surla, D.; Stokic, D. *Biped Locomotion: Dynamics, Stability, Control, and Application*; Springer: Berlin/Heidelberg, Germany, 1990.
2. Kajita, S.; Kanehiro, F.; Kaneko, K.; Fujiwara, K.; Harada, K.; Yokoi, K.; Hirukawa, H. Biped Walking Pat-tern Generation by using Preview Control of Zero-Moment Point. In Proceedings of the 2003 IEEE International Conference on Robotics and Automation, Taipei, Taiwan, 4–19 September 2003.
3. Choi, Y.; Kim, D.; Oh, Y.; You, B. Posture/Walking Control for Humanoid Robot Based on Kinematic Resolution of CoM Jacobian With Embedded Motion. *IEEE Trans. Robot.* **2007**, *23*, 1285–1293. [CrossRef]
4. Menga, G.; Ghirardi, M. Modelling, Simulation and Control of the Walking of Biped Robotic Devices—Part I: Modelling and Simulation Using Autolev. *Inventions* **2016**, *1*, 6. [CrossRef]
5. Menga, G.; Ghirardi, M. Modeling, Simulation and Control of the Walking of Biped Robotic Devices, Part II: Rectilinear Walking. *Inventions* **2016**, *1*, 7. [CrossRef]
6. Menga, G.; Ghirardi, M. Modeling, Simulation and Control of the Walking of Biped Robotic Devices—Part III: Turning while Walking. *Inventions* **2016**, *1*, 6. [CrossRef]
7. Menga, G.; Ghirardi, M. Lower Limb Exoskeleton for Rehabilitation with Improved Postural Equilibrium. *Robotics* **2018**, *7*, 28. [CrossRef]
8. Menga, G.; Ghirardi, M. Control of the sit-to-stand transfer of a biped robotic device for postural rehabilitation. *Robotics* **2019**, *8*, 91.
9. Espiau, B.; Boulic, R. On the Computation and control of the mass center of articulated chains. In *Research Report INRIA, n.3479*; INRIA: Rocquencourt, France, 1998. Available online: https://hal.inria.fr/inria-00073210/document (accessed on 10 June 2019).
10. Kajita, S.; Morisawa, M.; Miura, K.; Nakaoka, S.; Harada, K.; Kaneko, K.; Kanehiro, F.; Yokoi, K. Biped Walking Stabilization Based on Linear Inverted Pendulum Tracking. In Proceedings of the 2010 IEEE/RSJ International Conference on Intelligent Robots and Systems, Taipei, Taiwan, 18–22 October 2010.
11. Kasaei, M.; Lau, N.; Pereira, A. Comparison Study of Well-Known Inverted Pendulum Models for Balance Recovery in Humanoid Robot. *arXiv* **2019**, arXiv:1906.01936.

12. Levenberg, K. A Method for the Solution of Certain Non-linear Problems in Least Squares. *Q. Appl. Math.* **1944**, *2*, 164–168. [CrossRef]
13. Marquardt, D. An Algorithm for the Least-Squares Estimation of Nonlinear Parameters. *SIAM J. Appl. Math.* **1963**, *11*, 431–441. [CrossRef]
14. Colaneri, P.; Geromel, C.; Locatelli, A. *Control Theory and Design: An RH2 and RHinf Viewpoint*; Academic Pressl: San Diego, CA, USA, 1997.
15. Zhou, K.; Doyle, J.; Glover, K. *Robust and Optimal Control*; Prentice Hall: Upper Saddle River, NJ, USA, 1996.
16. Cotton, S.; Vanoncini, M.; Fraisse, P.; Ramdani, N.; Demircan, E.; Murray, A.; Keller, T. Estimation of the centre of mass from motion capture and force plate recordings: A study on the elderly. *Appl. Bionics Biomech.* **2011**, *8*, 67–84. [CrossRef]
17. Cotton, S.; Murray, A.; Fraisse, P. Estimation of the Center of Mass: From Humanoid Robots to Human Beings. *IEEE/ASME Trans. Mechatron.* **2009**, *14*, 707–712. [CrossRef]
18. Menga, G. Automatic Control. DAUIN—Politecnico di Torino. Available online: https://www.dropbox.com/s/jhkai3pabok0t8r/Book.pdf (accessed on 21 October 2019).
19. Menga, G. Automatic Control Software. DAUIN—Politecnico di Torino. Available online: https://www.dropbox.com/s/xs5mwy4hvzqigcn/controldesign3.zip (accessed on 21 October 2019).

robotics

MDPI

Article

Control of the Sit-To-Stand Transfer of a Biped Robotic Device for Postural Rehabilitation

Giuseppe Menga [1,*,†] **and Marco Ghirardi** [2,†]

1 Department of Control and Computer Engineering, Politecnico di Torino, Corso Duca degli Abruzzi 24, 10129 Torino, Italy
2 Department of Management and Production Engineering, Politecnico di Torino, Corso Duca degli Abruzzi 24, 10129 Torino, Italy; marco.ghirardi@polito.it
* Correspondence: giuseppe.menga@formerfaculty.polito.it; Tel.: +39-011 0907261
† These authors contributed equally to this work.

Received: 17 July 2019; Accepted: 21 October 2019; Published: 24 October 2019

Abstract: This paper deals with the control of the sit-to-stand transfer of a biped robotic device (either an autonomous biped robot or a haptic assistive exoskeleton for postural rehabilitation). The control has been synthesized, instead of considering the physiology, analyzing the basic laws of dynamics. The transfer of a human from sitting on a chair to an erect posture is an interesting case study, because it treats biped balance in a two-phase dynamic setting, with an external force disturbance (the chair–pelvis contact) affecting the center of pressure under the feet. At the beginning, a body is sitting, with a fixed pelvis moving with the hips going toward the supporting feet and, contemporaneously, releasing the load from the chair with ankles and knee torques. Then, after lift-off, it reaches and maintains an erect posture. The paper objectives are threefold: identifying the major dynamical determinants of the exercise; sythesizing an automatic control for an autonomous device; proposing an innovative approach for the rehabilitation process with an exoskeleton. For this last objective, the paper extends the idea of the authors of a haptic exoskeleton for rehabilitation. It is driven to control the joints by electromiographical signals from the patient. The two spaces, cartesian (world) and joint, where, respectively, the automatic control and the patient operate, are considered and a technique to blend the two actions is proposed. The exoskeleton is programed to perform the exercise autonomously. Then, during the evolution of the phases of rehabilitation, we postulated to seamlessly move the control from one space (purely autonomous) to another (completely driven by the patient), choosing and keeping the postural tasks and joints (heaps, knees, or ankles) on which to apply each one of the two actions without interaction.

Keywords: exoskeleton; haptics; rehabilitation; postural control; postural balance; multi-chain dynamical systems

1. Introduction

Robotic rehabilitation started with passive exoskeletons [1], or devices that impose (on the interested limbs) a forced motion. An early example for lower limbs is Lokomat [2]. It evolved into systems that create force tunnels to address the motion generated by the patient [3]. These were further extended with the introduction of a feedback from the patient, to offer cooperative controls (also called hybrid control) not only to guide but also to contribute to the efforts of the patient [4]. An extensive state-of-the-art of cooperative exoskeletons for rehabilitation is contained in [5]. Following this line of approach, we proposed a haptic exoskeleton where the joints are actuated using admittance control based on the patient's Electromiographical (EMG) signals [6].

One classical exercise for postural rehabilitation performed in a fixed position is the "sit-to-stand". Then, a haptic exoskeleton able to guide the patient to perform this exercise is highly desirable.

The study of the motion of the body during this apparently simple, but in reality not so simple, exercise has attracted interest for a long time [7], not only to understand the human physiological behavior, but also to mimic the control for autonomous biped robots or for actively cooperating exoskeletons.

The majority of available studies are related to the analysis of the human physiological behavior [8,9], but also examples of synthesis of the control based on optimization are available [10,11]. A recent comprehensive review can be found in [12]. However, none discussed the key determinants at the root of the exercise. Here, we follow a different approach. Recognizing that the human motion in performing the exercise is the direct consequence of the respect of physical laws of dynamics, these laws are analyzed and a feedback control based on them is synthesized. This also offers an explanation of well know physiological results such as the "Alexander STS technique" [11].

The exercise is composed of two dynamical phases: phase 1, when still sitting on the chair, the trunk, through the hips, is moved forward to gain balance on the feet, and phase 2, when the balance is maintained moving from the chair to an erect posture. In both phases, postural balance plays a key role, however, in phase 1, the coordination between the motion of the trunk and the torques on ankles and knees to release the load from the chair are also important. From the understanding of these dynamics, an automatic control can be synthesized. However, in the case of a lower limb exoskeleton for rehabilitation, such as in [6], where an automatic postural feedback operates in the *Cartesian* space and the patient controls the joints in the *joint* space, the interaction between the two players has to be considered. This paper proposes to program the exoskeleton to perform the exercise autonomously, then, with an innovative approach, to blend the two actions, moving seamlessy during the evolution of the rehabilitation, under the direction of the physiotherapist, from purely automatic to completely under the control of the patient. Moreover, according to the needs of the rehabilitation, some of the components of the coordinates of the *Cartesian* space, indicated here as elemental postural tasks, can be actuated by the automatic control and the remaining components, through selected joints by the patient, keeping the two groups separat from postural tasks without interfering between each other. Section 2 contains a background on the dynamics of the exercise. Section 3 presents the problem, describes the general adopted model and the autonomous control. Details of control during both phases are in Section 3.3. Section 4 applies the approach to a haptic exoskeleton. The results of a simulation are discussed in Section 5. The conclusions, mentioning future ongoing researches, in Section 6 complete the paper. Details of the control algorithms are contained in Appendices A–C.

2. Background on the Dynamical Behaviour of the Exercise

Two main aspects characterize the exercise: balance and coordination, briefly introduced in the following subsections.

2.1. Balance

The dynamic of a linearized inverted pendulum is a fundamental element to understand the balance of biped systems. It was introduced by Vukobratovic [13] for controlling exoskeletons, but actually has been widely exploited in autonomous biped robotics. He argued that balance is guaranteed if the center of pressure (*CoP*) of the reaction forces exerted by the ground on the feet and is maintained in the convex hull containing the surface of the feet. This point is coincident, on a flat horizontal surface, to a point he called *ZMP* (zero moment point-where the reaction from the constraint is a pure force with zero moment), a result from classical equivalence and replacement [14] in mechanics. Moreover, a linearized inverted pendulum is a really good approximation of the more complex kinematic chain of a biped, but also, adopting this approximation, and *ZMP* motion is linked by a linear relationship to position and acceleration. In a simplified 2D environment, (the paper considers motion only on the sagittal plane) the relationship is

$$ZMP_x = COG_x - COG_z/g \cdot \ddot{COG}_x \tag{1}$$

where ZMP_x and COG_x are the motion coordinates of ZMP and COG on the ground, COG_z is the height of the barycenter and g the gravity acceleration.

Hence, to perform a transition, the COG must be moved controlling the joint angles to which it is algebraically linked, having as objective the ZMP position. As this requires a certain degree of anticipation, usually it is achieved by tracking some special pre-computed references of the COG (preview control [15]). Choi in [16] showed that a closed loop feedback from the measures of COG and ZMP position is able to track a preview reference signal. This feedback has been improved in [17] by introducing in the loop a state estimator of $COG - ZMP$, extended to external disturbance forces.

2.2. Coordination

The biped in the sit-to-stand exercise is a dynamical system with changing non-holonomic constraints: the contacts of ground–feet are always present and the chair–pelvis is only present in the first phase only. This has consequences on the number of degrees of freedom of the multibody chain, and in the role of torque on the joints, that are defined in [18,19] as *Position* and *Auxiliary*. The former contributes to the motion, and the latter only to the reaction forces on the constraints. The balance with tracking of a preview reference of the $COG - ZMP$, based on the measures of the center of pressure under the feet and the pelvis–chair contact, and the coordination of *Position* and *Auxiliary* torques for motion and constraint force control are the elements of a "sit-to-stand" transition.

3. The Process, the Model and the Control

3.1. The Process

The exoskeleton is joined to the patient, moving on the sagittal plane with joint motion of the pairs of ankles, knees and hips of the two legs; it can be represented as a four-link chain: feet, legs, thighs and trunk. The trunk, comprising head and arms, is also often called *hat*. The hips are coincident with the pelvis. In a sitting position, as is represented in Figure 1, the reaction forces transferred from the chair are applied for simplicity at the pelvis point, without exchange of torque.

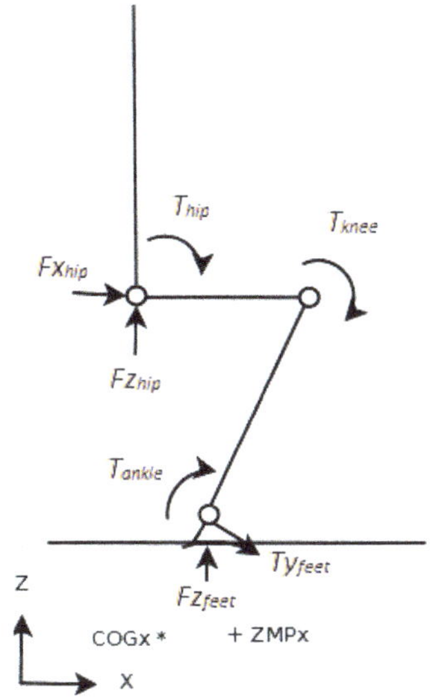

Figure 1. A biped sitting on a chair.

In a stationary posture, with relaxed muscles controlling ankles and knees, and the trunk attitude maintained by controlling the hips, COG_x and ZMP_x do not coincide as stated by the classical relationship (1), due to the presence of the external disturbance (with respect to the feet) represented by the chair reaction forces. COG_x position is well behind the soles of the feet, close to the chair support, while the ZMP_x alone is under the feet. The "sit-to-stand" transition starts by moving the COG_x from its actual position from the chair toward the feet, tracking a preview signal which will be followed, also, during the second phase of the exercise. A typical behavior of this signal, based on Equation (1), imposing a step without overshoot of the ZMP_x is given in Figure 2. Note that it depends on only one design parameter, i.e., the value of COG_z.

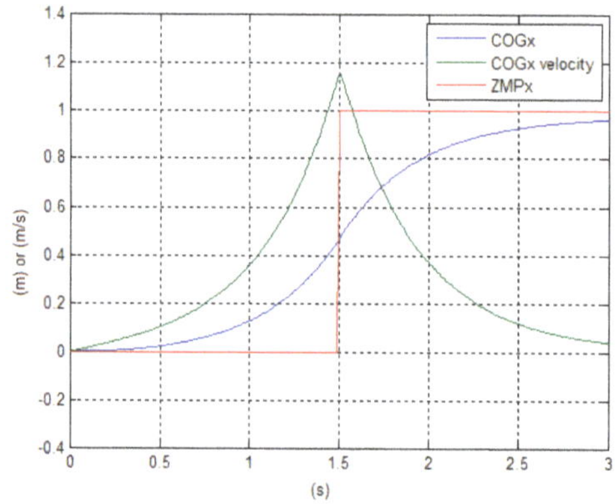

Figure 2. A preview signal for the COG_x position and velocity imposing a step forward to the ZMP_x.

At the same time during phase 1, two torques are applied to the ankles and the knees to bring the reaction force under the pelvis to zero. As these torques influence the reaction forces/torques at the constraints, it is easy to see that this action, without trunk motion, would move the ZMP_x back away from the feet (Figure 3), with obvious consequences on the balance.

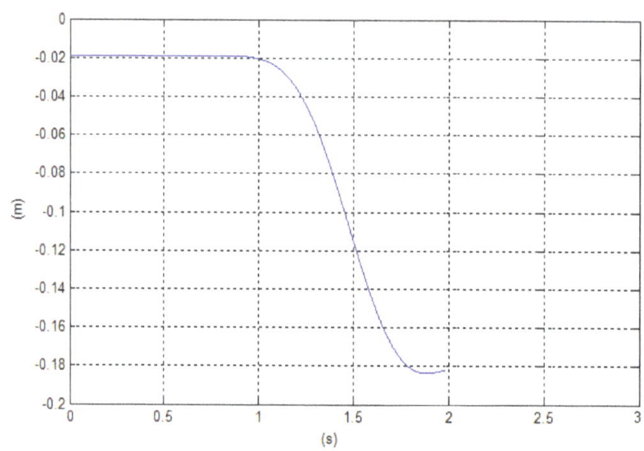

Figure 3. The ZMP_x on the feet releasing the load from the chair, without moving forward the trunk.

The physiological behavior through proprioception, sensing the reaction forces under the feet and the pelvis, coordinates the two actions, so that during phase 1 while the reaction forces under the pelvis are zeroed, moving the trunk forward the ZMP_x is approximately maintained in the original position under the feet guarateeing balance for the next lift-off. Comparing Figures 2 and 3, (the comparison is only qualitative, as Figure 2 is generic), with transition from phase 1 to phase 2 at time 1.5 s, the behaviour of the inverted pendulum and the role of the preview can be better understood: the motion of the COG must start well in advance of any other action in order to compensate the negative effect on the ZMP_x caused by releasing the load from the chair, moreover, dynamically postural equilibrium can be reached at the halfway through the transition period, even before the COG reaches the feet.

The choice of the design parameter COG_z for the preview transition of the COG is not irrelevant; in fact, it can be seen in Figure 4 how the behavior of $COG_x, \dot{COG}_x, ZMP_x$ changes if the design parameter is chosen below (red plot-with an overshoot) or above (blue plot-smoother response) the actual value (green plot) of the inverted pendulum. At last, the choice of this design parameter influences the rotational velocity of the trunk at the time of lift-off from the chair when moving to phase 2, with consequences on the kinetic energy spent and on efficiency. We argue that it is the main parameter explaining the Alexander STS technique frequently cited in the STS literature [11].

Phase 1 ends when the reaction forces from the chair become zero. At this point, the biped is in postural dynamical balance and all three joints now contribute to the motion in phase 2 to complete the exercise and reach an erect standing posture.

Obviously, these and the following are simplified observations. In reality, physiologically, through neural plasticity, almost any person who does not know anything about preview control and inverted pendulum, learns how to control the COG and the torques on the joints during the exercise optimizing, contemporaneously, the expenditure of energy.

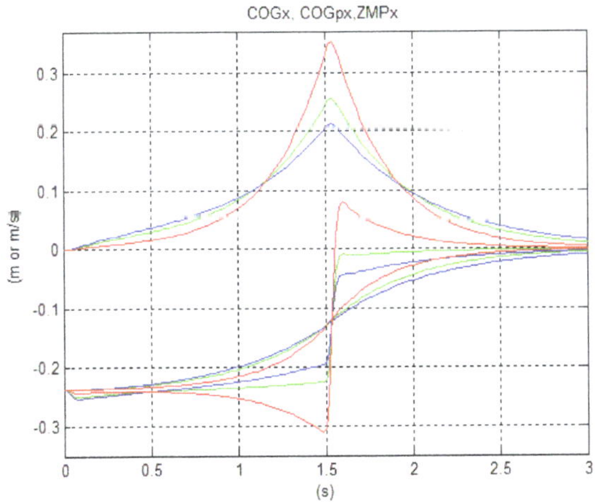

Figure 4. Responses of COG_x, $\dot{COG}_x$ and ZMP_x for previews with different parameter COG_z.

3.2. The Model

The model of the biped device in a standing up position is given in Figure 5. The kinematical chain has three degrees of freedom (DOF) with configuration variables of angles θ_{ankle}, θ_{knee}, θ_{hip}, representing the *joint* space, with control torques T_{ankle}, T_{knee}, T_{hip}. The three coordinates, target variables representing three elemental postura tasks, in the *Cartesian* space, as the problem is dealing with body balance and posture, COG_x, the height of the pelvis and the attitude θ_{trunk} of the trunk. For simplicity, the height of the pelvis is substituted by the knee angle θ_{knee}, to which, in balance, it is strictly related. This choice also avoids singularity of the Jacobian when the knees are fully stretched. With reference to Figure 1, reaction forces from the ground to the feet, applied for convention to the vertical projection on the ground of the ankle $Ankle_x$, and torque are $F_{z_{feet}}$, $F_{x_{feet}}$ (compensated by the ground friction, is not further represented, as it has no roles) , $T_{y_{feet}}$. Forces from the chair to the pelvis (in the sitting position) are $F_{z_{hip}}$, $F_{x_{hip}}$.

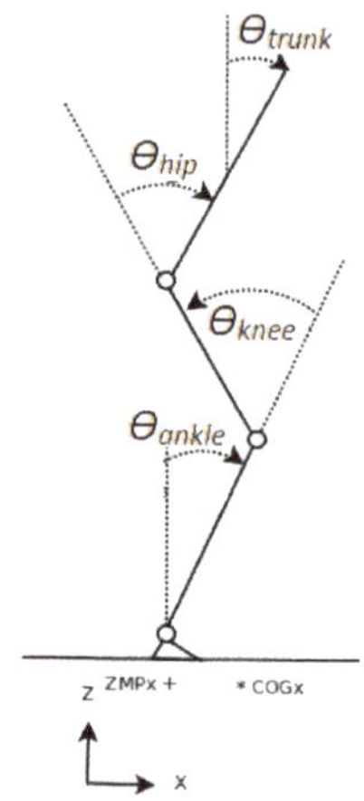

Figure 5. The model of the biped device.

From previous definitions, the displacement of ZMP_x, with respect to the position $Ankle_x$ of the conventional application of reaction forces is given by

$$\begin{gathered} \Delta ZMP_x = -T_{y_{feet}} / F_{z_{feet}} \\ \text{i.e.,} \\ ZMP_x = Ankle_x + \Delta ZMP_x \end{gathered} \tag{2}$$

This indicates that maintaining ZMP_x under the feet means keeping $T_{y_{feet}}$ close to zero.

During phase 1, in the presence of the non-holonomic constraint given by the chair–pelvis contact, the degrees of freedom of the kinematic chain are reduced: the only *Position* variable controlling the trunk attitude is the hip angle, while ankle and knee are *Auxiliary*. The latter torques control the constraint reaction forces and torques. Their mathematical expressions are composed of three terms: a nonlinear function of the angular speeds, a term generated by acceleration, and a linear function of the angular accelerations of the *Position* joints and of the control torques of the *Auxiliary* ones. Obviously, all coefficients are function of the angle values:

$$\begin{bmatrix} T_{y_{foot}} \\ F_{x_{hip}} \\ F_{z_{hip}} \end{bmatrix} = NL(\theta_{hip}, \dot{\theta}_{hip}) + G(g) + \begin{bmatrix} M_{11} M_{12} M_{13} \\ M_{21} M_{22} M_{23} \\ M_{31} M_{32} M_{33} \end{bmatrix} \cdot \begin{bmatrix} T_{ankle} \\ T_{knee} \\ \ddot{\theta}_{hip} \end{bmatrix} \tag{3}$$

In the right hand side of expression (3), only the linear relationships of the control torques are of interest, which coefficients M_{ij} are basically Jacobians. Target velocities in the *Cartesian* space are linked to the joint angle velocities by the Jacobian matrix $\boldsymbol{J}$:

$$\begin{bmatrix} C\dot{O}G_x \\ \dot{\theta}_{knee} \\ \dot{\theta}_{trunk} \end{bmatrix} = \boldsymbol{J} \cdot \begin{bmatrix} \dot{\theta}_{ankle} \\ \dot{\theta}_{knee} \\ \dot{\theta}_{hip} \end{bmatrix} \tag{4}$$

with J given by

$$J = \begin{bmatrix} J_{11} & J_{12} & J_{13} \\ 0 & 1 & 0 \\ 1 & 1 & 1 \end{bmatrix}. \tag{5}$$

The Jacobians of COG_x, present in the Jacobian matrix with embedded knee and trunk motion (5) and the coefficients M_{ij} in (3) can be generated in real time from the joint angles using expressions, which in our case, were obtained by the symbolic computational environment AutoLev [20] implementing Kane's method [14], when the kinematics is known, or better, for the COG_x, they can be derived from a SESC model [21] identified by a priori experiments. For details see [17].

3.3. The Control

Depending on the transition phase, the joint motor drivers are configured to be controlled either in velocity or in torque. The control is obtained by closing the loop using the measures of the center of pressure under the feet and the reaction forces under the pelvis, obtained from appropiate loading cells mounted on the chair and on the floor. Moreover, the position/velocities of the joints angles from the motor drivers, torques on the joints, and inertial datas from the trunk are available.

3.3.1. Velocity Control

When the joints are configured for velocity control, the inverse of the Jacobian matrix (4) of the previous section is used to generate the inputs of the motor drivers in the *joint* space to track the desired reference preview signals of the COG_x, of the trunk attitude and of the knee angle in the *Cartesian* space. These, during phase 2, are

$$\dot{\boldsymbol{\theta}}_{ref}(t) = \boldsymbol{J}^{-1} \cdot \boldsymbol{u}(t) \tag{6}$$

where

$$\dot{\boldsymbol{\theta}}_{ref}(t) = \begin{bmatrix} \dot{\theta}_{ankle_{ref}} \\ \dot{\theta}_{knee_{ref}} \\ \dot{\theta}_{hip_{ref}} \end{bmatrix} \tag{7}$$

is the vector of speed reference in the *joint* space and

$$\boldsymbol{u}(t) = \begin{bmatrix} u_{COG}(t) \\ u_{hat}(t) \\ u_{knee}(t) \end{bmatrix} \tag{8}$$

is the vector of the return difference signals of the feedback in the *Cartesian* space derived in the Appendix A. During phase 1, as only the hip is controlled in velocity, (6) simplifies to

$$\dot{\theta}_{hip_{ref}}(t) = J_{13}{}^{-1} \cdot u_{COG}(t). \tag{9}$$

3.3.2. Torque Control

For the drivers of ankles and knees configured to torque control during phase 1, Equation (3) are used to bring to zero the chair reaction forces, based on a recursive weighted least square scheme (described in details in Appendix B):

$$\begin{gathered} \boldsymbol{u_T}(t+\Delta) = \boldsymbol{u_T}(t) + \\ \alpha \boldsymbol{M}^{+} \cdot (\boldsymbol{F}_{react_{ref}}(t) - \boldsymbol{F}_{react}(t)) \end{gathered} \tag{10}$$

where Δ is the sampling time, $u_T(t)$ is the vector of reference inputs to the torque control drivers of the motors

$$\boldsymbol{u_T}(t) = \begin{bmatrix} T_{ankle_{ref}}(t) \\ T_{knee_{ref}}(t) \end{bmatrix} \tag{11}$$

α is a coefficient to guarantee convergence, $\boldsymbol{M}^+$ is the solution of the least square problem (A5), and $\boldsymbol{F}_{react_{ref}}(t)$ and $\boldsymbol{F}_{react}(t)$ are the vectors of reference to track and measured reaction force/torques :

$$\boldsymbol{F}_{react_{ref}}(t) = \begin{bmatrix} T_{y_{foot_{ref}}} \\ F_{x_{hip_{ref}}} \\ F_{z_{hip_{ref}}} \end{bmatrix}, \boldsymbol{F}_{react}(t) = \begin{bmatrix} T_{y_{foot}} \\ F_{x_{hip}} \\ F_{z_{hip}} \end{bmatrix} \tag{12}$$

$F_{x_{hip_{ref}}}, F_{z_{hip_{ref}}}$ are chosen to send the reaction forces on the chair from their starting value to zero, and $T_{y_{foot_{ref}}}$ is kept equal to 0. The measure of $T_{y_{foot}}$ is not avilable, however it can be inferred by the distance of ZMP_x from $Ankle_x$.

The choice of the weighted least square allows to blend torque transition with the need to control the ZMP_x. In fact, with $T_{y_{foot_{ref}}}$ close to zero ZMP_x becomes fairly insensitive to the choice of the other references signals.

3.3.3. Lift off Transition

The control switches from phase 1 to phase 2 when the module of the reaction forces reaches a neighbourhood of zero. At that moment, a smooth transition must be imposed on the control of knees and ankles to avoid torque spikes. The simplest approach, adopted here, is to maintain (for a fraction of time at the joints) the value of torques that is reached at the end of phase 1 while the velocity control of phase 2 goes into action with a loop gain increasing from zero to the design value.

4. A Haptic Exoskeleton

The control described in the previous sections refers to an autonomous behavior, and it can be used to program the exercise into a biped robot, or applied in an exoskeleton in the first phase of the reabilitation when a patient is completely unable to operate.

In the case of a haptic exoskeleton partially or totally controlled by the patient, two different aspects have to be considered: the joints controlled in torque and the jonts controlled in velocity. For both aspects, EMG signals that are measured on the appropriate muscles offer approximate information of the torques applied by the patient to his joints. For the torque control, the contribution to the patient's effort is achieved with a classical technique as described in [22].

For the motion, admittance control is adopted as described in [6]. The EMG signals, processed by an admittance filter, are translated into motion information that is used as reference velocity of the corresponding joint speed drivers of the exoskeleton.

Torque control applies to phase 1 of the exercise, the contribution of the exoskeleton is to help the patient to coordinate the two actions; *motion* with the rotation of the trunk and contemporaneously *torque* for the release of the load from the chair. This can be achieved by plotting on a display in front of the patient the position of the center of pressure in relation to the feet, with total or partial automatic support of the exoskeleton. Two options are available: automatic motion tracking a preview (different previews can be tested), and torques supported by the patient, or vice versa.

After transition to phase 2, only motion is involved, with three postural tasks, described in (4), to execute, and three joints to control. Let indicate the motion references driven by the patient's efforts, as output of the admittance filters, for ankle, knee and hip as

$$\dot{\theta}_p(t) = \begin{bmatrix} \dot{\theta}_{ankle_p}(t) \\ \dot{\theta}_{knee_p}(t) \\ \dot{\theta}_{hip_p}(t) \end{bmatrix}. \tag{13}$$

In a training program, it is desirable that the patient progressively takes control, in number and strength, of the joints of the exoskeleton, and in so doing assumes responsibility of one or more of the postural tasks, without affecting the remaining tasks performed autonomously by the feedback. For this, the concept of *tutoring-cooperation-coordination* is introduced, where some of the elemental postural tasks, under the control of the patient, and the complementary tasks, under the automatic postural loop, can be completely decupled. This is a variant of whole body coordination (WBC) discussed in [16] and introduced in [23]. *Tutoring* means that all tasks are performed fully automatically, without intervention of the patient that is completely tutored, it is equivalent to the WBC. *Coordination* is when the patient takes complete control, through some joints, of one or more elemental postural tasks of the exercise, without interfering with the complementary tasks controlled by the automatic postural loop and, hence, he must perform coordinate motions. In between of the two extreme situations with *cooperation*, the two players can operate jointly on some tasks and their actions are overlapped. Through a modulation parameter β the patient takes partial control and must cooperate with the automatic tasks, as depicted in Figure 6. The approach is based on two aspects: the separation of the elemental postural tasks operated by the two players, and for the tasks operated jointly, on the modulation of the two actions.

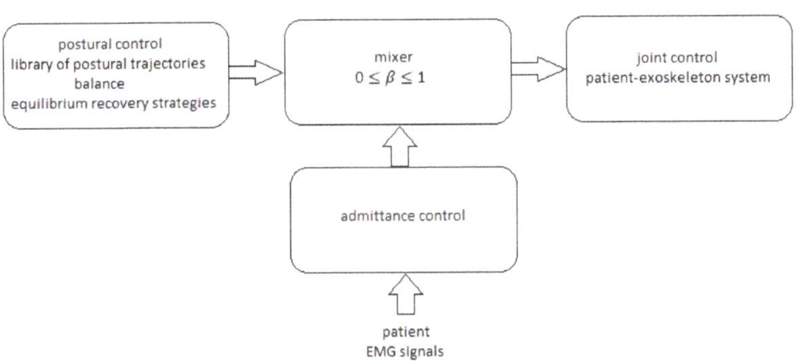

Figure 6. Mixing cartesian postural and patient joint controls.

Let R_u represents the 3×3 selection matrix, with ones and zeros on the diagonal covering the range of the elemental tasks performed exclusively by the autonomous feedback and N_u its nullspace with tasks controlled partially or totally by the patient through the joints in the range of R_p. As an example, when autonomous processes control COG_x and trunk attitude and the patient controls the height of the pelvis through the knee, the matrices R_u and R_p are:

$$R_u = \begin{bmatrix} 1 & 0 & 0 \\ 0 & 0 & 0 \\ 0 & 0 & 1 \end{bmatrix}, R_p = \begin{bmatrix} 0 & 0 & 0 \\ 0 & 1 & 0 \\ 0 & 0 & 0 \end{bmatrix}, \tag{14}$$

or, the patient, through knees and hips, controls height and attitude of the trunk, with the balance, through COG_x, guarateed automatically, the matrices R_u and R_p are:

$$R_u = \begin{bmatrix} 1 & 0 & 0 \\ 0 & 0 & 0 \\ 0 & 0 & 0 \end{bmatrix}, R_p = \begin{bmatrix} 0 & 0 & 0 \\ 0 & 1 & 0 \\ 0 & 0 & 1 \end{bmatrix}. \tag{15}$$

The postural tasks in the range of R_u are completely controlled by the postural feedback, while the complementary tasks in its null-space N_u are jointly controlled by the feedback and by the patient, through the joints in the range of R_p, according to the value of a coefficient $0 \leqslant \beta \leqslant 1$. β indicates the level of the patient's involvement in the control: *tutoring* is when $\beta = 0$, *coordination* is when $\beta = 1$, *cooperation* is when $\beta > 0$ *and* $\beta < 1$. As the coefficient β modifies the admittance filter gain, it has, also, influence on the compliance felt by the patient on the joints in the range of R_p (in fact, the joints in its null-space, not being controlled by the patient, don't offer any compliance): $\beta = 0$ completely stiff joints, $\beta = 1$ fully compliant.

The general expression for the references of the speed control of the actuators, merging automatic postural control and patient action through EMG signals, according to *tutoring-cooperation-coordination* is:

$$\begin{aligned} \dot{\boldsymbol{\theta}}_{ref}(t) &= \boldsymbol{J}^{-1} \cdot \boldsymbol{u}_p(t) + R_p \cdot \dot{\boldsymbol{\theta}}_p(t) \cdot \beta \\ \boldsymbol{u}_p &= (R_u + N_u \cdot (1-\beta)) \cdot \boldsymbol{u} - F \cdot R_p \cdot \dot{\boldsymbol{\theta}}_p \cdot \beta \end{aligned} \tag{16}$$

where $F = R_u \cdot \boldsymbol{J}$, and with the condition

$$range(N_u) \subset range(\boldsymbol{J} \cdot R_p). \tag{17}$$

The proof of Equation (16) is presented in Appendix C.

During phase 1, while the torque control operates on ankle and knee, the *COG* tracking (9) is modified as follows

$$\dot{\theta}_{hip_{ref}}(t) = J_{13}{}^{-1} \cdot u_{COG}(t) \cdot (1-\beta) + \dot{\theta}_{hip_p}(t) \cdot \beta. \tag{18}$$

Moving β from 0 to 1, the physiotherapist increases the admittance of the joints, so that the patient gains more and more control of them. Changing R_u to the zero matrix, and R_p to the identity, with $\beta = 1$, the patient assumes complete control of the exoskeleton. In this condition, the autonomous control can continue to monitor the postural balance, and eventually, it inhibits incorrect patient postures, by automatically returning β to 0.

5. A Simulation Example and Comparisons

In this section the simulation of the automatic control is offered. The test of the complete exercise on the exoskeleton of Figure 7 has not been performed yet. However, in [6] several examples (phase 2 only) of the transition with the cooperation of the patient using this control have been reported.

The simulation uses the same data of a patient of 75 kg, 1.8 m tall wearing an exoskeleton of 35 kg presented in [6]. The model considers only one leg and the weight of the trunk (hat) is divided by 2. So that the magnitudes of forces (N) and torques (Nm) are referred to each leg. The exercise lasts 3 s. Approximately at 1.5 s the transition from phase 1 to phase 2 occurs, with the lift off. The *COM* preview for a step transiton of the *ZMP* of Figure 2 has been chosen, testing different values of the parameter COG_z. Different experiments, with different degrees of difficulty, obtained with the feet in different positions with respect to the initial value of COG_x, i.e., to the chair, have been performed. Each position is characterized by the angle of the ankles from 10° to 45° when sitting, (i.e., feet far away from the chair and exercise more difficult, in the first case, or feet under the chair and easier exercise, in the second case). The lift-off time and transition from phase 1 to phase 2 is triggered when the module of the reaction forces on the pelvis enter into a neighbour of few Newtons of the zero. During phase 2, simple transition functions drive knees and trunk attitude angle positions and velocities from their initial values, left from phase 1, to zero in 1.5 s. The next figures represent the most difficult experiment (experiment 1) that can be performed using this control with ankle angle of 10°. With a lower angle, e.g., 5°, the *CoP* cannot be guaranteed to remain inside the foot print. The behavior of reaction forces and torques are shown in Figure 8.

Figure 7. The prototype exoskeleton for the sit-to-stand exercise.

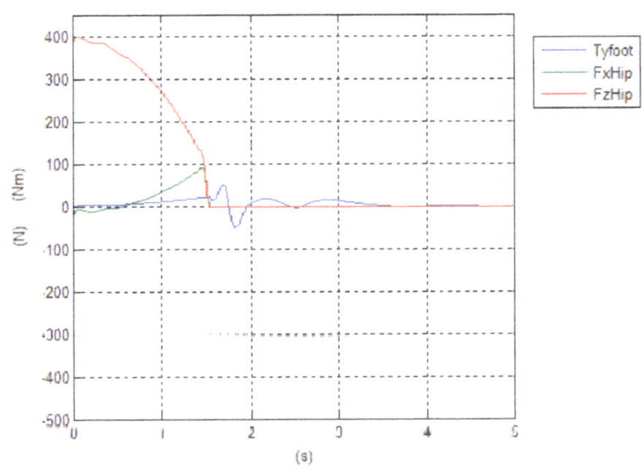

Figure 8. The behavior of reaction force on the hip and torque on the foot: experiment 1.

The behavior of COG_x and ZMP_x during the transition is given in Figure 9. The position of $Ankle_x$ is 0.

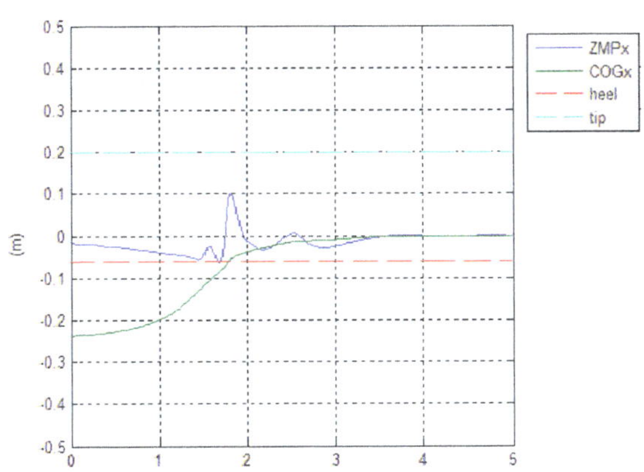

Figure 9. The behavior of COG_x and ZMP_x during the transition: experiment 1.

The torques at the joints are in Figure 10.

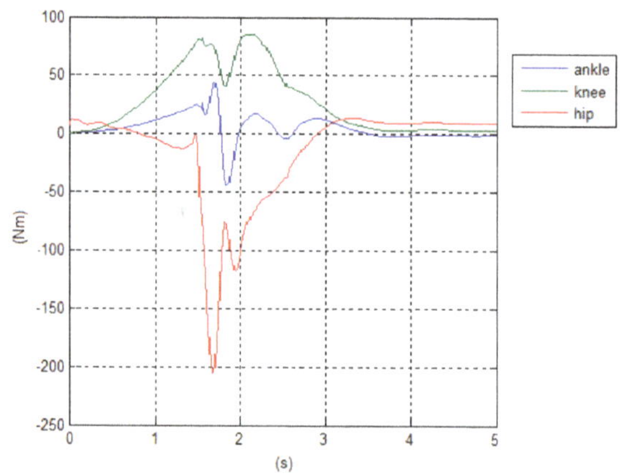

Figure 10. The torques at the joints: experiment 1.

Figure 11 shows the animation of the exercise.

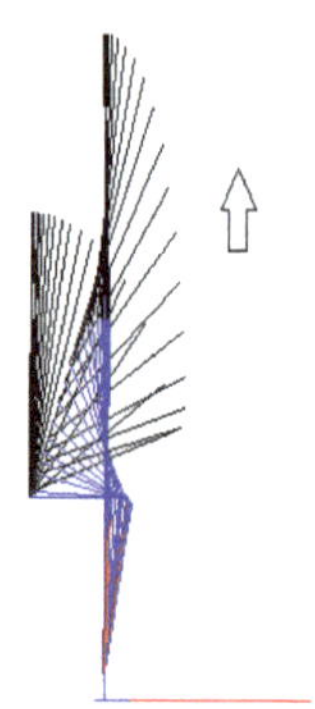

Figure 11. The animation of the exercise: experiment 1.

Figure 12 shows how the hat and knee angles track the references.

It is interesting to see the estimates $\hat{COG}_x$, $Z\hat{M}P_x$ theoretical assuming no disturbances and $Z\hat{M}P_x$ actual (i.e., the *CoP*), along with the estimate of the disturbances, i.e., the reaction forces on the chair, obtained by the extended estimator [17]. These are given in Figure 13.

The performance of the exercise, expressed as the ratio between the change in potential energy and the total work consumed to complete the motion, as defined in [11] is 60%, and the maximum angle reached by the trunk is 79°. As oppose, the remaining figures, from Figures 14–18, show a relatively easy exercise (experiment 2) with an ankle angle of 45°. The performance here is 95% with a maximum trunk angle value of 11°. To show the robustness of the control this exercise was performed with the identical control parameters of the previous experiment.

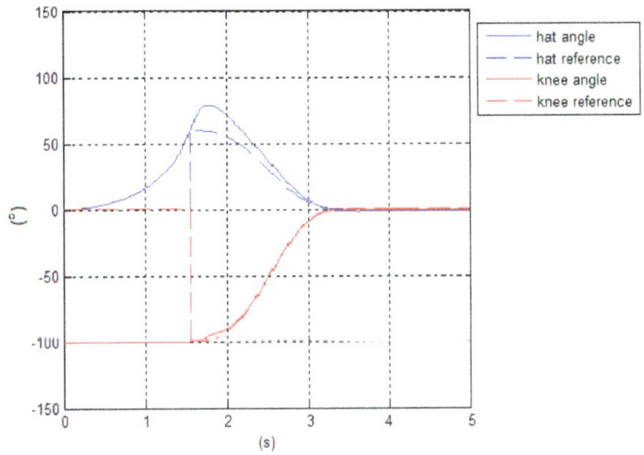

Figure 12. The tracking of hat and knee angles: experiment 1.

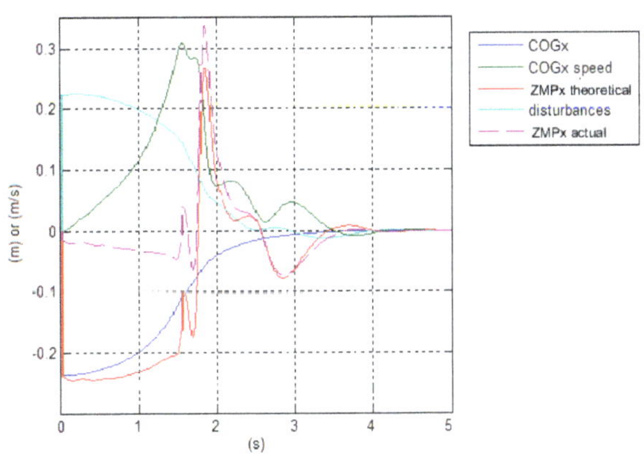

Figure 13. Estimates of COG and ZMP resulting from the extended estimator accounting for the pelvis-chair disturbances: experiment 1.

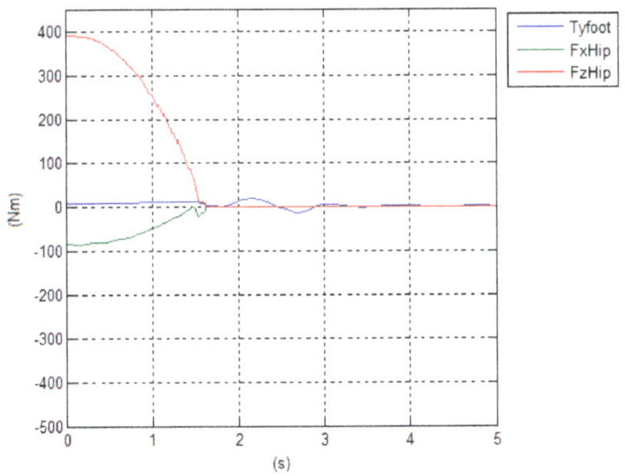

Figure 14. The behavior of reaction force on the hip and torque on the foot: experiment 2.

The behavior of COG_x and ZMP_x during the transition is given in Figure 15.

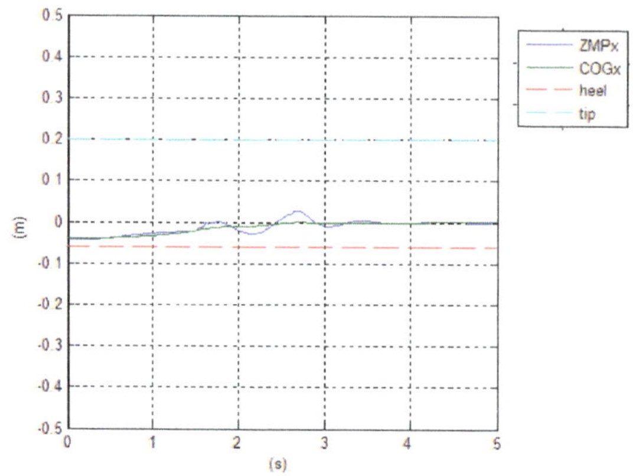

Figure 15. The behavior of COG_x and ZMP_x during the transition: experiment 2.

The torques at the joints are in Figure 16.

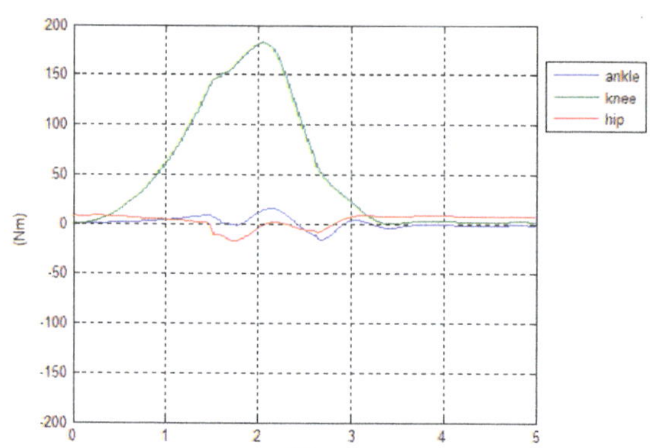

Figure 16. The Torques at the joints: experiment 2.

Figure 17 shows the animation of the exercise, while Figure 18 shows how the hat and knee angles track the references.

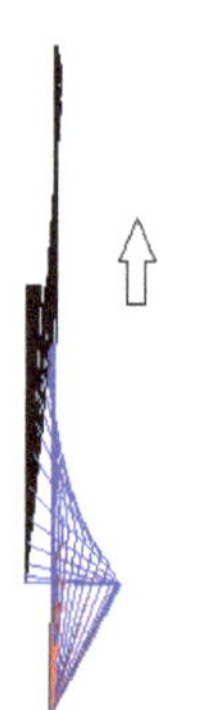

Figure 17. The animation of the exercise: experiment 2.

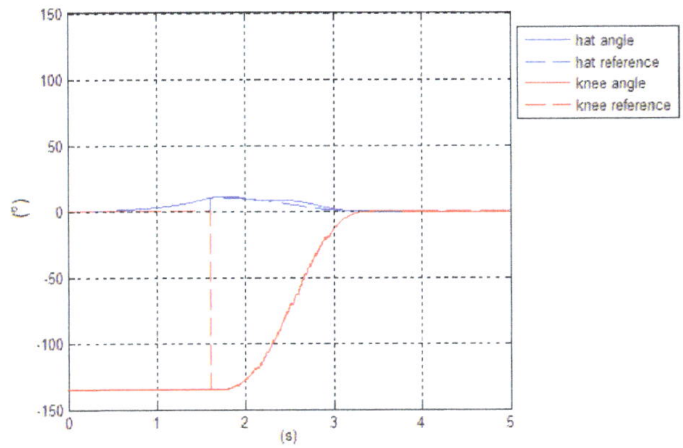

Figure 18. The tracking of hat and knee angles: experiment 2.

6. Conclusions

The paper approaches the sit-to-stand exercise under two aspects: analysis of the dynamics of the exercise/design of the feedback control of an autonomous biped device, and its application to an haptic exoskeleton with the cooperation of a patient.

In regard to the autonomous execution of the exercise, several authors have argued that two components are involved in solving the control problem in the humanoid STS motion: (1) phase and trajectory planning, and (2) feedback control.

With respect to the former, the most significant are the COM preview to follow in the transition, and the time coordination between motion of the hips and torques on ankles and knees in phase 1. From the experiments, the chair reaction forces, the knee angle and trunk attitude to reach an erect posture are not critical. The COM preview can be synthesized by exploiting the linearized inverted pendulum model, with basis the step transition of the ZMP, and selecting the appropiate parameter COG_z. The value of this parameter is not necessarily identical to the real height of the COG of the biped, but it is chosen according to the difficulty of the exercise. This difficulty here is represented by the distance of COG_x, i.e., the chair, from the feet when sitting. The design parameter COG_z can explain the Alexander STS technique. If its value is lower or equal than the height of the COG, the transition is steep, the lift off is early, the biped will be in dynamical balance, with the COG_x still far from the feet, and with a relatively high trunk angular velocity to be conveyed in the successive lift motion, on the contrary if the transistion is slower, the balance will be almost statical with the COG_x already under the feet, and the trunk attitude almost stationary. Based on these elements, the final trajectories can easily be planned to reduce maximum torque on the joints and energy expenditure. In the simulated experiments, two examples are considered: one is difficult with a performance, expressed as the ratio between the change in potential energy and the total work consumed to complete the motion of 60%, with the trunk reaching a maximum angle of 79°; the other is relatively easy with a performance of 95% and a maximum trunk angle of 11°.

With respect to the latter, exploiting the inverse or pseudo-inverse of Jacobian matrices and a linearized inverted pendulum model of a simple linear feedback, based on feet and chair pressure sensors, and position/velocity measures of the joint angles, can be obtained either for joint velocity or torque controls. This feedback requires a limited number of model parameters and design gains without the need to perform cumbersome inverse kinematics or dynamics. Symbolic environments such as Autolev can offer efficient expressions for the necessary Jacobians that can be computed in real time.

When the postural control is applied to an exoskeleton interacting with a patient, the paper proposes an innovative approach to blend the patient's actions with the automatic execution of the exercise. Several options are offered by the approach for proposing protocols of physiotherapy tracking the evolution of the process of rehabilitation, in terms of strenght of the support offered by the exoskeleton, and specific joints and postural tasks to be controlled by the patient. The process of rehabilitation can start with a complete *tutoring* of the patient, evolving to a *cooperation* between automatic control and patient on the same postural tasks and jonts, e.g., reaching an erect posture moving the knees, or a *coordination* between the two players operating on different tasks, e.g., the patient raises the pelvis and rotates the trunk through knees and hips while an automatic balance is guaranteed with a feedback on the ankles.

An exoskeleton for this specific exercise has been built, and some preliminary tests of phase 2 have been documented in a previous paper. However, more work is needed in future research: testing a complete exercise using the exoskeleton, not only during phase 2; decoupling the interaction not only from joints to elemental postural tasks, but also vice versa; generalizing the admittance filters from single joints and related muscles to multivariable filters processing the vector of the available EMG signals to collectively control all joints. This will be obtained exploiting muscle synergies, and training artificial neural networks that directly link EMG patterns to motion.

Author Contributions: Methodology, G.M. and M.G.; writing—original draft, G.M.; writing—review & editing, M.G.

Funding: This research has been partially supported by MIUR the Italian Ministry of Instruction, University and Research and the Piedmont Region.

Conflicts of Interest: The authors declare no conflict of interest.

Appendix A. Derivation of the Velocity Controls of Section 3.3.1

The COG tracking requires a preliminary disussion of the $COG - ZMP$ estimator. Evaluations of COG_x and $\dot{COG}_x$ are obtained from the joint angle measures through a SESC model [21], while ZMP_x is given by measuring the center of pressure under the feet (CoP). These outputs, using the state estimator based on an extended inverted pendulum described in the parent paper [17], generate in real time the estimates needed to close the loop: $\hat{COG}_x$, $\hat{\dot{COG}}_x$, $Z\hat{M}P_x$ (the theoretical value in the assumption of no disturbances) and $\hat{\delta}$, the difference between ZMP_x and CoP, needed to take into account the chair-pelvis contact. Then, the feedback return difference signals, at each sampling time, is:

$$\boldsymbol{u}(t) = \begin{bmatrix} u_{COG}(t) \\ u_{hat}(t) \\ u_{knee}(t) \end{bmatrix} \tag{A1}$$

where $u_{COG}(t)$ refers to to the COG tracking

$$\begin{aligned} u_{COG}(t) &= \dot{COG}_{x_{ref}}(t) + \\ kp_1 \cdot (COG_{x_{ref}}(t) &- \hat{COG}_x(t) - \hat{\delta}) \\ +kv_1 \cdot (\dot{COG}_{x_{ref}}(t) &- \hat{\dot{COG}}_x(t)) \\ -kz \cdot (ZMP_{x_{ref}}(t) &- Z\hat{M}P_x(y) - \hat{\delta}) \end{aligned} \tag{A2}$$

$u_{hat}(t)$ refers to to the trunk attitude tracking

$$\begin{aligned} u_{hat}(t) = \dot{\theta}_{hat_{ref}}(t) &+ kp_2 \cdot (\theta_{hat_{ref}}(t) - \hat{\theta}_{hat}(t)) \\ &+ kv_2 \cdot (\dot{\theta}_{hat_{ref}}(t) - \hat{\dot{\theta}}_{hat}(t)) \end{aligned} \tag{A3}$$

and $u_{knee}(t)$ refers to the knee angle tracking

$$u_{knee}(t) = \dot{\theta}_{knee_{ref}}(t) + kp_3 \cdot (\theta_{knee_{ref}}(t) - \hat{\theta}_{knee}(t)) \\ + kv_3 \cdot (\dot{\theta}_{knee_{ref}}(t) - \hat{\dot{\theta}}_{knee}(t)) \tag{A4}$$

Equation (A2) was originally suggested by Choi in [16] and extended in [17], while (A3) and (A4) are basically a proportional-derivative feedback with a feedforward contribution.

During both phase 1 and 2 a preview $COG_{x_{ref}}$ of the type of those in Figure 4 has been chosen, specifically the one of Figure A1, to transfer COG_x from the starting position during sitting to $Ankle_x$.

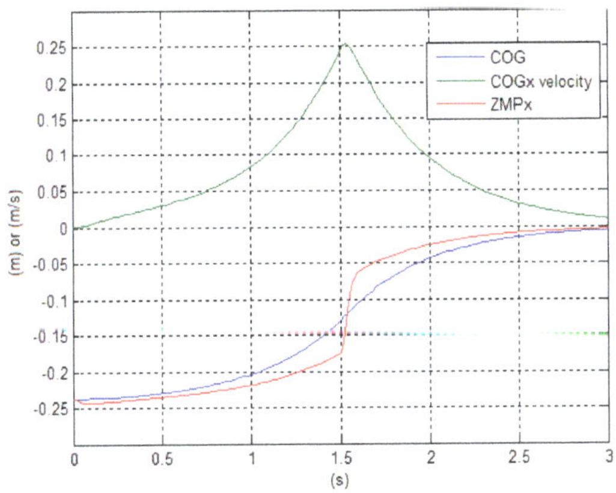

Figure A1. The COG preview signal for the sit to stand transition adopted in the examples.

For phase 2 $\theta_{hat_{ref}}, \theta_{knee_{ref}}$ are the references to bring the patient to a standing posture. $\theta_{knee_{ref}}$ has no special characteristics, it simply imposes a transition to the knee angle during the desired period from sitting to standing, vice versa, $\theta_{hat_{ref}}$ is chosen to take trunk attitude position and velocity from lift-off values at the end of phase 1 to zero (Figure 12).

$k_{p_i}, k_{v_i}, i = 1 - 3, k_z$ are the feedback gains.

Appendix B. Derivation of Torque Controls of Section 3.3.2

For torque control during phase 1, Equations (3) are used to solve the following weighted least square scheme:

$$\Lambda \cdot \begin{bmatrix} T_{y_{foot_{ref}}} - T_{y_{foot}} \\ F_{x_{hip_{ref}}} - F_{x_{hip}} \\ F_{z_{hip_{ref}}} - F_{z_{hip}} \end{bmatrix} \approx \Lambda \cdot M \cdot \begin{bmatrix} T_{ankle} \\ T_{knee} \end{bmatrix} \tag{A5}$$

and

$$M = \begin{bmatrix} M_{11} M_{12} \\ M_{21} M_{22} \\ M_{31} M_{32} \end{bmatrix} \tag{A6}$$

where Λ is a diagonal matrix of weighting factors, with solution

$$\begin{bmatrix} T^*_{ankle} \\ T^*_{knee} \end{bmatrix} = \boldsymbol{M}^+ \cdot \begin{bmatrix} T_{y_{foot_{ref}}} - T_{y_{foot}} \\ F_{x_{hip_{ref}}} - F_{x_{hip}} \\ F_{z_{hip_{ref}}} - F_{z_{hip}} \end{bmatrix} \tag{A7}$$

$$\boldsymbol{M}^+ = (M^T \Lambda^2 M)^{-1} \Lambda^2 M^T, \tag{A8}$$

Appendix C. Proof of Equation (16)

Consider the building block diagram Figure A2. It implements Equation (16), and shows the generation of the $\dot{\boldsymbol{\theta}}_{ref}$ with both autonomous control and patient's contribution. With block diagram manipulations advance the patient's contribution $\dot{\boldsymbol{\theta}}_p$ before $\boldsymbol{J}^{-1}$ from *joint* to *Cartesian* space.

The speed signals in the Cartesian space result as

$$(R_u + N_u \cdot (1-\beta)) \cdot \boldsymbol{u} + (\boldsymbol{I} - R_u) \cdot \boldsymbol{J} \cdot R_p \cdot \dot{\boldsymbol{\theta}}_p \cdot \beta \tag{A9}$$

It should be noted that $\boldsymbol{I} - R_u = N_u$, then

$$(R_u + N_u \cdot (1-\beta)) \cdot \boldsymbol{u} + N_u \cdot \boldsymbol{J} \cdot R_p \cdot \dot{\boldsymbol{\theta}}_p \cdot \beta \tag{A10}$$

Condition (17) assures that the action of the patient on the joints in the range of R_p encompasses all postural tasks not covered by the automatic control.

Equation (A10) can be interpreted as follows:

- If $\beta = 0$, all tasks are controlled by the feedback loop;
- If $\beta = 1$, tasks in R_u are controlled by the feedback loop, while tasks in N_u are controlled by the patient, that, obviously must coordinate his action with the other tasks;
- If $0 < \beta < 1$, tasks in R_u are controlled by the feedback loop, while tasks in N_u are jointly controlled by the feedback loop and by the patient, blended by the coefficient β.

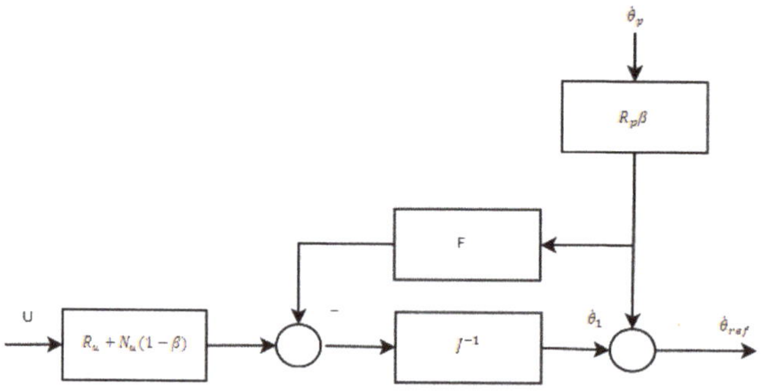

Figure A2. Integrating autonomuos control and patient action in controlling the joints of the exoskeleton.

References

1. Agrawal, S.K.; Banala, S.K.; Fattah, A.; Sangwan, V.; Krishnamoorthy, V.; Scholz, J.P.; Hsu, W.L. Assessment of Motion of a Swing Leg and Gait Rehabilitation with a Gravity Balancing Exoskeleton. *IEEE Trans. Neural Syst. Rehabil. Eng.* **2007**, *15*, 410–420. [CrossRef] [PubMed]
2. Tansey, K. The Use of the Lokomat System in Clinical Research. In Proceedings of the International Neurorehabilitation Symposium: Zurich, Switzerland, 12–14 February 2009.
3. Banala, S.; Seok, H.K.; Agrawal, S.; Scholz, J. Robot Assisted Gait Training with Active Leg Exoskeleton (ALEX). *IEEE Trans. Neural Syst. Rehabil. Eng.* **2000**, *17*, 2–8. [CrossRef] [PubMed]
4. Kawamoto, H.; Suwoong, L.; Kanbe, S.; Sankai, Y. Power assist method for HAL-3 using EMG-based feedback controller. In Proceedings of the 2003 IEEE International Conference on Systems, Man and Cybernetics, Washington, DC, USA, 8 October 2003 .
5. Duschau-Wicke, A.; von Zitzewitz, J.; Caprez, A.; Lunenburger, L.; Riener, R. Path Control: A Method for Patient-Cooperative Robot-AidedGait Rehabilitation. *IEEE Trans. Neural Syst. Rehabil. Eng.* **2010**, *18*, 38–48. [CrossRef] [PubMed]
6. Menga, G.; Ghirardi, M. Lower Limb Exoskeleton for Rehabilitation with Improved Postural Equilibrium. *Robotics* **2018**, *7*, 28. [CrossRef]
7. Shepherd, R.; Carr, J. Reflections on physiotherapy and the emerging science of movement rehabilitation. *Aust. J. Physiother.* **1994**, *40*, 39–47. [CrossRef]
8. Mistry, M.; Murai, A.; Yamane, K.; Hodgins, J. Sit-to-Stand Task on a Humanoid Robot from Human Demonstration. In Proceedings of the 10th IEEE-RAS International Conference on Humanoid Robots (Humanoids), Nashville, TN, USA, 6–8 December 2010.
9. Andani, M.E.; Bahrami, F.; Maralani, P.J. A Biologically Inspired Modular Structure to Control the Sit-to-Stand Transfer of a Biped Robot. In Proceedings of the 29th Annual International Conference of the IEEE EMBS, Lyon, France, 23–26 August 2007.
10. Mughal, A.M.; Iqbal, K. Bipedal Modeling and Decupled Optimal Control Design of Biomechanical Sit-to-Stand Transfer. In Proceedings of the IEEE International Workshop on Robotics and Sensor Environment, Ottawa, ON, Canada, 17–18 October 2008; pp. 46–51.
11. Wang, F.C.; Yu, C.H.; Lin, Y.L.; Tsai, C.E. Optimization of the Sit-to-Stand Motion. In Proceedings of the IEEE/ICME International Conference on Comple Medical Engineering, Shenzhen, China, 22–26 November 2007, pp. 1240–1253.
12. Bahar, B.; Miskon, M.F.; Bakar, N.A.; Shukor, A.Z.; Ali, F. Path generation of Sit to Stand Motion using Humanoid Robot. *Aust. J. Basic Appl. Sci.* **2014**, *8*, 168–182.
13. Vukobratovic, M. Humanoids Robots: Past, Present State, Future. In *SISY—Serbian-Hungarian Joint Symposium on Intelligent Systems*; Director Robotics Center, Mihailo Pupin Institute: Belgrade, Serbia, 2006.
14. Kane, T.R.; Levinson, D.A. *Dynamics: Theory and Applications*; McGraw-Hill: New York, NY, USA, 1985.
15. Kajita, S.; Kanehiro, F.; Kaneko, K.; Fujiwara, K.; Harada, K.; Yokoi, K.; Hirukawa, H. Biped Walking Pattern Generation by using Preview Control of Zero-Moment Point. In Proceedings of the 2003 IEEE International Conference on Robotics and Automation, Taipei, Taiwan, 14–19 September 2003.
16. Choi, Y.; Kim, D.; Oh, Y.; You, B. Posture/Walking Control for Humanoid Robot Based on Kinematic Resolution of CoM Jacobian with Embedded Motion. *IEEE Trans. Robot.* **2007**, *23*, 1285–1293. [CrossRef]
17. Menga, G.; Ghirardi, M. Estimation and Closed Loop Control of *COG/ZMP* in Biped Devices Blending *CoP* Measures and Kinematic Information. *Robotics* **2019**, *8*, 89. [CrossRef]
18. Menga, G.; Ghirardi, M. Modelling, Simulation and Control of the Walking of Biped Robotic Devices—Part I: Modelling and Simulation Using Autolev. *Inventions* **2016**, *1*, 6. [CrossRef]
19. Menga, G.; Ghirardi, M. Modeling, Simulation and Control of the Walking of Biped Robotic Devices, Part II: Rectilinear Walking. *Inventions* **2016**, *1*, 7. [CrossRef]
20. Mitiguy, P. MotionGenesis: Advanced Solutions for Forces, Motion, and Code-Generation. Available online: http://www.motiongenesis.com/ (accessed on 22 October 2019).
21. Espiau, B.; Boulic, R. *On the Computation and Control of the Mass Center of Articulated Chains*; Research Report INRIA, n.3479; INRIA: Rocquencourt, France, 1998. Available online: https://hal.inria.fr/inria-00073210/document (accessed on 22 October 2019).

22. Fleischer, C.; Hommel, G. Calibration of an EMG-Based Body Model with six Muscles to control a Leg Exoskeleton. In Proceedings of the 2007 IEEE International Conference on Robotics and Automation, Rome, Italy, 14–15 April 2007; pp. 2514–2519.
23. Sentis, L.; Khatib, O. Synthesis of whole-body behaviors through hierarchical control of behavioral primitives. *Int. J. Human. Rob.* **2005**, *2*, 505–518. [CrossRef]

robotics

MDPI

Article

Quantifying Age-Related Differences of Ankle Mechanical Properties Using a Robotic Device

Francesca Martelli [1,*], Juri Taborri [2], Zaccaria Del Prete [1], Eduardo Palermo [1] and Stefano Rossi [2]

1 Department of Mechanical and Aerospace Engineering, Sapienza University of Rome, 00184 Rome, Italy; zaccaria.delprete@uniroma1.it (Z.D.P.); eduardo.palermo@uniroma1.it (E.P.)
2 Department of Economics, Engineering, Society and Business Organization (DEIM), University of Tuscia, 01100 Viterbo, Italy; juri.taborri@unitus.it (J.T.); stefano.rossi@unitus.it (S.R.)
* Correspondence: francesca.martelli@uniroma1.it; Tel.: +39-064-458-5585

Received: 30 September 2019; Accepted: 11 November 2019; Published: 13 November 2019

Abstract: A deep analysis of ankle mechanical properties is a fundamental step in the design of an exoskeleton, especially if it is to be suitable for both adults and children. This study aims at assessing age-related differences of ankle properties using pediAnklebot. To achieve this aim, we enrolled 16 young adults and 10 children in an experimental protocol that consisted of the evaluation of ankle mechanical impedance and kinematic performance. Ankle impedance was measured by imposing stochastic torque perturbations in dorsi-plantarflexion and inversion-eversion directions. Kinematic performance was assessed by asking participants to perform a goal-directed task. Magnitude and anisotropy of impedance were computed using a multiple-input multiple-output system. Kinematic performance was quantified by computing indices of accuracy, smoothness, and timing. Adults showed greater magnitude of ankle impedance in both directions and for all frequencies, while the anisotropy was higher in children. By analyzing kinematics, children performed movements with lower accuracy and higher smoothness, while no differences were found for the duration of the movement. In addition, adults showed a greater ability to stop the movement when hitting the target. These findings can be useful to a proper development of robotic devices, as well as for implementation of specific training programs.

Keywords: ankle impedance; kinematic performance; pediAnklebot; robotics; measurements

1. Introduction

In recent years, advanced technologies have allowed robot-mediated therapy to become a prominent solution for rehabilitation, as an alternative and/or a supporting solution to traditional rehabilitative programs [1,2]. Robotic devices permit intensive, controlled, and tailored rehabilitation, as well as reducing the therapist's burden [3]. Since the ability of locomotion is fundamental to avoid the worsening of the quality of life [4], one of the main challenges in the robotic field is the design and development of robots for ankle rehabilitation [5]; in fact it is well-known that the ankle joint plays essential roles during walking, such as shock absorption, propulsion, lower limb coordination, adaptation to different environments, and maintenance of stability [6]. From this perspective, an appropriate design of robotic devices for the ankle joint is required for: (i) rehabilitating people affected by neuromuscular diseases [7,8]; (ii) restoring athletes after injuries [9]; and, (iii) augmenting human strength and endurance in industrial and military applications [10,11]. Through the aim of ankle robotic device development, a full insight into kinematic performance and dynamic characterization of ankle appears to be mandatory in order to design robots that operate in accordance with human behavior, leading to a stable and effective physical human–robot interaction [12].

As regards the kinematic performance, several experimental protocols have been developed for quantifying kinematic indices using a robotic device. Among the protocols developed for kinematic performance evaluation, goal-directed movements through the use of serious games represent the most commonly adopted approach in clinical settings [13–16]. Generally, goal-directed tasks are used to understand how the central nervous system optimizes kinematic parameters, such as movement accuracy, smoothness, and speed, when a dynamic task is required [14]. These factors can be also considered the most relevant to be pursued for a proper design of a robotic ankle [12]. However, few studies have been conducted to evaluate ankle kinematic performance during goal-directed tasks. Michmizos and Krebs evaluated the relationship between the speed and the accuracy in both dorsi-plantar (DP) and inversion-eversion (IE) movements performed by adults, assessing the possibility to describe this relation with Fitt's law [17]. The same authors, in [18], compared several models of speed profile in ankle pointing movements, finding that the best fitting models were those already used for upper limbs during pointing movements.

By moving to the dynamic characterization, ankle impedance represents one of the main properties to monitor during rehabilitation programs, as it is one of the most important mechanical components involved in lower body stability during locomotion [19], providing fundamental information for designing robotic devices physically interacting with human lower extremities [20]. In addition, it has been already demonstrated that neurological diseases lead to a significant deterioration of ankle impedance, with respect to healthy subjects [21]. The application of dynamic perturbations to the examined anatomical joint and successive analysis of torque vs. angle graphs is currently the most widespread methodology for the measurement of the dynamic joint mechanical impedance [20,22–24]. Different studies have been proposed in the literature for the objective measurement of ankle impedance in adult subjects. More specifically, Lee et al. validated a stochastic methodology for the quantification of ankle impedance, considering dumping, stiffness, and further dynamic aspects [25,26]. The innovative aspect proposed by the authors is related to the feasibility of impedance evaluation in multiple directions, overcoming the limits of the previously proposed approaches [27–29]. Following a similar approach, Dallali et al. evaluated ankle impedance in the external-internal direction by analyzing the lower limb muscle activation and by applying an artificial neural network that achieved accuracy of 85% in the impedance estimation [22]. Conversely, in the literature, a limited number of studies have focused on ankle impedance evaluation in children. Alhusaini et al. [30] assessed ankle impedance by analyzing the responses to imposed movements in dorsi-plantarflexion in children with cerebral palsy (CP), while Martelli and colleagues [23] applied the methodology proposed by Lee [25] to quantify the effects of botulinum toxin on dynamic ankle impedance.

Currently, the design of robotic devices for the rehabilitation of children is an increasingly appealing and challenging field [16,31–34]. However, properly scaling robotic devices designed for adults to match the characteristics of children, remains an existing challenge, and is generally recognized as an important goal to achieve in the robotic field [35]. Considering this aspect, the quantification of age-related differences in terms of kinematic and dynamic performance clearly represents the starting point. However, no studies, to the best of the authors' knowledge, have been conducted for investigating the age-related differences in terms of ankle impedance between adults and children, as well as regarding kinematic performance, such as accuracy and smoothness, during goal-directed movements. Thus, this study aims at providing full insight into the ankle properties' maturation by comparing kinematic and dynamic performance indices in both DP and IE directions related to healthy young adults and healthy children using a robotic device. The outcomes of the study could offer important guidelines for the correct design and development of robotic devices and rehabilitation protocols addressed for adults and children, as well as serving as a starting point for solving the issue related to the scalability of robotic devices.

2. Materials and Methods

2.1. Subjects

Sixteen healthy adults aged from 22 to 30 years old and ten healthy children aged from 5 to 9 years old were enrolled in the study. The inclusion criteria were: (i) absence of neurological and visual deficits, (ii) physiological range of motion (ROM) for ankle, (iii) adequate anthropometric measures in order to freely move the ankle in the robot workspace, and (iv) right footedness. The dominant leg was established by asking them to kick a ball [36].

Written informed consent was obtained from all subjects. The protocol was compliant with the ethical standards outlined in the Declaration of Helsinki.

2.2. Experimental Setup and Procedure

All measurements were conducted by means of the pediAnklebot [31]. Subjects were seated in front of a monitor with knee flexed at 45°. They wore a knee brace fixed to the limb by means of Velcro straps, and a shoe of proper size, firmly tightened to the foot with shoelaces to prevent foot slippage. The main body of the robot was attached to the knee brace and the end-effectors were connected to the bracket attached to the bottom of the shoe. The calf of the subjects leaned against an aluminum support, covered with foam rubber and linked to the chair. The robot was laterally attached to the chair to ensure collected data were free from the weight of the robot and to improve repeatability.

The robot was equipped with two linear encoders and two load cells to acquire displacements and forces of each end-effector at 200 Hz. From the acquired data, rotations and moments of the ankle were obtained as reported in [29,37]. The robot was mainly designed for children, but can be used by adults in the sitting configuration, by changing the dimensions of shoe and knee brace, and using a new ad-hoc developed linkage between the brace and chair. Different anthropometric features across enrolled subjects do not affect moment arm, which only depend on robot design. In particular, torque was computed from the sensors as follows [29]:

$$\tau_{DP} = \left(F_{right} + F_{left}\right)x_{length} \tag{1}$$

$$\tau_{IE} = \left(F_{right} - F_{left}\right)x_{width} \tag{2}$$

where τ_{DP} and τ_{IE} are the dorsi-plantar and inversion-eversion net torques at the ankle joint; F_{right} and F_{left} are the forces measured by the right and left force sensors; and x_{length} and x_{width} are the distances between the line of action of the actuator force and the point of attachment between the ankle and robot in the sagittal and the frontal planes, respectively. The experimental setup is shown in Figure 1.

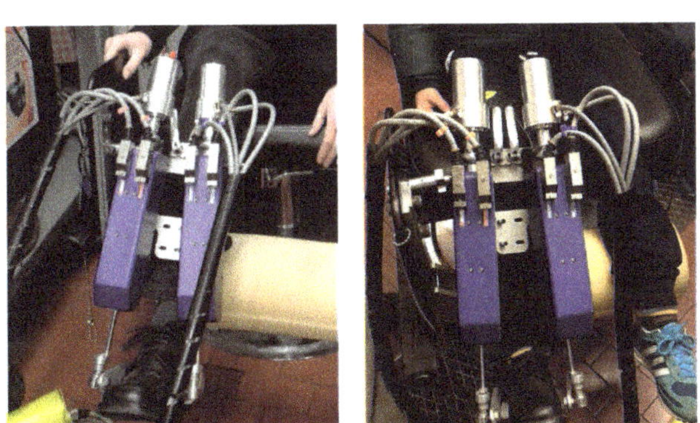

(a) (b)

Figure 1. Experimental setup: an adult (**a**) and a child (**b**) wearing the robot.

The experimental protocol consisted of two different phases: the first aimed at the evaluation of the ankle mechanical impedance, while the second aimed at the evaluation of ankle kinematic performance in a goal-directed task. Before starting the experimental procedure, the initial and reference positions were set by positioning the foot at 90° relative to the shank, for both the experimental phases. In addition, each participant performed a familiarization session that lasted until participants felt familiar with the equipment and the tasks. All subjects performed the entire protocol with the dominant limb.

2.2.1. Dynamic Ankle Impedance Evaluation

During the first experimental phase, the robot applied random stochastic torque perturbations to the ankle, both in IE and DP directions for 60 s, with a random white noise signal characterized by no periodicity in time and a flat spectrum in a frequency range of 0–100 Hz. In particular, the torque perturbation ranged between ±12.82 Nm in the IE direction and ±21.11 Nm in the DP direction, for both adults and children. The voluntary reactions of the participant were minimized due to the random nature of the chosen perturbation [38]. The commanded torques and corresponding angular displacements at the ankle were recorded at 200 Hz. Recorded data were expressed in the IE-DP space, which defined the joint coordinates.

2.2.2. Ankle Motor Performance Evaluation

In the second experimental phase, subjects carried out a goal-directed task, playing an in-home-developed serious game [14], based on the one proposed by Michmizos et al. [17]. In particular, subjects were asked to move a pointer controlled through ankle rotations in both sagittal (DP direction) and frontal (IE direction) planes, in order to hit targets appearing on the game scene alternatively at the top and at the bottom of the game scene. Subjects were instructed to reach the targets as fast as possible without stopping during the movement, to stop the ankle movement as soon as the target was reached, and to wait until a new target appeared on the scene. The game scenario is shown in Figure 2. The pointer (yellow dot) could be moved in every part of the scene by means of ankle rotations both in the frontal plane, i.e., inversion-eversion (IE), and in the sagittal plane, i.e., dorsi-plantarflexion (DP). Then, each point of the game scenario was described by 2 coordinates (x and y) that corresponded to the inversion and plantarflexion angles, respectively. The coordinates of the two target centers reported in the ankle reference system were: 0° in inversion-eversion and 10° in dorsiflexion for the up-target, and 0° in inversion-eversion and −10° in plantarflexion for the down-target. One second after hitting the target, a new target appeared in the opposite position. This experimental phase consisted of 2 blocks of 40 targets, 20 up (up-targets) and 20 down (down-targets), making a total of 80 goal-directed movements for each subject.

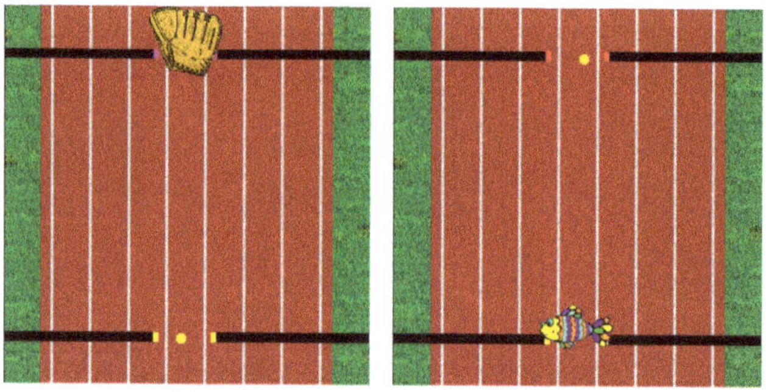
(a) (b)

Figure 2. Scenario of the game with up-(**a**) and down-targets (**b**) for the ankle motor performance evaluation.

2.3. Data Analysis

Data acquired by the sensors of the robot were processed offline. Regarding the evaluation of dynamic ankle impedance, multiple-input multiple-output (MIMO) system identification was used on the recorded signals, in accordance with Lee et al. [39], whereas, for the characterization of ankle motor performance in the goal-directed task, a set of kinematic indices was computed.

2.3.1. Dynamic Ankle Impedance Evaluation

A brief description of the MIMO system identification methodology is here reported for clarity. The MIMO system is characterized by 2 input and 2 output signals that are the commanded torques ($\boldsymbol{\tau}$) and the corresponding angular displacements ($\boldsymbol{\vartheta}$) at the ankle, respectively.

$$\boldsymbol{\tau} = (\tau_{IE},\ \tau_{DP}) \tag{3}$$

$$\boldsymbol{\vartheta} = (\vartheta_{IE},\ \vartheta_{DP}) \tag{4}$$

The first 200 samples of each signal were ignored to remove mechanical noise due to motor activation.

To obtain ankle impedance ($\mathbf{Z_{Ankle}}$), mechanical admittance ($\mathbf{Y_{CL}}$) of the closed-loop system was firstly identified:

$$\boldsymbol{\vartheta} = \mathbf{Y_{CL}}\boldsymbol{\tau} \tag{5}$$

$$\begin{bmatrix} \vartheta_{IE} \\ \vartheta_{DP} \end{bmatrix} = \begin{bmatrix} Y_{IE,\,IE}(f) & Y_{IE,\,DP}(f) \\ Y_{DP,\,IE}(f) & Y_{DP,\,DP}(f) \end{bmatrix} \begin{bmatrix} \tau_{IE} \\ \tau_{DP} \end{bmatrix} \tag{6}$$

Finally, mechanical impedance of the closed loop system ($\mathbf{Z_{CL}}$) was obtained from the inverse of mechanical admittance ($\mathbf{Y_{CL}}$):

$$\boldsymbol{\tau} = \mathbf{Y_{CL}}^{-1}\boldsymbol{\vartheta} = \mathbf{Z_{CL}}\boldsymbol{\vartheta} \tag{7}$$

$$\mathbf{Z_{CL}} = \begin{bmatrix} Z_{IE,\,IE}(f) & Z_{IE,\,DP}(f) \\ Z_{DP,\,IE}(f) & Z_{DP,\,DP}(f) \end{bmatrix} \tag{8}$$

Since the ankle and the robot shared the same displacement, the impedance of the closed loop system is the parallel between robot and ankle impedances. Thus, to obtain $\mathbf{Z_{Ankle}}$, the impedance component due to the robot dynamics was subtracted from $\mathbf{Z_{CL}}$. Bode plots of ankle impedance $\mathbf{Z_{Ankle}}$ in both IE and DP direction were determined.

Power spectral density of $\mathbf{Y_{CL}}$ was estimated using Welch's periodogram approach. The number of fast Fourier transform points was set at 1024, thus obtaining a spectral resolution of 0.19 Hz, and a periodic Hamming window was used, with a 50% overlap of the window size. These parameters were chosen considering a trade-off among spectral resolution, bias error, and variance of estimation [38].

Directional variation (anisotropy) of ankle impedance in two DOFs (Degree of Freedom) was evaluated applying rotations from 0° to 90° (step of 1°) to the original joint coordinates, thus obtaining the impedance magnitude for each rotation in the space defined by IE and DP directions. Then, the ankle impedance was represented through a polar plot, to obtain the Direction Depend Map (DDM) (Figure 3). Ankle impedance was analyzed as a function of frequency, by dividing the analyzed range of frequencies into three bands: low ($0\ \text{Hz} \le f \le 2\ \text{Hz}$), mid ($2\ \text{Hz} < f \le 5\ \text{Hz}$), and high ($5\ \text{Hz} < f \le 8\ \text{Hz}$) frequency ranges. These ranges are named, hereafter, LF, MF, and HF, respectively. The ranges were selected from the results of Bode plot analysis. In particular, Bode plots were consistent with a second order system, characterized by a stiffness dominated region below 2 Hz, whereas inertial component was not negligible starting from 5 Hz. Beyond approximately 10 Hz, Bode plots showed oscillations that can be ascribed to vibrational modes of shoe bracket and of the chair that the robot was attached to [23]. Impedance $\mathbf{Z_{Ankle}}$ was averaged in each range and the following parameters were evaluated: Z_{IE}, Z_{DP}, Z_{Ratio}, and Φ. Specifically, Z_{IE} and Z_{DP} represent ankle impedance magnitude in IE and

DP directions, respectively; and Z_{Ratio} and Φ were defined as shape and orientation index of DDM, respectively. Z_{Ratio} represents the ratio between minimum (Z_{min}) and maximum (Z_{max}) values of ankle impedance and quantifies how accentuated is the "peanut" shape of DDM: the closer Z_{Ratio} is to 1, the more the peanut shape is transformed into a circumference. Φ is the absolute value of the angle between Z_{DP} and the direction individuated by Z_{max}, and quantifies how the DDM is tilted with respect to the DP direction.

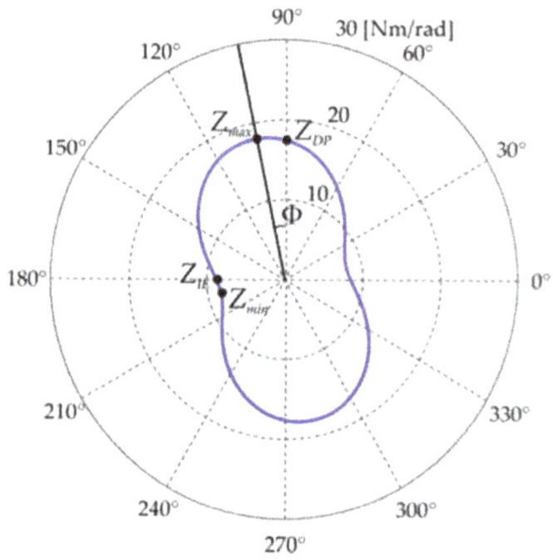

Figure 3. Example of direction dependent map (blue line) and the evaluated parameters (Z_{IE}, Z_{DP}, Z_{max}, Z_{min}, and Φ).

2.3.2. Ankle Motor Performance Evaluation

The recorded position of the pointer was filtered with a 6th order, zero phase shift low-pass Butterworth filter, with a cut-off frequency of 10 Hz, and then differentiated, with a two-point differentiation, to obtain speed, acceleration, and jerk. The global data obtained from each subject were divided into single goal-directed movements and grouped into plantarflexion (down-target) and dorsiflexion (up-target) movements. Each movement was assumed to start when the speed magnitude became greater than 10% of the peak speed; the movement was assumed to end when the speed dropped and remained below 10% of the peak speed [40].

Trajectories were accurately screened and discarded if one of following cases occurred: (i) the movement began before the new target appearance, i.e., the initial velocity was not equal to zero, and (ii) when the subject stopped the ankle movement before reaching the target. The number of discarded trials was lower than 5% for each subject. For the characterization of movement kinematics, a set of indices was computed and grouped into: (i) accuracy and smoothness indices, (ii) temporal indices, and (iii) stopping indices.

The first group comprises the lateral deviation (*LD*) and the normalized jerk (*NJ*), as accuracy and smoothness indexes, respectively. In particular, *LD* is defined as the highest deviation from the straight line connecting the starting and the target position. The *LD* value decreases when the movement accuracy increases. *NJ* is the normalized jerk, as proposed by Teulings et al. [41]. Lower values of *NJ* indicate smoother movements.

The set of temporal indices is constituted by the duration of movement (*T*), the time position symmetry (*TPS*), and the time velocity symmetry (*TVS*). *T* is the time between the movement onset and the movement termination, which was evaluated according to the speed threshold. The remaining two indices quantify the temporal symmetry of kinematic parameters of the trajectory. In particular, *TPS* is defined as follows:

$$TPS = \frac{\Delta t_E}{T} \tag{9}$$

TPS represents the temporal duration (Δt_E) of the eversion rotations with respect to the total duration of the trajectory. In fact, when the foot is moved upward or downward, the corresponding rotation should be a pure dorsiflexion or plantarflexion, but actually a component of inversion or eversion is always present. *TPS* ranges from 0 to 1. If the *TPS* value is equal to 0.5, it means an equal temporal duration of the inversion and eversion rotations during dorsi-plantarflexion movements. A value of *TPS* higher than 0.5 means a prevalence of eversion with respect to inversion rotations, and vice versa, a value lower than 0.5 means a prevalence of inversion. Thus, *TPS* can be considered a measure of the contribution of the ankle rotation around the secondary movement axis, i.e., inversion/eversion, in dorsiflexion/plantarflexion. *TVS* is defined as the time t_v in which the velocity peak occurred, normalized to the duration the trajectory.

$$TVS = \frac{t_v}{T} \tag{10}$$

TVS ranges from 0 to 1. The closer *TVS* is to 1, the more the velocity peak occurs close to the end of the trajectory. Thus, a *TVS* value close to 0.5 represents a perfect bell-shaped trajectory, in which the peak speed is close to the middle of the trajectory.

The stopping indices are delay (ΔT) and dispersion (σ_{trj}). They evaluate the ability of the subject to stop the movement once the target is reached [13]. ΔT represents the temporal delay, normalized to T, between the time in which the subject hits the target and the end of the movement evaluated accordingly to the velocity threshold. A ΔT value close to zero indicates that the subject rapidly stops the ankle movement after hitting the target. σ_{trj} is defined as:

$$\sigma_{trj} = std\sqrt{(d_{IE} - C_{IE})^2 + (d_{DP} - C_{DP})^2} \tag{11}$$

where d_{IE} and d_{DP} are the coordinates of the trajectory performed after hitting the target, and C_{IE} and C_{DP} are the coordinates of the target center position. Thus, σ_{trj} is a measure of the dispersion of the trajectory travelled after hitting the target. Low values of σ_{trj} imply a trajectory confined in a small area.

All the aforementioned indices were evaluated for both dorsiflexion and plantarflexion movements.

2.4. Statistical Analysis

A two-way repeated measures ANOVA was performed on Z_{IE}, Z_{DP}, Z_{Ratio}, and Φ considering, as independent variables, the age (adults vs. children) and the frequency range (low, mid, and high frequency). If the interaction effects were significant, the interactions were broken down by comparing each age at each frequency with a one-way repeated measures ANOVA and, vice versa, with an unpaired t-test. A Bonferroni test for multiple comparisons was performed when statistical differences were found. All data were tested for normality with the Shapiro–Wilk test and sphericity was checked. If sphericity was violated, Greenhouse–Gasser correction was applied.

A two-way repeated measures ANOVA was performed on all the kinematic indices to find differences between age and movement direction. If the interaction effects were significant, the interactions were broken down by comparing each age at each movement direction with an unpaired *t*-test and, vice versa, with a paired *t*-test. All data were tested for normality with a Shapiro–Wilk test.

The significance level was set to 0.05 for all performed tests.

3. Results

3.1. Ankle Impedance

Mean and standard deviation values of Z_{IE}, Z_{DP}, Z_{Ratio}, and Φ averaged across subjects for the tree frequency range (LF, MF, HF) are reported in Figure 4.

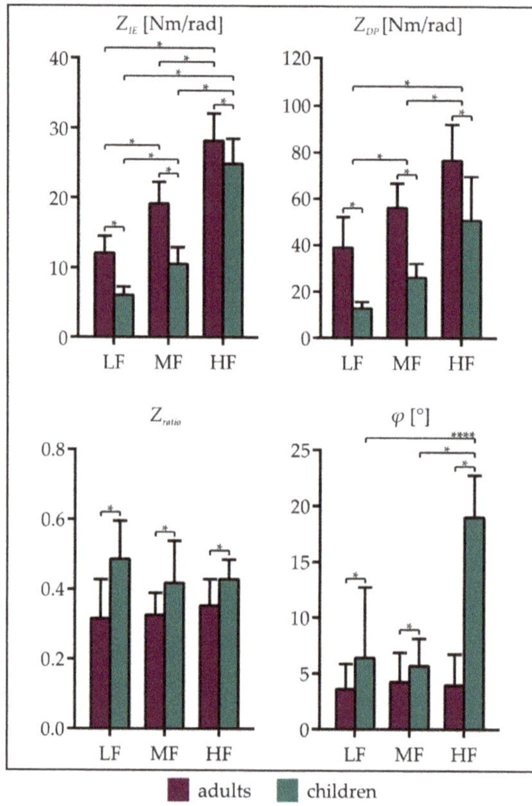

Figure 4. Means and standard deviations of the ankle impedance parameters Z_{IE}, Z_{DP}, Z_{Ratio}, and Φ averaged across subjects related to adults and children for all frequency ranges (LF, MF, and HF). Asterisks denote statistical differences ($p < 0.05$).

Z_{IE} was higher for adults with respect to children for all the frequency ranges ($p < 0.01$). Moreover, it was higher in HF with respect to both MF and LF, and in MF with respect to LF, for both adults and children (adults: HF vs. MF, HF vs. LF, and MF vs. LF: $p < 0.01$; children: HF vs. MF, HF vs. LF, and MF vs. LF: $p < 0.01$). As regards Z_{DP}, it showed higher values for adults than children, in all frequency ranges ($p < 0.01$). Furthermore, Z_{DP} was higher in HF with respect to both MF and LF, and in MF with respect to LF ($p < 0.01$ for all the comparisons).

Considering Z_{Ratio}, it showed lower values for adult subjects than children in LF ($p < 0.01$), MF ($p = 0.02$) and HF ($p = 0.01$), whereas no differences were found between frequency ranges.

Finally, Φ was lower in the adult group for all frequency ranges. Moreover, considering the children group, Φ was higher in HF with respect to both MF ($p < 0.01$) and LF ($p < 0.01$). No differences were found among frequency ranges for adult subjects.

The representation of DDM as a function of frequency is reported in Figure 5, through two representative examples related to adults (Figure 5a) and children (Figure 5b). It emerged that the characteristic peanut shape remains unchanged for all the frequencies in both adults and children, whereas the impedance magnitude is different between the two cohorts in the whole DP-IE space and for all frequencies. Finally, the orientation of DMM with respect to the DP axis remains unaltered in the adult whereas it visibly rotates in the child.

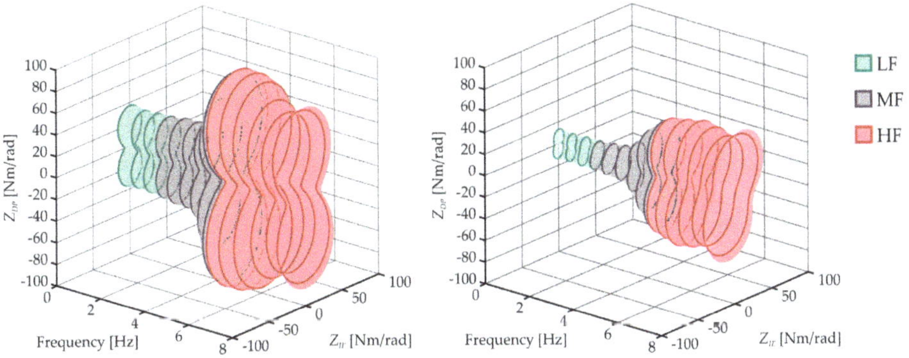

(a) (b)

Figure 5. Example of directional variation (anisotropy) of ankle mechanical impedance as a function of frequency related to a representative subject of the adult group (**a**) and of the child group (**b**).

3.2. Kinematic Indices

Mean and standard deviation of all the kinematic indices averaged across subjects for both plantarflexion and dorsiflexion movements are reported in Figure 6.

For all the indices, except *NJ*, the interaction factor was significant. In particular, as regards the accuracy index (*LD*), no differences were found between plantarflexion and dorsiflexion movements related to the same group. Instead, *LD* was higher in children than adults for both plantarflexion ($p < 0.01$) and dorsiflexion ($p < 0.01$) movements. The opposite occurred for the smoothness index (*NJ*), which showed higher values in the adult group for both plantarflexion and dorsiflexion movements ($p = 0.04$).

As regards the temporal indices, no differences were found for the duration of the movement (*T*), neither between age or movement direction. Considering *TPS*, it showed higher values in adults for plantarflexion movements ($p = 0.05$), whereas the opposite occurred for dorsiflexion movements ($p < 0.01$). Moreover, *TPS* was higher in dorsiflexion than in plantarflexion movements ($p < 0.01$) related to the child group, and no differences were found between dorsiflexion and plantarflexion in the adult group. Finally, *TVS* showed lower values in the adult group, for both plantarflexion ($p < 0.01$) and dorsiflexion ($p = 0.03$) movements. Furthermore, *TPS* was higher in plantarflexion with respect to dorsiflexion in the child group ($p < 0.01$).

Considering the stopping indices, ΔT was higher in children for both plantarflexion ($p < 0.01$) and dorsiflexion ($p = 0.01$) movements. Additionally, ΔT was higher in plantarflexion than in dorsiflexion for both adults ($p < 0.01$) and children ($p < 0.01$). A similar behavior was found for σ_{trj}, which showed lower values in the adult group, for both plantarflexion ($p < 0.01$) and dorsiflexion ($p < 0.01$) movements. Moreover, σ_{trj} was lower in dorsiflexion than in plantarflexion for both adults ($p < 0.01$) and children ($p < 0.01$).

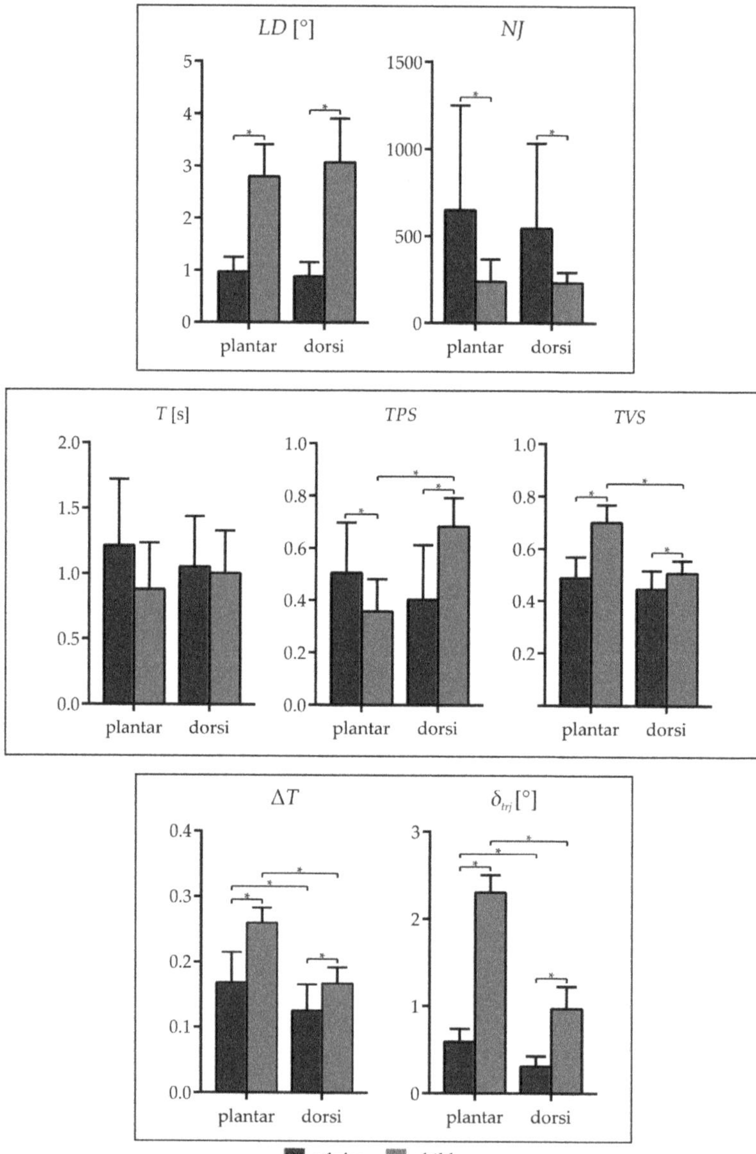

Figure 6. Means and standard deviations of the kinematic parameters averaged across subjects related to adults and children, grouped in plantarflexion (plantar) and dorsiflexion (dorsi) movements. (**Top row**): accuracy and smoothness indices, (**mid row**): temporal indices, (**bottom row**): stopping indices. Asterisks denote statistical differences ($p < 0.05$).

4. Discussion

With the aim of investigating age-related differences of ankle mechanical properties, we compared ankle mechanical behaviors, in terms of ankle impedance and kinematic performance, of children and young adult subjects. The experimental protocol was performed using pediAnklebot, a robotic device for the ankle joint.

4.1. Age-Related Differences of Ankle Impedance

The results of both groups confirm that ankle impedance in DP and IE directions assumes a peanut shape in the diagram of DP vs. IE, as also reported in [23,25]. This finding indicates that the two examined populations are characterized by a similar behavior, i.e., a greater value of impedance in DP rather than in IE direction, even though all the examined children were less than 14 years, which is recognized as the age of complete maturation of the ankle joint [42].

Despite the similar shape, statistical differences were found for the magnitude values of impedance in both directions. These differences can be ascribed to the incomplete maturation of elastic properties of tendon structures that lead to an incorrect transmission of force exerted by muscle to the bone [43]. In addition, our outcomes are in accordance with [44], confirming that ankle stiffness increases with age and stature. The lowest values of ankle impedance found for children can be also considered one of the main causes of the dynamic instability of children when walking [33,42]. In fact, a complete development of lower limb joints guarantees a natural interaction between limbs and the environment, and a proper value of ankle impedance allows correct regulating and controlling movements [45].

Differences among frequencies in both groups are in line with the concept that ankle impedance is characterized by different behaviors according to the frequency range [25]. More specifically, the size of the peanut shape rapidly increases with the frequency, indicating that the contribution to the ankle impedance of the inertia is dominant at high frequencies, whereas, at low and medium frequencies, stiffness and viscosity represent the main contributions to the impedance magnitude value, since the size of the peanut remains stable [39].

By moving to the analysis of shape parameters, we can state that ankle impedance was highly direction dependent, being weak in IE direction, in all examined range frequencies. This outcome can be considered as one of the main sources of the greater prevalence of ankle injuries in IE direction in both adults and children [46]. In addition, this anisotropy is more evident in children rather than in adults, as confirmed by the statistical differences found between the two groups. This outcome can be justified by considering the greater joint and ligament laxity typical of children younger than 10 years old [47]. This greater laxity, also known as hypermobility, is caused by an incomplete development of muscle elasticity [47]. The excessive anisotropy could be considered a causal factor that leads to the high rate of instability in children, leading to a greater prevalence of ankle sprains, especially in IE direction [48]. However, the anisotropy is not influenced by frequencies in both examined groups; thus, we can speculate that children have the capability to maintain constant differences between DP and IE directions at different frequencies as adults, even considering the above-mentioned factors of incomplete joint development. This finding could suggest that children can also perform highly dynamic activities without increasing the risk of injuries. Furthermore, the parameter related to the orientation of the peanut shape is frequency-invariant in adults, while the highest values were obtained when analyzing the high frequencies in children. This implies that the maximum value of impedance in children is not in correspondence of the DP direction and can be ascribed to the incomplete development of muscles. This finding indicates that children are not able to provide maximum impedance during movements that mainly involve the dorsi-plantarflexion of the ankle, such as walking; thus, it could be one of the main reasons for the well-known higher stride-to-stride variability found for children [33].

These findings should be taken into account for proper development and design of ankle robotic devices, especially when they are addressed to recover the functionality of the ankle in both degrees of freedom.

4.2. Age-Related Differences of Ankle Kinematic Performance

Results related to accuracy and smoothness indices highlighted that, independently of movement direction, adults performed trajectories characterized by a higher accuracy, whereas children performed smoother trajectories. Additionally, these differences in terms of accuracy and smoothness were not reflected in the duration of the movement, which revealed no age-related differences. The lower accuracy characterizing children movements can be ascribed to a lower proprioceptive ability, as already found for the upper limbs in a comparison between children and adolescent movements [49]. In addition, it has been demonstrated that children performed movement favoring smoothness with respect to accuracy [50].

The *TPS* index can be considered a measure of the coupling between ankle rotations performed in sagittal and frontal planes. In fact, ankle rotations are usually described as movements around fixed mutually perpendicular axes, but they actually change their position depending on ankle rotation [51]. This complexity yields to a significant coupling between the two DOFs: when the foot is moved upward or downward, the resulting rotation is always a combination of plantar-dorsiflexion and inversion-eversion. Considering the child group, there is a significance difference of *TPS* when comparing dorsiflexion and plantarflexion movements. In particular, plantarflexion movements are characterized by the prevalence of inversion rotations, whereas dorsiflexion movements show a prevalence of eversion rotations. This behavior is not present in the adult group, which revealed no differences between movement directions, showing a *TPS* value always close to 0.5 for both plantarflexion and dorsiflexion. These results indicate that adults are able to balance the contribution of rotations around the secondary axis and that this ability is not completely developed by children in the examined age range. Thus, it implies a higher mobility of ankle movement axes in children, which is probably due to the greater joint and ligament laxity typical of children younger than 10 years old [47], as already discussed for the ankle impedance. Moreover, results from *TPS* are consistent with the findings related to the accuracy and smoothness indices. Children, in fact, present less accurate trajectories, i.e., highly deviating from a straight line, which are then characterized by a significant contribution of inversion or eversion movements. On the contrary, the smoothness of the movement was higher in the child group. Thus, it can be speculated that, during the motor development, subjects prefer to optimize the movement accuracy in order to obtain a better-balanced trajectory even though it implies performing fewer smooth movements.

Focusing on the bell-shaped profile of velocity, results from *TVS* highlighted that, in the child group, the velocity peak occurred early in dorsiflexion with respect to plantarflexion. This finding could be explained by considering that subjects needed to counter gravity when the foot is moved upwards, exerting a greater force at the beginning of the movement, which allowed reaching the maximum speed earlier in dorsiflexion than in plantarflexion. On the contrary, adults showed a perfect bell-shaped velocity profile and, consequently, they are not influenced by gravity. These results are in line with [18], in which authors established that, in adults subjects, the velocity profile is not affected by gravity. Thus, we can speculate that the differences between children and adults could be due to higher force levels that can be exerted by adults at the ankle, implying a negligible effect of gravity.

Considering the stopping indices, it emerged that both adults and children can stop the movement more easily during dorsiflexion than plantarflexion. This result can be explained by considering that plantarflexion is a movement performed in the same direction of gravity. Thus, after hitting the target, subjects have to counter gravity to stop ankle movement. Another important consideration emerging from the stopping indices is that, independently of the movement direction, children showed higher difficulty in stopping the ankle movement as soon as the target was reached. This characteristic can be ascribed to the lower magnitude of ankle impedance of children that emerged from the impedance evaluation.

From an overall analysis of the kinematic indices it seems that the complete maturation of the ankle joint leads to a loss in the smoothness of the movement but, at the same time, to an increase in the accuracy. Moreover, the incomplete maturation of the ankle joint leads to differences in plantarflexion and dorsiflexion movements in terms of symmetry of the trajectory evaluated by means of *TPS* and *TVS*. These differences, in fact, were only found in the child group.

Results related to the kinematic parameters should be considered together with those reported for the ankle impedance for a full understanding of ankle properties, leading to a more suitable development of robots.

5. Conclusions

Age-related differences in terms of dynamic properties and kinematic performance related to the ankle joint were assessed by means of a robotic device.

Outcomes revealed for children were characterized by a lower magnitude of ankle impedance in both dorsi-plantarflexion and inversion-eversion directions and by a greater anisotropy. By considering a goal-directed task, adults performed more accurate and less smooth movements, and a lower symmetry of movement in both directions was found in children. Moreover, a greater ability to stop the movement after hitting the target was found in adults.

These findings provide fundamental information usable as guidelines for researcher groups involved in the design of robotic devices and training protocols for rehabilitation purposes. Furthermore, they could represent a starting point for solving the issues related to the scalability of robots suitable for adults and children.

Author Contributions: Conceptualization, F.M., J.T. and S.R.; methodology, F.M., J.T. and S.R.; software, F.M. and S.R.; formal analysis, F.M., J.T. and S.R.; data curation, F.M., J.T. and S.R.; writing—original draft preparation, F.M., J.T. and S.R.; writing—review and editing, F.M., J.T., Z.D.P., E.P. and S.R.; supervision, Z.D.P., E.P. and S.R.; project administration, F.M., J.T., E.P. and S.R.

Conflicts of Interest: The authors declare no conflict of interest.

References

1. Marchal-Crespo, L.; Reinkensmeyer, D.J. Review of control strategies for robotic movement training after neurologic injury. *J. Neuroeng. Rehabil.* **2009**, *6*, 20. [CrossRef]
2. Meyer-Heim, A.; Ammann-Reiffer, C.; Schmartz, A.; Schäfer, J.; Sennhauser, F.H.; Heinen, F.; Knecht, B.; Dabrowski, E.; Borggraefe, I. Improvement of walking abilities after robotic-assisted locomotion training in children with cerebral palsy. *Arch. Dis. Child.* **2009**, *94*, 615–620. [CrossRef]
3. Adiputra, D.; Nazmi, N.; Bahiuddin, I.; Ubaidillah, U.; Imaduddin, F.; Rahman, M.A.A.; Mazlan, S.A.; Zamzuri, H. A Review on the Control of the Mechanical Properties of Ankle Foot Orthosis for Gait Assistance. *Actuators* **2019**, *8*, 10. [CrossRef]
4. Edwards, M.K.; Rosenbaum, S.; Loprinzi, P.D. Differential Experimental Effects of a Short Bout of Walking, Meditation, or Combination of Walking and Meditation on State Anxiety Among Young Adults. *Am. J. Health Promot.* **2018**, *32*, 949–958. [CrossRef]
5. Alvarez-Perez, M.G.; Garcia-Murillo, M.A.; Cervantes-Sánchez, J.J. Robot-assisted ankle rehabilitation: A review. *Disabil. Rehabil. Assist. Technol.* **2019**, 1–15. [CrossRef]
6. Trevino, J.; Lee, H. Sex Differences in 2-DOF Human Ankle Stiffness in Relaxed and Contracted Muscles. *Ann. Biomed. Eng.* **2018**, *46*, 2048–2056. [CrossRef]
7. Sawicki, G.S.; Ferris, D.P. Powered ankle exoskeletons reveal the metabolic cost of plantar flexor mechanical work during walking with longer steps at constant step frequency. *J. Exp. Biol.* **2009**, *212*, 21–31. [CrossRef]
8. Contreras-Vidal, J.L.; A Bhagat, N.; Brantley, J.; Cruz-Garza, J.G.; He, Y.; Manley, Q.; Nakagome, S.; Nathan, K.; Tan, S.H.; Zhu, F.; et al. Powered exoskeletons for bipedal locomotion after spinal cord injury. *J. Neural Eng.* **2016**, *13*, 031001. [CrossRef] [PubMed]
9. Jamwal, P.K.; Xie, S.Q.; Hussain, S.; Parsons, J.G. An Adaptive Wearable Parallel Robot for the Treatment of Ankle Injuries. *IEEE/ASME Trans. Mechatron.* **2014**, *19*, 64–75. [CrossRef]
10. Zoss, A.B.; Kazerooni, H.; Chu, A. Biomechanical design of the Berkeley lower extremity exoskeleton (BLEEX). *IEEE/ASME Trans. Mechatron.* **2006**, *11*, 128–138. [CrossRef]
11. Walsh, C.J.; Eond, K.; Herr, H. A quasi-passive leg exoskeleton for load-carrying augmentation. *Int. J. Humanoid Robot.* **2007**, *4*, 487–506. [CrossRef]
12. Miao, Q.; Zhang, M.; Wang, C.; Li, H. Towards Optimal Platform-Based Robot Design for Ankle Rehabilitation: The State of the Art and Future Prospects. *J. Healthc. Eng.* **2018**, *2018*, 1–9. [CrossRef]
13. Martelli, F.; Palermo, E.; Rossi, S. A novel protocol to evaluate ankle movements during reaching tasks using pediAnklebot. In Proceedings of the 2017 International Conference on Rehabilitation Robotics (ICORR), London, UK, 17–20 July 2017; Volume 2017, pp. 326–331.

14. Martelli, F.; Bertini, E.; del Prete, Z.; Palermo, E.; Rossi, S. Submovement changes in goal-directed and non-goal-directed ankle movements using pediAnklebot. In Proceedings of the 2018 IEEE International Symposium on Medical Measurements and Applications (MeMeA), Rome, Italy, 11–13 June 2018; pp. 1–5.
15. Perez-Ibarra, J.C.; Gomez-Pena, G.; Siqueira, A.A.G. Estimation of Patient Participation during Therapeutic Movements Using the Anklebot: Technique and Preliminary Results. In Proceedings of the 2015 12th Latin American Robotics Symposium and 2015 3rd Brazilian Symposium on Robotics (LARS-SBR), Uberlândia, Brazil, 28 October–1 November 2015; pp. 198–203.
16. Krebs, H.I.; Ladenheim, B.; Hippolyte, C.; Monterroso, L.; Mast, J. Robot-assisted task-specific training in cerebral palsy. *Dev. Med. Child Neurol.* **2009**, *51*, 140–145. [CrossRef]
17. Michmizos, K.P.; Krebs, H.I. Pointing with the ankle: The speed-accuracy trade-off. *Exp. Brain Res.* **2014**, *232*, 647–657. [CrossRef]
18. Michmizos, K.P.; Vaisman, L.; Krebs, H.I. A comparative analysis of speed profile models for ankle pointing movements: Evidence that lower and upper extremity discrete movements are controlled by a single invariant strategy. *Front. Hum. Neurosci.* **2014**, *8*, 1–15. [CrossRef]
19. Hansen, A.; Childress, D.; Miff, S.; Gard, S. The human ankle during walking: Implications for design of biomimetic ankle prostheses. *J. Biomech.* **2004**, *37*, 1467–1474. [CrossRef]
20. Nalam, V.; Lee, H. Development of a Two-Axis Robotic Platform for the Characterization of Two-Dimensional Ankle Mechanics. *IEEE/ASME Trans. Mechatron.* **2019**, *24*, 459–470. [CrossRef]
21. Lee, H.; Patterson, T.; Ahn, J.; Klenk, D.; Lo, A.; Krebs, H.I.; Hogan, N. Static ankle impedance in stroke and multiple sclerosis: A feasibility study. In Proceedings of the 2011 Annual International Conference of the IEEE Engineering in Medicine and Biology Society EMBS, Boston, MA, USA, 30 August–3 September 2011; pp. 8523–8526.
22. Dallali, H.; Knop, L.; Castelino, L.; Ficanha, E.; Rastgaar, M. Using lower extremity muscle activity to obtain human ankle impedance in the external–internal direction. *Int. J. Intell. Robot. Appl.* **2018**, *2*, 29–42. [CrossRef]
23. Martelli, F.; Rossi, S.; Frascarelli, F.; Germanotta, M.; Petrarca, M.; Castelli, E.; Cappa, P. Estimation of multivariable dynamic ankle impedance after botulinum toxin injection in children with cerebral palsy. In Proceedings of the 2016 IEEE International Symposium on Medical Measurements and Applications (MeMeA), Benevento, Italy, 15–18 May 2016; pp. 1–6.
24. Dai, C.; Martel, S.; Martel, F.; Rancourt, D.; Clancy, E.A. Single-trial estimation of quasi-static EMG-to-joint-mechanical-impedance relationship over a range of joint torques. *J. Electromyogr. Kinesiol.* **2019**, *45*, 18–25. [CrossRef]
25. Lee, H.; Krebs, H.I.; Hogan, N. Multivariable dynamic ankle mechanical impedance with active muscles. *IEEE Trans. Neural Syst. Rehabil. Eng.* **2014**, *22*, 971–981. [CrossRef]
26. Rastgaar, M.A.; Ho, P.; Lee, H.; Krebs, H.I.; Hogan, N. Stochastic Estimation of the Multi-Variable Mechanical Impedance of the Human Ankle with Active Muscles. In Proceedings of the ASME 2010 Dynamic Systems and Control Conference, Cambridge, MA, USA, 12–15 September 2010; pp. 1–3.
27. Hunter, I.W.; Kearney, R.E. Dynamics of human ankle stiffness: Variation with mean ankle torque. *J. Biomech.* **1982**, *15*, 747–752. [CrossRef]
28. Kearney, R.E.; Hunter, I.W. Dynamics of human ankle stiffness: Variation with displacement amplitude. *J. Biomech.* **1982**, *15*, 753–756. [CrossRef]
29. Roy, A.; Krebs, H.I.; Williams, D.J.; Bever, C.T.; Forrester, L.W.; Macko, R.M.; Hogan, N. Robot-aided neurorehabilitation: A novel robot for ankle rehabilitation. *IEEE Trans. Robot.* **2009**, *25*, 569–582. [CrossRef]
30. Alhusaini, A.A.; Crosbie, J.; Shepherd, R.B.; Dean, C.M.; Scheinberg, A. Mechanical properties of the plantarflexor musculotendinous unit during passive dorsiflexion in children with cerebral palsy compared with typically developing children. *Dev. Med. Child Neurol.* **2010**, *52*, 101–106. [CrossRef] [PubMed]
31. Michmizos, K.; Rossi, S.; Castelli, E.; Cappa, P.; Krebs, H. Robot-Aided Neurorehabilitation: A Pediatric Robot for Ankle Rehabilitation. *IEEE Trans. Neural Syst. Rehabil. Eng.* **2015**, *23*, 1056–1067. [CrossRef]
32. Patanè, F.; Rossi, S.; del Sette, F.; Taborri, J.; Cappa, P. WAKE-Up Exoskeleton to Assist Children with Cerebral Palsy: Design and Preliminary Evaluation in Level Walking. *IEEE Trans. Neural Syst. Rehabil. Eng.* **2017**, *25*, 906–916. [CrossRef]

33. Mileti, I.; Taborri, J.; Rossi, S.; Petrarca, M.; Patanè, F.; Cappa, P. Evaluation of the effects on stride-to-stride variability and gait asymmetry in children with Cerebral Palsy wearing the WAKE-up ankle module. In Proceedings of the 2016 IEEE International Symposium on Medical Measurements and Applications (MeMeA), Benevento, Italy, 15–18 May 2016.
34. Germanotta, M.; Taborri, J.; Rossi, S.; Frascarelli, F.; Palermo, E.; Cappa, P.; Castelli, E.; Petrarca, M. Spasticity Measurement Based on Tonic Stretch Reflex Threshold in Children with Cerebral Palsy Using the PediAnklebot. *Front. Hum. Neurosci.* **2017**, *11*, 277. [CrossRef]
35. Kardofaki, M.; Tabti, N.; Alfayad, S.; Ouezdou, F.B.; Chitour, Y.; Dychus, E. Mechanical Development of a Scalable Structure for Adolescent Exoskeletons. In Proceedings of the 2019 IEEE 16th International Conference on Rehabilitation Robotics (ICORR), Toronto, ON, Canada, 24–28 June 2019; Volume 2019, pp. 323–330.
36. Taborri, J.; Palermo, E.; Masiello, D.; Rossi, S. Factorization of EMG via muscle synergies in walking task: Evaluation of intra-subject and inter-subject variability. In Proceedings of the I2MTC 2017—2017 IEEE International Instrumentation and Measurement Technology Conference, Turin, Italy, 22–25 May 2017.
37. Krebs, H.I.; Michmizos, K.P.; Monterosso, L.; Mast, J. Pediatric anklebot: pilot clinical trial. In Proceedings of the 2016 6th IEEE International Conference on Biomedical Robotics and Biomechatronics (BioRob), Singapore, 26–29 June 2016.
38. Perreault, E.J.; Kirsch, R.F.; Acosta, A.M. Multiple-input, multiple-output system identification for characterization of limb stiffness dynamics. *Biol. Cybern.* **1999**, *80*, 327–337. [CrossRef]
39. Lee, H.; Krebs, H.; Hogan, N. Multivariable dynamic ankle mechanical impedance with relaxed muscles. *IEEE Trans. Neural Syst. Rehabil. Eng.* **2014**, *22*, 1104–1114. [CrossRef]
40. Emken, J.L.; Benitez, R.; Sideris, A.; Bobrow, J.E.; Reinkensmeyer, D.J. Motor Adaptation as a Greedy Optimization of Error and Effort. *J. Neurophysiol.* **2007**, *97*, 3997–4006. [CrossRef]
41. Teulings, H.L.; Contreras-Vidal, J.L.; Stelmach, G.E.; Adler, C.H. Parkinsonism reduces coordination of fingers, wrist, and arm in fine motor control. *Exp. Neurol.* **1997**, *146*, 159–170. [CrossRef]
42. Hausdorff, J.M.; Zemany, L.; Peng, C.; Goldberger, A.L. Maturation of gait dynamics: Stride-to-stride variability and its temporal organization in children. *J. Appl. Physiol.* **1999**, *86*, 1040–1047. [CrossRef] [PubMed]
43. Kubo, K.; Kanehisa, H.; Kawakami, Y.; Fukanaga, T. Growth Changes in the Elastic Properties of Human Tendon Structures. *Int. J. Sports Med.* **2001**, *22*, 138–143. [CrossRef] [PubMed]
44. Lambertz, D.; Mora, I.; Grosset, J.-F.; Pérot, C. Evaluation of musculotendinous stiffness in prepubertal children and adults, taking into account muscle activity. *J. Appl. Physiol.* **2003**, *95*, 64–72. [CrossRef] [PubMed]
45. Hogan, N. The mechanics of multi-joint posture and movement control. *Biol. Cybern.* **1985**, *52*, 315–331. [CrossRef]
46. Baumhauer, J.F.; Alosa, D.M.; Renström, A.F.; Trevino, S.; Beynnon, B. A Prospective Study of Ankle Injury Risk Factors. *Am. J. Sports Med.* **1995**, *23*, 564–570. [CrossRef]
47. Clinch, J.; Deere, K.; Sayers, A.; Palmer, S.; Riddoch, C.; Tobias, J.H.; Clark, E.M. Epidemiology of generalized joint laxity (hypermobility) in fourteen-year-old children from the UK: A population-based evaluation. *Arthritis Rheum.* **2011**, *63*, 2819–2827. [CrossRef]
48. Mandarakas, M.; Pourkazemi, F.; Sman, A.; Burns, J.; Hiller, C.E. Systematic review of chronic ankle instability in children. *J. Foot Ankle Res.* **2014**, *7*, 21. [CrossRef]
49. Goble, D.J.; Lewis, C.A.; Hurvitz, E.A.; Brown, S.H. Development of upper limb proprioceptive accuracy in children and adolescents. *Hum. Mov. Sci.* **2005**, *24*, 155–170. [CrossRef]
50. Bo, J.; Contreras-Vidal, J.L.; Kagerer, F.A.; Clark, J.E. Effects of increased complexity of visuo-motor transformations on children's arm movements. *Hum. Mov. Sci.* **2006**, *25*, 553–567. [CrossRef]
51. Nordin, M.; Frankel, V.H. *Basic Biomechanics of the Musculoskeletal System*; Lippincott Williams & Wilkins: Philadelphia, PA, USA, 2001.

robotics

MDPI

Review

Human–Robot Collaboration in Manufacturing Applications: A Review

Eloise Matheson [1], Riccardo Minto [2,*], Emanuele G. G. Zampieri [2], Maurizio Faccio [3] and Giulio Rosati [2]

1. Department of Mechanical Engineering, Imperial College London, London SW7 2AZ, UK; e.matheson@imperial.ac.uk
2. Department of Industrial Engineering, University of Padova, 35131 Padova, Italy; emanuele.zampieri@unipd.it (E.G.G.Z.); giulio.rosati@unipd.it (G.R.)
3. Department of Management and Engineering, University of Padova, 36100 Vicenza, Italy; maurizio.faccio@unipd.it

* Correspondence: riccardo.minto@unipd.it; Tel.: +39-049-827-6810

Received: 30 September 2019; Accepted: 3 December 2019; Published: 6 December 2019

Abstract: This paper provides an overview of collaborative robotics towards manufacturing applications. Over the last decade, the market has seen the introduction of a new category of robots—collaborative robots (or "cobots")—designed to physically interact with humans in a shared environment, without the typical barriers or protective cages used in traditional robotics systems. Their potential is undisputed, especially regarding their flexible ability to make simple, quick, and cheap layout changes; however, it is necessary to have adequate knowledge of their correct uses and characteristics to obtain the advantages of this form of robotics, which can be a barrier for industry uptake. The paper starts with an introduction of human–robot collaboration, presenting the related standards and modes of operation. An extensive literature review of works published in this area is undertaken, with particular attention to the main industrial cases of application. The paper concludes with an analysis of the future trends in human–robot collaboration as determined by the authors.

Keywords: collaborative robotics; cobots; human–robot interaction; literature review

1. Introduction

Traditional industrial robotic systems require heavy fence guarding and peripheral safety equipment that reduce flexibility while increasing costs and required space. The current market, however, asks for reduced lead times and mass customization, thus imposing flexible and multi-purpose assembly systems [1]. These needs are particularly common for small- and medium-sized enterprises (SMEs). Collaborative robots (or cobots [2]) represent a natural evolution that can solve existing challenges in manufacturing and assembly tasks, as they allow for a physical interaction with humans in a shared workspace; moreover, they are designed to be easily reprogrammed even by non-experts in order to be repurposed for different roles in a continuously evolving workflow [3]. Collaboration between humans and cobots is seen as a promising way to achieve increases in productivity while decreasing production costs, as it combines the ability of a human to judge, react, and plan with the repeatability and strength of a robot.

Several years have passed since the introduction of collaborative robots in industry, and cobots have now been applied in several different applications; furthermore, collaboration with traditional robots is considered in research, as it takes advantage of the devices' power and performance. Therefore, we believe that it is the proper time to review the state of the art in this area, with a particular focus on industrial case studies and the economic convenience of these systems. A literature review is considered a suitable approach to identify the modern approaches towards Human–Robot Collaboration (HRC), in order to better understand the capabilities of the collaborative systems and highlight the possible existing gap on the basis of the presented future works.

The paper is organized as follows: After a brief overview of HRC methods, Section 2 provides an overview of the economic advantages of the collaborative systems, with a brief comparison with traditional systems. Our literature review analysis is presented in Section 3, and Section 4 contains a discussion of the collected data. Lastly, Section 5 concludes the work.

Background

Despite their relatively recent spread, the concept of cobots was invented in 1996 by J. Edward Colgate and Michael Pashkin [2,4]. These devices were passive and operated by humans, and are quite different from modern cobots that are more represented by the likes of lightweight robots such as KUKA LBR iiwa, developed since the 1990s by KUKA Roboter GmbH and the Institute of Robotics and Mechatronics at the German Aerospace Center (DLR) [5], or the first commercial collaborative robot sold in 2008, which was a UR5 model produced by the Danish company Universal Robots [6].

First of all, we believe that it is important to distinguish the different ways of collaboration, since the term collaboration often generates misunderstandings in its definition.

Müller et al. [7] proposed a classification for the different methodologies in which humans and cobots can work together, as summarized in Figure 1, where the final state shows a collaborative environment.

- Coexistence, when the human operator and cobot are in the same environment but generally do not interact with each other.
- Synchronised, when the human operator and cobot work in the same workspace, but at different times.
- Cooperation, when the human operator and cobot work in the same workspace at the same time, though each focuses on separate tasks.
- Collaboration, when the human operator and the cobot must execute a task together; the action of the one has immediate consequences on the other, thanks to special sensors and vision systems.

It should be noted that neither this classification nor the terminology used are unique, and others may be found in the literature [8–11].

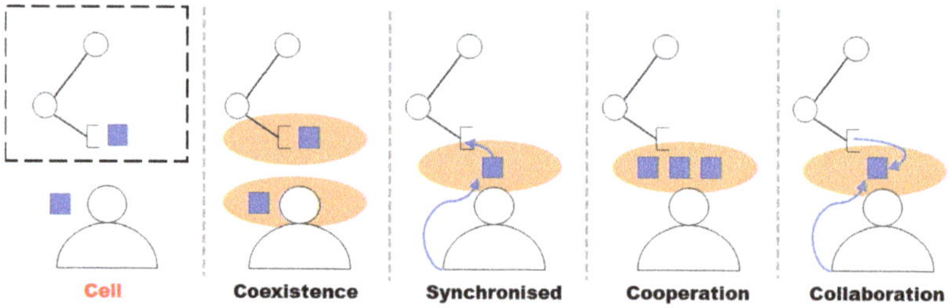

Figure 1. Types of use of a collaborative robot.

To provide definitions and guidelines for the safe and practical use of cobots in industry, several standards have been proposed. Collaborative applications are part of the general scope of machinery safety regulated by the Machinery Directive, which defines the RESS (Essential Health and Safety Requirements). For further documentation, we refer to [12].

The reference standards as reported in the Machinery Directive are:

- UNI EN ISO 12100:2010 "Machine safety, general design principles, risk assessment, and risk reduction".
- UNI EN ISO 10218-2:2011 "Robots and equipment for robots, Safety requirements for industrial robots, Part 2: Systems and integration of robots".

- UNI EN ISO 10218-1:2012 "Robots and equipment for robots, Safety requirements for industrial robots, Part 1: Robots".

In an international setting, the technical specification ISO/TS 15066:2016 "Robots and robotic devices, Collaborative Robots" is dedicated to the safety requirements of the collaborative methods envisaged by the Technical Standard UNI EN ISO 10218-2:2011.

According to the international standard UNI EN ISO 10218 1 and 2, and more widely explained in ISO/TS 15066:2016, four classes of safety requirements are defined for collaborative robots:

- Safety-rated monitored stop (SMS) is used to cease robot motion in the collaborative workspace before an operator enters the collaborative workspace to interact with the robot system and complete a task. This mode is typically used when the cobot mostly works alone, but occasionally a human operator can enter its workspace.
- Hand-guiding (HG), where an operator uses a hand-operated device, located at or near the robot end-effector, to transmit motion commands to the robot system.
- Speed and separation monitoring (SSM), where the robot system and operator may move concurrently in the collaborative workspace. Risk reduction is achieved by maintaining at least the protective separation distance between operator and robot at all times. During robot motion, the robot system never gets closer to the operator than the protective separation distance. When the separation distance decreases to a value below the protective separation distance, the robot system stops. When the operator moves away from the robot system, the robot system can resume motion automatically according to the requirements of this clause. When the robot system reduces its speed, the protective separation distance decreases correspondingly.
- Power and force limiting (PFL), where the robot system shall be designed to adequately reduce risks to an operator by not exceeding the applicable threshold limit values for quasi-static and transient contacts, as defined by the risk assessment.

Collaborative modes can be adopted even when using traditional industrial robots; however, several safety devices, e.g., laser sensors and vision systems, or controller alterations are required. Thus, a commercial cobot that does not require further hardware costs and setup can be a more attractive solution for industry.

Lastly, cobots are designed with particular features that distinguish them considerably from traditional robots, defined by Michalos et al. [13] as technological and ergonomic requirements. Furthermore, they should be equipped with additional features with respect to traditional robots, such as force and torque sensors, force limits, vision systems (cameras), laser systems, anti-collision systems, recognition of voice commands, and/or systems to coordinate the actions of human operators with their motion. For a more complete overview, we refer to [8,13]. Table A1 shows the characteristics of some of the most popular cobots, with a brief overview of some kinematic schemes in Table 2.

2. Convenience of Collaborative Robotics

The choice towards human–robot collaborative systems is mainly dictated by economic motivations, occupational health (ergonomics and human factors), and efficient use of factory space. Another advantage is the simplification in the robot programming for the actions necessary to perform a task [14]. In addition, learning by demonstration is a popular feature [15].

Furthermore, the greater convenience of collaborative systems is their flexibility: Theoretically, since collaborative cells do not require rigid safety systems, they could be allocated in other parts of plants more easily and more quickly; therefore, they could adapt well to those cases in which the production layout needs to change continuously [16]. However, it should be noted that high-risk applications have to be constrained as in any other traditional system, thus restricting the flexibility.

Collaborative systems can also achieve lower direct unit production costs: [17] observed that a higher degree of collaboration, called $c_{\%}$, has a high impact on throughput; moreover, depending on the assembly process considered, the throughput can be higher than in traditional systems.

Table 1 provides a comparison between collaborative and traditional systems for four different jobs: assembly (the act of attaching two or more components), placement (the act of positioning each part in the proper position), handling (the manipulation of the picked part), and picking (the act of taking from the feeding point). In order to adapt to market needs, a manual assembly system could be used, though this can lead to a decrease in productivity due to variations in quality and fluctuations in labor rates [18]. Comparing the human operator capabilities to automated systems, it is clear that the performance of manual assembly is greatly influenced by ergonomic factors, which restrict the product weight and the accuracy of the human operator [19]. Therefore, these restrictions limit the capabilities of human operators in the handling and picking tasks of heavy/bulky parts. These components can be manipulated with handling systems such as jib cranes: These devices could be considered as large workspace-serving robots [20], used for automated transportation of heavy parts. However, to the authors' knowledge, there are no commercial end-effectors that allow these systems to carry out complex tasks, such as assembly or precise placing, since they are quite limited in terms of efficiency and precision [21].

Traditional robotic systems [22] bridge the presented gap, presenting manipulators with both high payload (e.g., FANUC M-2000 series with a payload of 2.3 t [23]) and high repeatability. However, the flexibility and dexterity required for complex assembly tasks could be too expensive, or even impossible, to achieve with traditional robotic systems [24]. This gap can be closed by collaborative systems, since they combine the capabilities of a traditional robot with the dexterity and flexibility of the human operator. Collaborative robots are especially advantageous for assembly tasks, particularly if the task is executed with a human operator. They are also suitable for pick and place applications, though the adoption of a traditional robot or a handling system can offer better results in terms of speed, precision, and payload.

Table 1. Qualitative evaluation of the most suitable solutions for the main industry tasks.

	Human Operator	**Collaborative Systems**	**Traditional Robot**	**Handling Systems**
Assembly	High dexterity and flexibility	Combines human dexterity with robot capabilities [24]	Dexterity/flexibility could be unreachable [24]	No complex tasks with commercial end-effectors [21]
Placement	High dexterity	Commercial cobots have lower repeatability	High repeatability and payload	High payload
Handling	Product weight restricted [19]	Typical cobots have low payload	High payload and speed [23]	High payload
Picking	Product weight restricted [19]	Typical cobots have low payload	High payload and repeatability [23]	Bin picking difficult due to size

3. Literature Review

This literature review analyses works from 2009–2018 that involved collaborative robots for manufacturing or assembly tasks. Reviewed papers needed to include a practical experiment involving a collaborative robot undertaking a manufacturing or assembly task; we ignored those that only considered the task in simulation. This criterion was implemented as, often, only practical experiments with real hardware can highlight both the challenges and advantages of cobots.

For this literature review, three search engines were used to collect papers over our time period that were selected using the following boolean string: ((collaborative AND robot) OR cobot OR cobotics) AND (manufacturing OR assembly). Our time period of 2009–2018 was chosen as the timeline for this literature review, as it is only in the last 10 years that we have seen the availability of collaborative robots in the market.

- ScienceDirect returned 124 results, from which 26 were found to fit our literature review criteria after reading the title and abstract.
- IEEExplore returned 234 results, from which 44 were found to fit our literature review criteria after reading the title and abstract.
- Web of Science returned 302 results, from which 62 were found to fit our literature review criteria after reading the title and the abstract.

Of all these relevant results, 16 were duplicated results, leaving us with 113 papers to analyze. Upon a complete read-through of the papers, 41 papers were found to fully fit our criteria and have been included in this review. It should be noted that in the analysis regarding industry use cases, only 35 papers are referenced, as 6 papers were focused on the same case study as others and did not add extra information to our review.

The following parameters were studied: The robot used, control system, application, objectives, key findings, and suggested future work for all these studies, as summarized in Table 3. These were chosen for the following reasons. The robot choice is important, as it highlights which systems are successfully implemented for collaborative applications. The control system is interesting to analyze, as it dictates both safety and performance considerations of the task. Furthermore, when a human is in the control loop, the control system choice is specific to the manner of human–machine interaction—by seeing which methods are more popular and successfully implemented, we can identify trends and future directions. We characterized control systems as vision systems (such as cameras and laser sensors), position systems (such as encoders which are typical of traditional industrial robots), impedance control systems (through haptic interfaces), admittance control (taking advantage of the cobot torque sensors or voltage measurement), audio systems (related to voice command and used for voice/speech recognition), and other systems (that were not easily classified, or that were introduced only in one instance).

The application represents the task given to the cobot, which we believe allows a better understanding to be made regarding the capabilities of collaborative robots. These tasks were divided into assembly (when the cobot collaborates with the operator in an assembly process), human assistance (when the cobot acts as an ergonomic support for the operator, e.g., movable fixtures, quality control, based on vision systems), and lastly, machine tending (when the cobot performs loading/unloading operations).

Furthermore, we divided the objectives into three main topics: Productivity, representing the studies focused on task allocation, quality increase, and reduction of cycle time; safety, which includes not only strictly safety-related topics such as collision avoidance, but also an increase in human ergonomics and reduction of mental stress; and HRI (Human–Robot Interaction), which is focused on the development of new HRI methodologies, e.g., voice recognition. It should be noted that in no way is the proposed subdivision univocal; an interesting example could be [25–27]. These works were considered as safety because, even if the proposed solutions keep a high level of productivity, they operate on HRC safety.

The key findings were not grouped, since we believe they depend on the specific study and are too varied; however, they have been summarized in Table 3. Key findings were useful to present the capabilities of the collaborative systems and what HRC studies have achieved. They were included in our analysis in order to identify common solutions. Future work has been grouped into: HRI (works that focus on increasing HRI knowledge and design), safety (works that focusing on increasing the operator safety when working with the cobot), productivity (works focusing on increasing the task productivity in some manner), task complexity (works that focus on increasing the complexity of the task for a particular application), applicability (works that focus on increasing the scope of the work to be used for other industrial applications), and method (works that focus on enhancing the method of HRI via modeling, using alternative robots, or applying general rules and criteria to the design and evaluation process). From these groupings, we can identify ongoing challenges that still need to be solved in the field; by seeing what researchers identify as future work for industrial uptake, we can find trends across the industry in the direction research on which is focused. Our analysis of these parameters is presented in Section 4.

4. Discussion: Trends and Future Perspective

4.1. Trends in the Literature

By examining the literature as summarized in Table 3, we can identify several trends in the use of cobots in industrial settings. It should be noted that, for some of the considered studies in Table 3, we could not identify all of our parameters as specified in Section 3; thus, they are not considered in this specific discussion.

Although early researches utilized traditional industrial robots (Figure 2), the subsequent spread of cobots led to several studies based on the DLR LWR-III (2011), followed by the upgraded KUKA iiwa (from 2016 to 2018), ABB YuMi (also called FRIDA) in 2017 and 2018, and Universal Robots from 2014 to 2018. Several researchers applied the collaborative methods to industrial robots, usually due to their increased performance and widespread availability; however, the disadvantage of this choice is the increase in cost and complexity due to the inclusion of several external sensors and the limited HRC methodologies available. A relationship between the kinematics of the cobot and the application was not explicitly considered, since we believe that other parameters, such as the presence of force sensors in each axis, influenced the cobot choice made in these papers. However, it should be noted that the kinematics—precisely, the number of axes—was a feature considered in [28], whereas future works are focused on verifying their findings with kinematically redundant robots [29] or utilizing the redundancy for achieving better stiffness in hand-guiding [30].

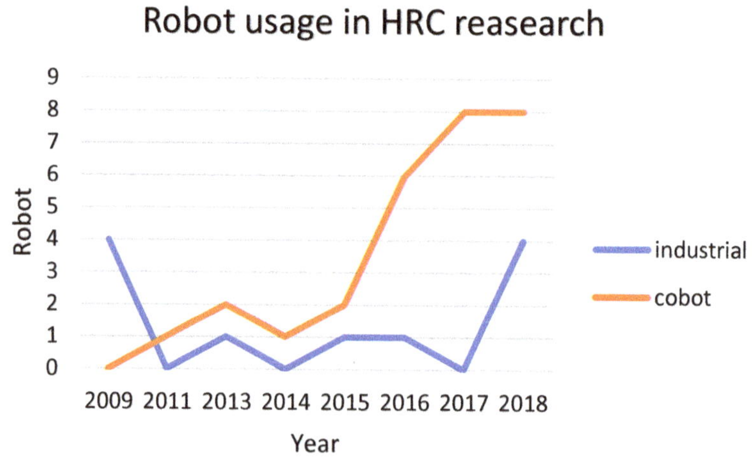

Figure 2. Robot usage in selected human–robot collaboration studies in the period 2009–2018.

Figure 3 presents the different control systems in the selected human–robot collaboration (HRC) studies. Position control systems were only used for traditional industrial robots, often using extra vision systems for safety reasons. Due to the inherent compliance of cobots, impedance control was more commonly chosen for these systems, though in many cases where an inherently compliant cobot was used, vision was also included for feedback [31–33]. Robot compliance can often be a trade-off with robot precision, so including a separate channel for feedback to monitor collisions and increase safety can be a useful method of maintaining manipulation performance. Vision is indeed the prevalent sensor used in HRC studies, also due to the flexibility and affordability of the systems, especially when using depth cameras such as Microsoft Kinect cameras. It is interesting to note that in recent years, Augmented Reality (AR) systems, such as the Microsoft Hololens, have been used more in HRC research, as they are able to provide information to the operator without obscuring their view of the assembly process. In one study, a sensitive skin was incorporated with the cobot to provide environmental information and maintain the operator's safety. As these skins become more widely studied and developed, we could see this feedback control input become more common, though challenges such as response time must still be solved [34].

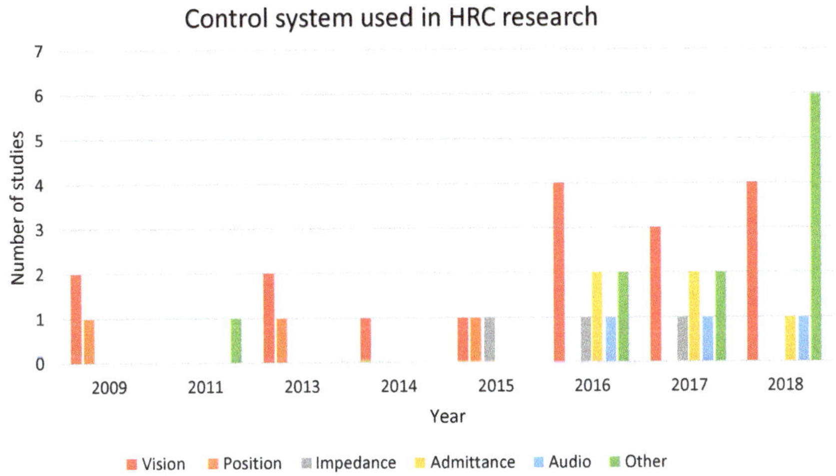

Figure 3. Control systems used in selected human–robot collaboration studies in the period 2009–2018: In red, the number of vision systems; in orange, position-controlled systems (used especially for traditional industrial robots); in gray, the cases for impedance control (e.g., through haptic interfaces); in yellow, admittance control (e.g., through torque sensors); in blue, audio systems (for voice/speech recognition); and green for other systems.

The considered studies used the aforementioned robots, both traditional industrial robots and cobots with different collaborative methodologies. Early studies were focused on SSM and PFL methodologies; we believe this focus is due to the need for safety and flexibility in traditional robotic systems and the early spread of cobots. Since 2016 and the introduction of ISO/TS 15066:2016, the considered research sample began to study other methodologies, especially the HG method, which, as shown in Figure 4, has become prevalent in recent years. The HG method is indeed a representative function of collaborative robots [30], since it allows even unskilled users to interact with and program the cobot, which can allow some degree of flexibility—even if the robot moves only on predefined directions—without the need for expensive algorithms [35]. It should be noted that the HG method could also be employed with traditional industrial robots, such as a COMAU NJ130 [36]: This allows one to take advantage of the robot's characteristics, such as high speed and power, and increase the system's flexibility.

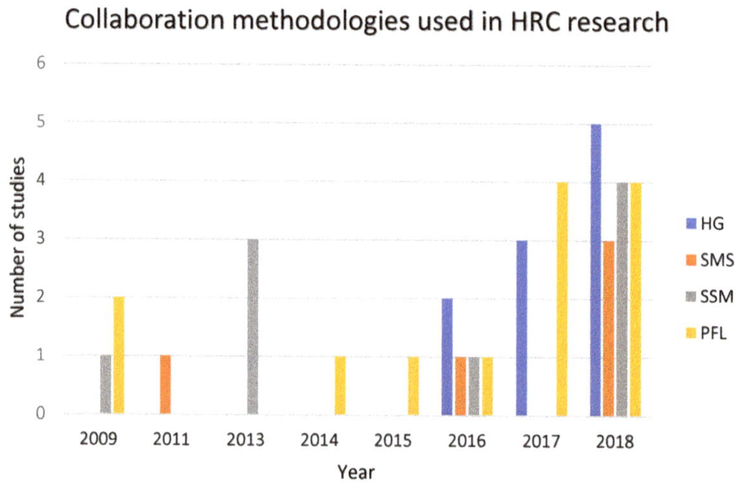

Figure 4. Collaboration methods used in selected human–robot collaboration in the period 2009–2018: In blue, hand guiding (HG); in orange, safety-rated monitored stop (SMS); in gray, speed and separation monitoring (SSM); in yellow, power and force limiting (PFL).

As stated previously, the collaborative mode depends on the considered application. Figure 5 depicts the considered tasks over the last decade. The most studied task is assembly, likely due to the required flexibility in the task, which makes traditional robotic systems too expensive or difficult to implement. However, the task of production also requires flexibility, and could greatly benefit from collaborative applications. Likely, until the fundamental challenges of setting up collaborative workcells are solved for the easier tasks of assembly, we will not see many case studies targeting production.

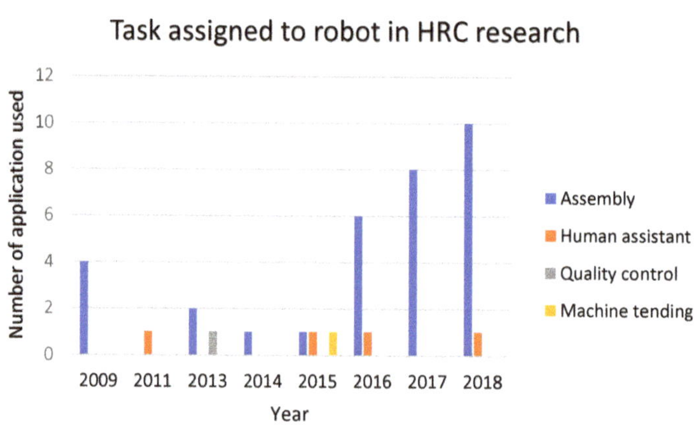

Figure 5. Tasks assigned to the robot in selected collaborative applications in research in the period 2009–2018: In blue, assembly tasks; in orange, the tasks used to assist the operator, e.g., handover of parts, quality control tasks, or machine tending, i.e., loading and/or unloading.

In our review, 35 papers presented unique case studies of industrial applications. Two industries seem to drive this research—the automotive industry accounted for 22.85% of studies, and the electronics industry a further 17.14%. Interestingly, research for the automotive industry only began after 2015, and will likely continue to drive research in this area.

HRC studies present several objectives that can be grouped into three main topics. Figure 6 depicts the focus of HRC studies in the last decade. It is interesting to note that the first phase of HRC study [37–41] was more focused on increasing the production and safety aspects of HRC, at least in a manufacturing context. As the research progressed, an increasing number of studies were focused on HRI methodologies, becoming a predominant objective in 2017. The ostensible reduction in 2018 should not mislead us to believe that HRI studies were abandoned in that year: As stated before, the presented classification is not univocal, thus studies such as [42–44] could also be considered HRI studies.

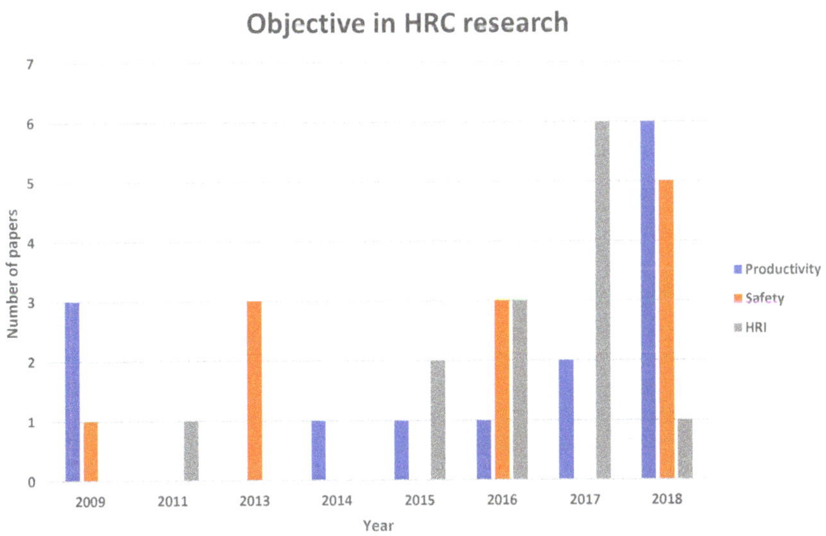

Figure 6. Main topics or objectives in HRC studies. The objectives were divided into productivity studies (blue), safety studies (orange), e.g., ergonomics and collision avoidance, HRI (Human–Robot Interaction) studies (gray), e.g., development or improvement of HRI methodologies.

The key findings of these studies highlight challenge areas that research has successfully addressed, or even solved, when cobots are used for industrial tasks. Multiple studies reported an increase in task performance—e.g., by reducing completion time and minimizing error [25,37,38,43]—as well as a better understanding of the operator space [29,31,32,41] and higher precision of workpiece manipulation [28,30,45]. Thematic areas of research intent can be identified, such as increasing and quantifying the trust of the operator in the robotic system [29,46,47], as well as improving safety by minimizing collisions [40].

The directions of future work identified in literature are summarized in Figure 7. Historically, researchers aimed to increase the HRI relevance of their work, also with a focus on higher safety requirements and more complex tasks. In recent years, the scope of future work has expanded, with researchers focusing on more complex methods that improve the performance of their systems—whether this is by applying their method to different application fields or more complex tasks. This is likely due to the prevalence of new cobots and sensing methodologies coming onto the market, maturing algorithms, and experience in designing collaborative workcells.

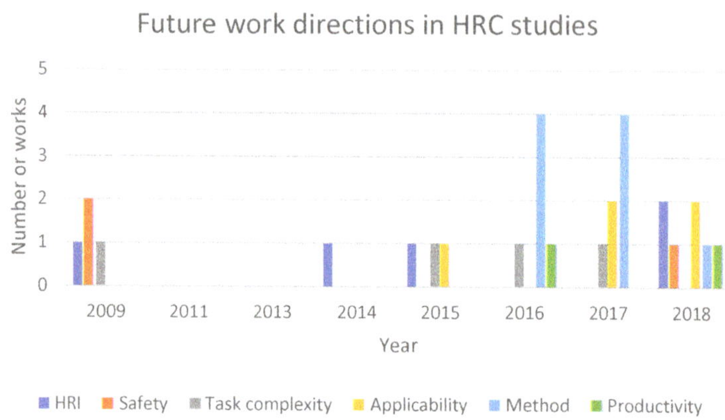

Figure 7. Future work topics from HRC studies. The work was divided into directions of HRI (dark blue), safety (orange), task complexity (gray), applicability (yellow), method (light blue), and productivity (green).

Many of the reviewed works highlight future work in terms of the method they used, whether it be by increasing the complexity of their modeling of the operator and/or environment [48], or using different metrics to evaluate performance [33,49,50] and task choice [51]. Others believe that expanding their research setup to other application areas is the next step [31,45,52]. In our view, these works can be achieved without any step change in existing technology or algorithms; rather, it requires more testing time. To increase safety, productivity, and task performance, researchers will need to improve planners, [39,53], environment and task understanding [28,40,54,55], operator intention understanding [38], and ergonomic cell setups [37,56]. To improve HRI systems, common future work focuses on increasing the robots' and operators' awareness of the task and environment by object recognition [44] and integrating multi-modal sensing in an intuitive manner for the operator [3,32,36].

In essence, this future direction focuses on having better understanding of the scene—whether this is what the operator intends to do, what is happening in the environment, or the status of the task. Researchers propose solving this by using more sensors and advanced algorithms, and fusing this information in a way that is easy to use and intuitive for the operator to understand. These systems will inherently lead to better safety, as unexpected motions will be minimized, leading consequently to more trust and uptake. We can expect that many of these advances can come from other areas of robotics research, such as learning by demonstration through hand-guiding or simulation techniques that make it easy to teach a robot a task, and advances in computer vision and machine learning for object recognition and semantic mapping. Other reviews, such as [8], identify similar trends, namely those of improved modeling and understanding, better task planning, and adaptive learning. It will be very interesting to see how this technology is incorporated into the industrial setting to take full advantage of the mechanics and control of cobots and the HRI methodologies of task collaboration.

4.2. Trend of the Market

We believe that the current market should also be presented in order to better place our literature review in the manufacturing context. According to [57], the overall collaborative robot market is estimated to grow from 710 million USD in 2018 to 12,303 million USD by 2025 at a compounded annual growth rate (CAGR) of 50.31% during the forecasted period. However, the International Federation of Robotics (IFR), acknowledging an increase in the robot adoption with over 66% of new sales in 2016, expects that market adoption may proceed at a somewhat slower pace over the forecasted timeframe [58]. However they suggest that the fall in robot prices [59] has led to a growing market for cobots, especially considering that small- and medium-sized enterprises (SMEs), which represent almost 70% of the global number of manufacturers [60] and could not afford robotic applications due to the high capital costs, are now adopting cobots, as they require less expertise and lower installation expenses, confirming a trend presented in scientific works [3].

Finally, [57] highlights that cobots, presenting different payloads, were preferred with up to 5 kg payload capacity; indeed, they held the largest market size in 2017, and a similar trend is expected to continue from 2018 to 2025. This preference of the market towards lightweight robots, which are safer but do not present the high speed and power typically connected with industrial robots [36,61], restrains the HRC possibilities in the current manufacturing scenario. However, we believe that without proper regulation, the current market will continue to mark a dividing line between heavy-duty tasks and HRC methods.

5. Conclusions

Human–robot collaboration is a new frontier for robotics, and the human–robot synergy will constitute a relevant factor in industry for improving production lines in terms of performances and flexibility. This will only be achieved with systems that are fundamentally safe for human operators, intuitive to use, and easy to set up. This paper has provided an overview of the current standards related to Human–Robot Collaboration, showing that it can be applied in a wide range of different modes. The state of the art was presented and the kinematics of several popular cobots were described. A literature analysis was carried out and 41 papers, presenting 35 unique industrial case studies, were reviewed.

Within the context of manufacturing applications, we focused on the control systems, the collaboration methodologies, and the tasks assigned to the cobots in HRC studies. From our analysis, we can identify that the research is largely driven by the electronics and automotive industries, but as cobots become cheaper and easier to integrate into workcells, we can expect SMEs from a wide range of industrial applications to lead their adoption. Objective, key findings and future research directions are also identified, the latter highlighting ongoing challenges that still need to be solved. We can expect that many of the advances needed in the identified directions could come from other areas of robotics research; how these will be incorporated into the industrial setting will lead to new challenges in the future.

Author Contributions: Conceptualization, E.M. and G.R.; Methodology, E.M. and G.R.; Formal analysis, E.M., R.M., and E.G.G.Z.; Investigation, E.M., R.M. and E.G.G.Z.; Data curation, E.M., R.M., and E.G.G.Z.; Writing—original draft preparation, E.M., R.M., and E.G.G.Z.; Writing—review and editing, M.F., E.M., R.M., G.R., and E.G.G.Z.; Supervision, M.F. and G.R.; Project administration, M.F. and G.R.; Funding acquisition, G.R.

Funding: This research was funded by University of Padua—Program BIRD 2018—Project no. BIRD187930, and by Regione Veneto FSE Grant 2105-55-11-2018.

Conflicts of Interest: The authors declare no conflict of interest.

Appendix A. Tables

Table A1. List of characteristics of some of the most used cobots for different kinematics.

Brand	Product Name	Payload (Range) [kg]	Reach (Range) [mm]	Repeatability [mm]	Internal Sensors & Control System	n° of Axis	Kinematics
ABB (CHE-SWE)	YuMi (dual arm IRB 14000, single arm IRB 14050)	0.5 for each arm	500–559	±0.02	Position control (encoder) and voltage measurement	7 for each arm	//
Comau (ITA)	AURA	170	2800	±0.1	Position control (encoder), sensitive skin, laser scanner (optional)	6	//
Fanuc (JPN)	CR-4iA, CR-7iA, CR-7iA/L, CR-14iA/L, CR-15iA, CR-35iA	4–35	550–1813	± 0.01–0.03	Force sensor (base)	6	Spherical wrist
Kawasaki (JPN)	duAro1, duAro2	2–3 for each arm	760–785	±0.05	Position controlled	4 for each arm	Dual arm scara (duAro1)
Kuka (DEU)	iisy, iiwa 7 R800, iiwa 14 R820	3–14	600–820	±0.1–0.15 (iisy not yet defined)	Torque sensor in each axis and position control (iiwa)	6 (iisy) 7 (iiwa)	Spherical wrist (and shoulder for iiwa)
Omron (JAP)/ Techman (TWN)	TM5 700, TM5 900, TM12, TM14	4–14	700–1300	±0.05 (TM5) ±0.1	2d wrist camera, position control (encoder), voltage measurement	6	Three parallel axes
Precise Automation (USA)	PP100, PF3400, PAVP6, PAVS6	2–7	432–770 or 685–1270	±0.02–0.1	Position control (encoder), voltage measurement	4 (cartesian, scara) 6 (spherical wrist)	Cartesian, scara, spherical wrist
Rethink Robotics (USA/DEU)	Baxter, Sawyer	4 (2 for each arm for Baxter)	1260 for each arm	±0.1	Torque sensors in each axis, position control	7 for each arm	//
Staübli	TX2-40, TX2(touch)-60, TX2(touch)-60L, TX2(touch)-90, TX2(touch)-90L, TX2touch-90XL	2.3–20	515–1450	±0.02–0.04	Position control, sensitive skin (touch)	6	Spherical wrist
Universal Robots (DNK)	UR3, UR3e, UR5, UR5e, UR10, UR10e, UR16e	3–16	500–1300	±0.03–0.1	Position control (encoder), voltage measuremen, Force torque multiaxis load cell (e-series)	6	Three parallel axes
Yaskawa (JPN)	HC10, HC10DT	10	1200	±0.1	Torque sensors in each axis, position control	6	//

Table 2. Denavit–Hartenberg parameters and singularity configurations for the considered kinematic schemes.

Kinematic Scheme	Denavit–Hartenberg Parameters	Singularity Configurations

Six axes with spherical wrist

T	α	a	θ	d
T_{10}	0	0	q_1	d_1
T_{21}	-90	a_1	$-90+q_2$	0
T_{32}	0	a_2	q_3	0
T_{43}	-90	a_3	q_4	d_4
T_{54}	90	0	q_5	0
T_{65}	-90	0	q_6	d_6

Wrist: J4 and J6 aligned 

Shoulder: Wrist aligned with J1 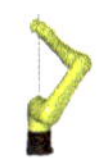

Elbow: Wrist coplanar with J2 and J3 

Six axes with three parallel axes

T	α	a	θ	d
T_{10}	0	0	q_1	d_1
T_{21}	-90	0	$-90+q_2$	d_2
T_{32}	0	a_2	q_3	$-d_3$
T_{43}	0	a_3	$-90+q_4$	d_4
T_{54}	-90	0	$180+q_5$	d_5
T_{65}	-90	0	q_6	d_6

Wrist: J6 // J4 

Shoulder: Intersection of J5 and J6 coplanar with J1 and J2 

Elbow: J2, J3, and J4 coplanar 

Six axes with offset wrist

T	α	a	θ	d
T_{10}	0	0	q_1	d_1
T_{21}	90	0	$90+q_2$	$-d_2$
T_{32}	180	a_2	$90+q_3$	$-d_3$
T_{43}	90	0	$180+q_4$	$-d_4$
T_{54}	90	0	$180+q_5$	$-d_5$
T_{65}	90	0	q_6	$-d_6$

Wrist: J5 $\approx 0°$ or $180° \pm 15°$ 

Shoulder: Wrist point near the yellow column (300 mm of radius) 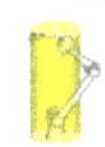

Elbow: J3 $\approx 0°$ or $180° \pm 15°$ 

Seven axes with spherical joints

T	α	a	θ	d
T_{10}	0	0	q_1	d_1
T_{21}	-90	0	q_2	0
T_{32}	90	0	q_3	d_3
T_{43}	90	0	q_4	0
T_{54}	-90	0	q_5	d_5
T_{65}	-90	0	q_6	0
T_{76}	90	0	q_7	d_7

Wrist motion: J6 = 0 & J4 = 90° 

Shoulder Motion: J2 = 0 & J3 = ± 90° 

Elbow Motion: J5 = ± 90° & J6 = 0 

Seven axes without spherical joints

T	α	a	θ	d
T_{10}	0	0	q_1	d_1
T_{21}	-90	$-a_1$	q_2	0
T_{32}	90	a_2	q_3	d_3
T_{43}	90	a_3	$90+q_4$	0
T_{54}	90	a_4	q_5	d_5
T_{65}	90	$-a_5$	q_6	0
T_{76}	-90	a_6	$180+q_7$	d_7

Wrist: J5 // J7 

Shoulder: Wrist point near J1 direction 

Elbow: J3 // J5 

Table 3. Literature review analysis.

Author	Year	Robot Used	Control System	Collaboration Methods	Application	Objective of the Study	Key Findings	Future Works
J. T. C. Tan et al. [37]	2009	Industrial robot			Assembly	**Productivity:** Information display for operator support	Development and implementation of an information support system, which leads to a reduction in assembly time.	**HRI**: Studies on the position of the LCD TV (information source) in order to improve information reception.
T. Arai et al. [38]	2009	Industrial robot	Vision	Power and Force limiting	Assembly	**Productivity:** Improve efficiency through devices that support the operator	HRC doubles productivity in comparison with manual assembly; reduction of human errors up to no defects	**Safety:** Monitoring operator's fatigue and intention
J. T. C. Tan et al. [39]	2009	Industrial robot	Vision	Power and Force limiting	Assembly	**Productivity:** Optimize working efficiency, improving quality and productivity	Safety analysis lead to an increased distance and a reduction in the robot speed; mental strain due to working with robots reduced	**Task complexity:** Analysis of the improvements obtained with HRC to order tasks.
C. Lenz et al. [40]	2009	Industrial robot	Position	Speed and separation monitoring	Assembly	**Safety:** HRC with traditional industrial robots, with focus on safety	Collision avoidance method based on restricted robot movements and virtual force fields	**Safety:** Estimation and tracking of human body pose to avoid collision without the use of any markers.
T. Ende et al. [41]	2011	DLR LWR-III 3	Gesture recognition		Human assistant	**Productivity:** Gather gestures for HRC from humans	Eleven gestures present recognition rate over 80%; recognition problem when torso position is part of the gesture	N/A

H. Ding et al. [25]	2013	ABB FRIDA (YuMi)	Vision	Speed and separation monitoring	Assembly	**Safety:** Collaborative behavior with operator safety and without productivity losses due to emergency stops	Speed reduction applied based on the distance between human arm and Kinect position avoids emergency stops	N/A
H. Ding et al. [26]	2013	ABB FRIDA (YuMi)	Vision	Speed and separation monitoring	Assembly	**Safety:** Multiple operators in collaborative behavior with operators safety concern and without productivity losses	Development of a finite state automaton; speed reduction improves the uptime while respecting safety constraints	N/A
A. M. Zanchettin et al. [27]	2013	Industrial robot	Position	Speed and separation monitoring	Quality control	**Safety:** Compromise between safety and productivity; Adaptable robot speed	Development of a safety-oriented path-constrained motion planning, tracking operator, and reducing robot speed	N/A
K. P. Hawkins et al. [32]	2014	Universal Robots UR10	Vision	Power and Force limiting	Assembly	**Productivity:** Robots need to anticipate human actions even with task or sensor ambiguity	Compromises between human wait times and confidence in the human action detection	**HRI:** impact of the system on the operator's sense of fluency, i.e., synchronization between cobot and operator
K. R. Guerin et al. [3]	2015	Universal Robots UR5	Impedance		Machine tending	**Productivity:** Robot assistant with set of capabilities for typical SMEs	Test of machine tending: 82% of parts taken from machine (due to bad weld or bad grasp)	**HRI:** Test ease of use and focus on HRC, gesture recognition for learning
I. D. Walker et al. [46]	2015	Industrial robot	Position		Human assistant	**HRI:** Trust not considered in handoffs; derive model for robot trust on operator	The robot pose changes accordingly to trust in human, reducing impact forces in case of low trust	**Applicability:** Effectiveness of the approach in SME scenarios

T. Hamabe et al. [54]	2015	Kawada HIRO	Vision	Power and Force limiting	Assembly	**HRI:** Learn task from human demonstration	Human and robot roles fixed, due to limits in robot's manipulation capabilities; changing task order increases time due to recognition system	**Task complexity:** Complete set of scenarios assumed, cobot should be able to recognize new tasks autonomously
S. M. M. Rahman et al. [29]	2016	Kinova MICO 2-finger	Vision		Assembly	**HRI:** Derive model for robot trust on human; trust-based motion planning for handover tasks	pHRI gets better with trust for contextual information transparent to human; increase: 20% safety, 30% handover success, 6.73% efficiency	**Method:** Apply the proposed method with kinematically redundant robot
S. M. M. Rahman et al. [49]	2016	Rethink Baxter	Vision		Assembly	**Productivity:** Autonomous error detection with human's intervention	Regret based method leads to improvement of fluency, due to an increase in synchronization, reduction in mean cognitive workload and increase in human trust compared to a Bayesian approach	**Method:** Different objective criteria for the regret-based approach to evaluate HRC performance
L. Rozo et al. [48]	2016	WAM robot/ KUKA LWR iiwa	Impedance/ admittance	Hand-guiding	Assembly	**HRI:** Robotic assistant needs to be easily reprogrammed, thus programming by demonstration	Model adapts to changes in starting and ending point, task, and control mode; with high compliance, the robot can not follow trajectory	**Method:** Estimation of the damping matrix for the spring damper model; study how interaction forces can change the robot behaviors
A. M. Zanchettin et al. [62]	2016	ABB FRIDA (YuMi)	Vision	Speed and separation monitoring	Assembly	**Safety:** Collision avoidance strategy: Decrease the speed of the cobot	Speed reduction method based on minimum distance; distance threshold adaptable to the programmed speed; continuous speed scaling	N/A

A. Cherubini et al. [63]	2016	KUKA LWR4+	Admittance	Safety rated monitored stop	Assembly	**Safety:** Collaborative human–robot application to assembly a car homokinetic joint	Framework that integrates many state-of-the-art robotics components, applied in real industrial scenarios	**Productivity:** Deploying the proposed methodologies on mobile manipulator robots to increase flexibility
H. Fakhruldeen et al. [51]	2016	Rethink Baxter	Vision and audio recognition	Power and Force limiting	Assembly	**HRI:** Implementation of a self-built planner in a cooperative task where the cobot actively collaborates	API development combining an object-oriented programming scheme with a Prolog meta interpreter to create these plans and execute them	**Method:** add cost to action evaluation, add non-productive actions, action completion percentage should be considered
S. Makris et al. [64]	2016	Industrial robot	AR system	Hand-guiding	Human assistant	**Safety:** Development of an AR system in aid of operators in human–robot collaborative environment to reduce mental stress	The proposed system minimizes the time required for the operator to access information and send feedback; it decreases the stoppage and enhances the training process	**Task complexity:** Application in other industrial environments
B. Whitsell et al. [28]	2017	KUKA LBR iiwa	Impedance / admittance	Hand-guiding/ Power and Force Limiting	Assembly	**HRI:** Cooperate in everyday environment; robots need to adapt to human; haptics should adapt to the operator ways	100% correct placement of a block in 1440 trials; robot can control a DOF if the operator does not control it (95.6%); lessening the human responsibility by letting the robot control an axis reduces the completion time	**Applicability:** Adapt variables to environment, e.g. task and robot coordinate system not aligned

J. Bös et al. [65]	2017	ABB YuMi	Admittance	Power and Force limiting	Assembly	**Productivity:** Increase assembly speed without reducing the flexibility or increasing contact forces using iterative learning control (ILC)	Increase acceleration by applying Dynamic Movement Primitives to an ILC, reduce contact forces by adding a learning controller. Stochastic disturbances do not have a long term effect; task duration decreases by 40%, required contact force by 50%	**Method:** Study and theoretical proof on stability in the long term
M. Wojtynek et al. [33]	2017	KUKA LBR iiwa	Vision	Hand-guiding	Assembly	**Productivity:** Create a modular and flexible system; abstraction of any equipment	Easy reconfiguration, without complex programming	**Method:** Introduce metrics for quantitative measurement of HRC
B. Sadrfaridpour et al. [47]	2017	Rethink Baxter	Human tracking system	Power and Force limiting	Assembly	**HRI:** Combination of pHRI and sHRI in order to predict human behavior and choose robot path and speed	Augmenting physical/social capabilities increases one subjective measure (trust, workload, usability); assembly time does not change	N/A
I. El Makrini et al. [56]	2017	Rethink Baxter	Vision	Power and Force limiting	Assembly	**HRI:** HRC based on natural communication; framework for the cobot to communicate	The framework is validated; more intuitive HRI	**Task Complexity:** Adapt robot to user; adjust parts position based on user's height
P. J. Koch et al. [52]	2017	KUKA LWR4+	Admittance		Screwing for maintenance	**HRI:** Cobot development: Focus on intuitive human–robot interface	HR interface, simple for user reconfiguration. Steps in order to transform a mobile manipulator into a cobot	**Applicability:** Expand to several industrial maintenance tasks

M. Haage et al. [66]	2017	ABB YuMi	Vision		Assembly	**HRI:** Reduce the time and required expertise to setup a robotized assembly station	A web-based HRI for assisting human instructors to teach assembly tasks in a straightforward and intuitive manner	N/A
P. Gustavsson et al. [50]	2017	Universal Robots UR3	Impedance and audio recognition	Hand-guiding/ Power and Force Limiting	Assembly	**HRI:** Joint Speech recognition and a haptic control in order to obtain an intuitive HRC	Developed a simplified HRI responsive to vocal commands, that guides the user in the progress of the task with haptics	**Method:** Test if haptic control can be used to move the robot with linear motions; an automatic way of logging the accuracy
M. Safeea et al. [30]	2017	KUKA LBR iiwa	Admittance	Hand-guiding/ Safety rated Monitored Stop	Assembly	**Safety:** Precise and intuitive hand guiding	Possible to hand-guide the robot with accuracy, with no vibration, and in a natural and intuitive way	**Method:** Utilizing the redundancy of iiwa to achieve better stiffness in hand-guiding
W. Wang et al. [45]	2018	Industrial robot	Gesture recognition	Hand-guiding/ Power and Force Limiting	Assembly	**Productivity:** Easier reconfiguration of the cobot using a teaching-by-demonstration model	95% of accuracy (higher than previous methods based on vision); Lower inefficient time	**Applicability:** Use multimodal information to investigate different applications
N. Mendes et al. [31]	2018	KUKA LBR iiwa	Vision	Speed and separation monitoring	Assembly	**HRI:** Need for a flexible system with simple and fast interface	Gesture is intuitive but delays process; constrained flexibility	**Applicability:** Expand use case to several industrial fields
K. Darvish et al. [42]	2018	Rethink Baxter	Vision	Power and Force limiting	Assembly	**Productivity:** Increase robot adaptability integrated in the FlexHRC architecture by an online task planner	Planning and task representation require little time (less than 1% of total); the simulation follows the real data very well	N/A

A. Zanchettin et al. [43]	2018	ABB YuMi	Vision	Power and Force limiting	Assembly	**Productivity:** Predict human behavior in order to increase the robot adaptability	Decrease of task time equal to 17%;	N/A
G. Pang et al. [34]	2018	ABB YuMi	Sensitive skin	Safety rated monitored stop/ Speed and separation monitoring	Test of collision	**Safety:** Cobot perceives stimulus only by its torque sensors, not guaranteeing collision avoidance	Integration on cobot of sensitive skin; delay in the system reaction	**Method:** Reduce contact area of the sensitive skin; test with multisensing systems
V. V. Unhelkar et al. [53]	2018	Universal Robots UR10	Vision	Speed and separation monitoring	Human assistant	**Productivity:** Cobots can be successful, but they have restricted range: Mobile cobots for delivering parts	Prediction of long motion (16 s) with a prediction time horizon up to 6 s; in simulation, reduced safety rated monitored stops, increasing task efficiency	**Safety:** Recognize unmodeled motion and incremental planners
V. Tlach et al. [55]	2018	Industrial robot	Admittance	Hand-guiding/ Safety rated Monitored Stop	Assembly	**Productivity:** Design of collaborative tasks in an application	The method is flexible to the type of product	**Productivity:** Improve methods for recognizing objects
S. Heydaryan et al. [35]	2018	KUKA LBR iiwa 14 R820	Vision / admittance	Hand-guiding/ Safety rated Monitored Stop	Assembly	**Safety:** Task allocation to ensure safety of the operator, increase productivity by increasing ergonomics	Assembly time of 203 s in SMS, but the robot obstructs the access to some screws; proposed a hand-guided solution (210s)	N/A
G. Michalos et al. [36]	2018	Industrial robot	Admittance	Hand-guiding	Assembly	**Safety:** Implementation of a robotic system for HRC assembly	Development of HRC assembly cell with high payload industrial robots and human operators.	**HRI:** Improve human immersion in the cell. Integrate all the sensing and interaction equipment

V. Gopinath et al. [61]	2018	Industrial robot	AR system		Assembly	**Safety:** Development of a collaborative assembly cell with large industrial robots	Development of two work stations	N/A
A. Blaga et al. [44]	2018	Rethink Baxter	Vision and audio recognition	Power and Force limiting	Assembly	**Productivity:** Improve the possibilities of the integration of Augmented Reality in collaborative tasks to shorten lead times	AR and HRC were integrated into an unitary system, meant to ease a worker's daily tasks regarding the visualization of the next possible assembly step	**HRI:** Using object recognition combined with 3D printing, along with the latest HMD devices

References

1. Barbazza, L.; Faccio, M.; Oscari, F.; Rosati, G. Agility in assembly systems: A comparison model. *Assem. Autom.* **2017**, *37*, 411–421. [CrossRef]
2. Colgate, J.E.; Edward, J.; Peshkin, M.A.; Wannasuphoprasit, W. Cobots: Robots for Collaboration with Human Operators. In Proceedings of the 1996 ASME International Mechanical Engineering Congress and Exposition, Atlanta, GA, USA, 17–22 November 1996; pp. 433–439
3. Guerin, K.R.; Lea, C.; Paxton, C.; Hager, G.D. A framework for end-user instruction of a robot assistant for manufacturing. In Proceedings of the 2015 IEEE International Conference on Robotics and Automation (ICRA), Seattle, WA, USA, 26–30 May 2015; pp. 6167–6174.
4. Peshkin, M.A.; Colgate, J.E.; Wannasuphoprasit, W.; Moore, C.A.; Gillespie, R.B.; Akella, P. Cobot architecture. *IEEE Trans. Robot. Autom.* **2001**, *17*, 377–390. [CrossRef]
5. DLR—Institute of Robotics and Mechatronics. History of the DLR LWR. Available online: https://www.dlr.de/rm/en/desktopdefault.aspx/tabid-12464/21732_read-44586/ (accessed on 30 November 2019).
6. Universal Robots. Low Cost and Easy Programming Made the UR5 a Winner. Available online: https://www.universal-robots.com/case-stories/linatex/ (accessed on 30 November 2019).
7. Müller, R.; Vette, M.; Geenen, A. Skill-based dynamic task allocation in Human-Robot-Cooperation with the example of welding application. *Procedia Manuf.* **2017**, *11*, 13–21. [CrossRef]
8. Wang, L.; Gao, R.; Váncza, J.; Krüger, J.; Wang, X.V.; Makris, S.; Chryssolouris, G. Symbiotic human-robot collaborative assembly. *CIRP Ann.* **2019**, *68*, 701–726. [CrossRef]
9. Müller, R.; Vette, M.; Mailahn, O. Process-oriented task assignment for assembly processes with human-robot interaction. *Procedia CIRP* **2016**, *44*, 210–215. [CrossRef]
10. Wang, X.V.; Kemény, Z.; Váncza, J.; Wang, L. Human–robot collaborative assembly in cyber-physical production: Classification framework and implementation. *CIRP Ann.* **2017**, *66*, 5–8. [CrossRef]
11. Krüger, J.; Lien, T.K.; Verl, A. Cooperation of human and machines in assembly lines. *CIRP Ann.* **2009**, *58*, 628–646.
12. Gaskill, S.; Went, S. Safety issues in modern applications of robots. *Reliab. Eng. Syst. Saf.* **1996**, *53*, 301–307. [CrossRef]
13. Michalos, G.; Makris, S.; Tsarouchi, P.; Guasch, T.; Kontovrakis, D.; Chryssolouris, G. Design considerations for safe human-robot collaborative workplaces. *Procedia CIrP* **2015**, *37*, 248–253. [CrossRef]
14. Gravel, D.P.; Newman, W.S. Flexible robotic assembly efforts at Ford Motor Company. In Proceeding of the 2001 IEEE International Symposium on Intelligent Control (ISIC'01) (Cat. No. 01CH37206), Mexico City, Mexico, 5–7 September 2001; pp. 173–182.
15. Zhu, Z.; Hu, H. Robot learning from demonstration in robotic assembly: A survey. *Robotics* **2018**, *7*, 17.
16. Fechter, M.; Foith-Förster, P.; Pfeiffer, M.S.; Bauernhansl, T. Axiomatic design approach for human-robot collaboration in flexibly linked assembly layouts. *Procedia CIRP* **2016**, *50*, 629–634. [CrossRef]
17. Faccio, M.; Bottin, M.; Rosati, G. Collaborative and traditional robotic assembly: A comparison model. *Int. J. Adv. Manuf. Technol.* **2019**, *102*, 1355–1372. [CrossRef]
18. Edmondson, N.; Redford, A. Generic flexible assembly system design. *Assem. Autom.* **2002**, *22*, 139–152. [CrossRef]
19. Battini, D.; Faccio, M.; Persona, A.; Sgarbossa, F. New methodological framework to improve productivity and ergonomics in assembly system design. *Int. J. Ind. Ergon.* **2011**, *41*, 30–42. [CrossRef]
20. Sawodny, O.; Aschemann, H.; Lahres, S. An automated gantry crane as a large workspace robot. *Control Eng. Pract.* **2002**, *10*, 1323–1338. [CrossRef]
21. Krüger, J.; Bernhardt, R.; Surdilovic, D.; Spur, G. Intelligent assist systems for flexible assembly. *CIRP Ann.* **2006**, *55*, 29–32. [CrossRef]
22. Rosati, G.; Faccio, M.; Carli, A.; Rossi, A. Fully flexible assembly systems (F-FAS): A new concept in flexible automation. *Assem. Autom.* **2013**, *33*, 8–21. [CrossRef]
23. FANUC Italia, S.r.l. M-2000—The Strongest Heavy Duty Industrial Robot in the Marker. Available online: https://www.fanuc.eu/it/en/robots/robot-filter-page/m-2000-series (accessed on 30 November 2019).
24. Hägele, M.; Schaaf, W.; Helms, E. Robot assistants at manual workplaces: Effective co-operation and safety aspects. In Proceedings of the 33rd ISR (International Symposium on Robotics), Stockholm, Sweden 7–11 October 2002; Volume 7.

25. Ding, H.; Heyn, J.; Matthias, B.; Staab, H. Structured collaborative behavior of industrial robots in mixed human-robot environments. In Proceedings of the 2013 IEEE International Conference on Automation Science and Engineering (CASE), Madison, WI, USA, 17–20 August 2013; pp. 1101–1106.
26. Ding, H.; Schipper, M.; Matthias, B. Collaborative behavior design of industrial robots for multiple human-robot collaboration. In Proceedings of the IEEE ISR 2013, Seoul, Korea, 24–26 October 2013; pp. 1–6.
27. Zanchettin, A.M.; Rocco, P. Path-consistent safety in mixed human-robot collaborative manufacturing environments. In Proceedings of the 2013 IEEE/RSJ International Conference on Intelligent Robots and Systems, Tokyo, Japan, 3–7 November 2013; pp. 1131–1136.
28. Whitsell, B.; Artemiadis, P. Physical human–robot interaction (pHRI) in 6 DOF with asymmetric cooperation. *IEEE Access* **2017**, *5*, 10834–10845. [CrossRef]
29. Rahman, S.M.; Wang, Y.; Walker, I.D.; Mears, L.; Pak, R.; Remy, S. Trust-based compliant robot-human handovers of payloads in collaborative assembly in flexible manufacturing. In Proceedings of the 2016 IEEE International Conference on Automation Science and Engineering (CASE), Fort Worth, TX, USA, 21–25 August 2016; pp. 355–360.
30. Safeea, M.; Bearee, R.; Neto, P. End-effector precise hand-guiding for collaborative robots. In *Iberian Robotics Conference*; Springer: Berlin, Germany, 2017; pp. 595–605.
31. Mendes, N.; Safeea, M.; Neto, P. Flexible programming and orchestration of collaborative robotic manufacturing systems. In Proceedings of the 2018 IEEE 16th International Conference on Industrial Informatics (INDIN), Porto, Portugal, 18–20 July 2018; pp. 913–918.
32. Hawkins, K.P.; Bansal, S.; Vo, N.N.; Bobick, A.F. Anticipating human actions for collaboration in the presence of task and sensor uncertainty. In Proceedings of the 2014 IEEE International Conference on Robotics and Automation (ICRA), Hong Kong, China, 31 May–7 June 2014; pp. 2215–2222.
33. Wojtynek, M.; Oestreich, H.; Beyer, O.; Wrede, S. Collaborative and robot-based plug & produce for rapid reconfiguration of modular production systems. In Proceedings of the 2017 IEEE/SICE International Symposium on System Integration (SII), Taipei, Taiwan, 11–14 December 2017; pp. 1067–1073.
34. Pang, G.; Deng, J.; Wang, F.; Zhang, J.; Pang, Z.; Yang, G. Development of flexible robot skin for safe and natural human–robot collaboration. *Micromachines* **2018**, *9*, 576. [CrossRef]
35. Heydaryan, S.; Suaza Bedolla, J.; Belingardi, G. Safety design and development of a human robot collaboration assembly process in the automotive industry. *Appl. Sci.* **2018**, *8*, 344. [CrossRef]
36. Michalos, G.; Kousi, N.; Karagiannis, P.; Gkournelos, C.; Dimoulas, K.; Koukas, S.; Mparis, K.; Papavasileiou, A.; Makris, S. Seamless human robot collaborative assembly—An automotive case study. *Mechatronics* **2018**, *55*, 194–211. [CrossRef]
37. Tan, J.T.C.; Zhang, Y.; Duan, F.; Watanabe, K.; Kato, R.; Arai, T. Human factors studies in information support development for human-robot collaborative cellular manufacturing system. In Proceedings of the RO-MAN 2009—The 18th IEEE International Symposium on Robot and Human Interactive Communication, Toyama, Japan, 27 September–2 October 2009; pp. 334–339.
38. Arai, T.; Duan, F.; Kato, R.; Tan, J.T.C.; Fujita, M.; Morioka, M.; Sakakibara, S. A new cell production assembly system with twin manipulators on mobile base. In Proceedings of the 2009 IEEE International Symposium on Assembly and Manufacturing, Suwon, Korea, 17–20 November 2009; pp. 149–154.
39. Tan, J.T.C.; Duan, F.; Zhang, Y.; Watanabe, K.; Kato, R.; Arai, T. Human-robot collaboration in cellular manufacturing: Design and development. In Proceedings of the 2009 IEEE/RSJ International Conference on Intelligent Robots and Systems, Saint Louis, MO, USA, 10–15 October 2009; pp. 29–34.
40. Lenz, C.; Rickert, M.; Panin, G.; Knoll, A. Constraint task-based control in industrial settings. In Proceedings of the 2009 IEEE/RSJ International Conference on Intelligent Robots and Systems, Saint Louis, MO, USA, 10–15 October 2009; pp. 3058–3063.
41. Ende, T.; Haddadin, S.; Parusel, S.; Wüsthoff, T.; Hassenzahl, M.; Albu-Schäffer, A. A human-centered approach to robot gesture based communication within collaborative working processes. In Proceedings of the 2011 IEEE/RSJ International Conference on Intelligent Robots and Systems, San Francisco, CA, USA, 25–30 September 2011; pp. 3367–3374.
42. Darvish, K.; Bruno, B.; Simetti, E.; Mastrogiovanni, F.; Casalino, G. Interleaved Online Task Planning, Simulation, Task Allocation and Motion Control for Flexible Human-Robot Cooperation. In Proceedings of the 2018 27th IEEE International Symposium on Robot and Human Interactive Communication (RO-MAN), Nanjing, China, 27–31 August 2018; pp. 58–65.

43. Zanchettin, A.; Casalino, A.; Piroddi, L.; Rocco, P. Prediction of human activity patterns for human-robot collaborative assembly tasks. *IEEE Trans. Ind. Inf.* **2018**, *15*, 3934–3942. [CrossRef]
44. Blaga, A.; Tamas, L. Augmented Reality for Digital Manufacturing. In Proceedings of the 2018 26th Mediterranean Conference on Control and Automation (MED), Akko, Israel, 1–4 July 2018; pp. 173–178.
45. Wang, W.; Li, R.; Diekel, Z.M.; Chen, Y.; Zhang, Z.; Jia, Y. Controlling Object Hand-Over in Human–Robot Collaboration Via Natural Wearable Sensing. *IEEE Trans. Human-Mach. Syst.* **2018**, *49*, 59–71. [CrossRef]
46. Walker, I.D.; Mears, L.; Mizanoor, R.S.; Pak, R.; Remy, S.; Wang, Y. Robot-human handovers based on trust. In Proceedings of the 2015 IEEE Second International Conference on Mathematics and Computers in Sciences and in Industry (MCSI), Sliema, Malta, 17 August 2015; pp. 119–124.
47. Sadrfaridpour, B.; Wang, Y. Collaborative assembly in hybrid manufacturing cells: An integrated framework for human–robot interaction. *IEEE Trans. Autom. Sci. Eng.* **2017**, *15*, 1178–1192. [CrossRef]
48. Rozo, L.; Calinon, S.; Caldwell, D.G.; Jimenez, P.; Torras, C. Learning physical collaborative robot behaviors from human demonstrations. *IEEE Trans. Robot.* **2016**, *32*, 513–527. [CrossRef]
49. Rahman, S.M.; Liao, Z.; Jiang, L.; Wang, Y. A regret-based autonomy allocation scheme for human-robot shared vision systems in collaborative assembly in manufacturing. In Proceedings of the 2016 IEEE International Conference on Automation Science and Engineering (CASE), Fort Worth, TX, USA, 21–25 August 2016; pp. 897–902.
50. Gustavsson, P.; Syberfeldt, A.; Brewster, R.; Wang, L. Human-robot collaboration demonstrator combining speech recognition and haptic control. *Procedia CIRP* **2017**, *63*, 396–401. [CrossRef]
51. Fakhruldeen, H.; Maheshwari, P.; Lenz, A.; Dailami, F.; Pipe, A.G. Human robot cooperation planner using plans embedded in objects. *IFAC-PapersOnLine* **2016**, *49*, 668–674. [CrossRef]
52. Koch, P.J.; van Amstel, M.K.; Dębska, P.; Thormann, M.A.; Tetzlaff, A.J.; Bøgh, S.; Chrysostomou, D. A skill-based robot co-worker for industrial maintenance tasks. *Procedia Manuf.* **2017**, *11*, 83–90. [CrossRef]
53. Unhelkar, V.V.; Lasota, P.A.; Tyroller, Q.; Buhai, R.D.; Marceau, L.; Deml, B.; Shah, J.A. Human-aware robotic assistant for collaborative assembly: Integrating human motion prediction with planning in time. *IEEE Robot. Autom. Lett.* **2018**, *3*, 2394–2401. [CrossRef]
54. Hamabe, T.; Goto, H.; Miura, J. A programming by demonstration system for human-robot collaborative assembly tasks. In Proceedings of the 2015 IEEE International Conference on Robotics and Biomimetics (ROBIO), Zhuhai, China, 6–9 December 2015; pp. 1195–1201.
55. Tlach, V.; Kuric, I.; Zajačko, I.; Kumičáková, D.; Rengevič, A. The design of method intended for implementation of collaborative assembly tasks. *Adv. Sci. Technol. Res. J.* **2018**, *12*, 244–250. [CrossRef]
56. El Makrini, I.; Merckaert, K.; Lefeber, D.; Vanderborght, B. Design of a collaborative architecture for human-robot assembly tasks. In Proceedings of the 2017 IEEE/RSJ International Conference on Intelligent Robots and Systems (IROS), Vancouver, BC, Canada, 24–28 September 2017; pp. 1624–1629.
57. MarketsandMarkets™ Research Private Ltd. Collaborative Robots Market by Payload Capacity (Up to 5 kg, Up to 10 kg, Above 10 kg), Industry (Automotive, Electronics, Metals & Machining, Plastics & Polymer, Food & Agriculture, Healthcare), Application, and Geography—Global Forecast to 2023. Available online: https://www.marketsandmarkets.com/Market-Reports/collaborative-robot-market-194541294.html (accessed on 30 November 2019).
58. International Federation of Robotics (IFR). Robots and the Workplace of the Future. 2018. Available online: https://ifr.org/papers (accessed on 30 November 2019).
59. Barclays Investment Bank. Technology's Mixed Blessing. 2017. Available online: https://www.investmentbank.barclays.com/our-insights/technologys-mixed-blessing.html (accessed on 30 November 2019).
60. Tobe, F. Why Co-Bots Will Be a Huge Innovation and Growth Driver for Robotics Industry. 2015. Available online: https://spectrum.ieee.org/automaton/robotics/industrial-robots/collaborative-robots-innovation-growth-driver (accessed on 30 November 2019).
61. Gopinath, V.; Ore, F.; Grahn, S.; Johansen, K. Safety-Focussed Design of Collaborative Assembly Station with Large Industrial Robots. *Procedia Manuf.* **2018**, *25*, 503–510. [CrossRef]
62. Zanchettin, A.M.; Ceriani, N.M.; Rocco, P.; Ding, H.; Matthias, B. Safety in human-robot collaborative manufacturing environments: Metrics and control. *IEEE Trans. Autom. Sci. Eng.* **2015**, *13*, 882–893. [CrossRef]
63. Cherubini, A.; Passama, R.; Crosnier, A.; Lasnier, A.; Fraisse, P. Collaborative manufacturing with physical human–robot interaction. *Robot. Comput. Integr. Manuf.* **2016**, *40*, 1–13. [CrossRef]

64. Makris, S.; Karagiannis, P.; Koukas, S.; Matthaiakis, A.S. Augmented reality system for operator support in human–robot collaborative assembly. *CIRP Ann.* **2016**, *65*, 61–64. [CrossRef]
65. Bös, J.; Wahrburg, A.; Listmann, K.D. Iteratively Learned and Temporally Scaled Force Control with application to robotic assembly in unstructured environments. In Proceedings of the 2017 IEEE International Conference on Robotics and Automation (ICRA), Singapore, 29 May–3 June 2017; pp. 3000–3007.
66. Haage, M.; Piperagkas, G.; Papadopoulos, C.; Mariolis, I.; Malec, J.; Bekiroglu, Y.; Hedelind, M.; Tzovaras, D. Teaching assembly by demonstration using advanced human robot interaction and a knowledge integration framework. *Procedia Manuf.* **2017**, *11*, 164–173. [CrossRef]

robotics

MDPI

Article

Trajectory Optimization of a Redundant Serial Robot Using Cartesian via Points and Kinematic Decoupling

Matteo Bottin * and Giulio Rosati

Department of Industrial Engineering, University of Padova, 35131 Padova, Italy; giulio.rosati@unipd.it
* Correspondence: matteo.bottin@unipd.it

Received: 1 October 2019; Accepted: 2 December 2019; Published: 9 December 2019

Abstract: Moving from a given position to another with an industrial robot can be a challenging problem when the task is redundant around the tool axis. In this case, there are infinite ways of choosing both the starting and the ending configurations, so that the movement between the given points is not uniquely defined. In this paper, an algorithm that calculates the suboptimal movement between two positions is proposed, which automatically generates a cloud of safe via points around the workpiece and then by exploiting such points finds the suboptimal safe path between the two positions that minimizes movement time. The proposed method, in which the search of the suboptimal path is based on graph theory and the Dijkstra algorithm, can iteratively evaluate a high number of starting and ending configurations in low computational time, allowing performing a reasonably wide search of the suboptimal path within the infinite possible motions between the given points.

Keywords: robot motion; redundancy; trajectory optimization; Dijkstra algorithm; graph

1. Introduction

To increase the productivity of an industrial robotic workcell, the cycle time has to be reduced. Sometimes, the time needed to perform the tasks is fixed, so it is necessary to reduce the robot movement times between the positions. This can be achieved by defining the best control strategy for the robot's motors and, if possible, by finding the optimal path that reduces movement time. However, the optimization of motor control is already implemented in the controllers of common industrial robots, so the only available option is to define an optimal path that the robot has to follow.

For example, in robotic deburring, the task time is fixed by the process, so it is not possible to reduce it. However, it is possible to perform the same task with different robot configurations by rotating around the spindle axis. Therefore, the movement time between working points can be modified by changing the spindle angles in such points, and as a result, the cycle time will vary. In such a case, we say the robot is functionally redundant, and this can lead to cycle times that can be very far from the optimal one if wrongly defined by an inexperienced human operator.

In addition, the movement between two working positions has to be completed without colliding with the workcell environment. The workpiece is not the only obstacle inside the workspace since there can be structures, equipment, conveyors, and barriers that must not be touched by the robot. To do so, in a generic movement, the robot can move through several via points between the starting and the ending points. For a given layout of the workcell and a given couple of starting and ending points, via points can be chosen in different ways, and their choice can widely affect cycle time.

The goal of our research is to define a novel path planning algorithm for functionally redundant robots, able to minimize locally the movement time between two given Cartesian working points. In the proposed approach, we generate automatically a set of Cartesian via points around the workpiece, then we seek for the minimum time motion path that: (a) is safe in the sense that no collision occurs between the robot and the surrounding objects along the path; (b) includes the given starting/ending points and a subset of the via points. The search is performed considering multiple tool angles both in the starting and in the ending points, to exploit robot functional redundancy.

The main novelty in our approach lies in the particular way we generate the set of via points, which allows simplifying and decoupling the path planning problem from a kinematic perspective. In fact, the motion path is generated with reference to the wrist center only, whereas wrist rotations at the chosen via points are left totally free. As a result, the rotations of the last three joints of the robot can be planned form the starting point to the ending point in the most convenient way, minimizing their rotations and movement time. In the paper, with reference to the case of a functionally redundant anthropomorphic robot performing a point to point motion, our method is compared to standard probabilistic roadmap path planning in terms of both robot movement time and of computational time.

2. Related Work

Trajectory planning is very important in reducing the cycle time of robotic workcells. Many algorithms have been proposed to reduce the overall cycle time [23]. Minimum kinematic parameters can be obtained by focusing on kinematics, dynamics, or minimum kinetic energy factors. However, this optimization is already implemented in the controllers of common industrial robots (even if it could be improved [9], reducing chattering effects). One of the parameters to be optimized is the path between the working points, avoiding the objects placed in the environment.

In the industrial workcells, the robotic workspace is cluttered with objects. To avoid collisions, the task can be simulated with off-line methods. Some of them were described in [7,20,32]. The objects within the workspace are usually described offline as patches. It is possible to detect the collision between multiple patches [24], but very wide patch collisions can be heavy to compute [30]. Since the geometry of the objects placed in the environment can be very complex, it is usually more convenient to encapsulate them inside bounding volumes [8]. These volumes can be simple spheres or cubes, Oriented Bounding Boxes (OBBs) [11], Sphere Swept Volumes (SSVs) [16], and many more. Usually, the tighter the volume is to the object, the slower is the collision detection algorithm [16]. This is due to the complexity (and the number) of the elements that define the volumes: to define tighter volumes, more components are required (e.g., a tighter Discretely Oriented Polytope (k-DOP) requires several bounding planes [15]). Rodriguez-Garavito et al. [26] used bounding boxes to encapsulate the robot and the objects in the environment. Then, the collision detection algorithm had to evaluate the intersection of the bounding rectangles to find out if a collision occurred.

It is important to notice that the slower is the collision detection algorithm, the lower the amount of function evaluations can be performed in real time. This becomes particularly crucial with the algorithms, called sampling-based methods [7], in which random samples are chosen within the configuration space as via points, and each configuration is then tested to see if the robot collides with the environment. Among these methods, the most known frameworks are Probabilistic Roadmaps (PRM) and Rapidly exploring Random Trees (RRTs) [17], but many more algorithms have been proposed [32]. These algorithms have been used both with kinematic and kinodynamic approaches [18] and can be used with robots with a high number of joints [13]. Usually, the Dijkstra algorithm [5] is used to find the optimal path.

Another possibility for defining a safe path is using collision avoidance algorithms: it is possible to move the robot away from a collision point [3] and iteratively create a safe path. A similar approach was shown for real-time planning in [12,19]. An interesting approach was proposed by Khatib in [14], where a collision avoidance method based on potential fields was presented: robots get repelled by the space zones in which the obstacles are placed, acting like repulsive forces. However, sometimes, the robot can get stuck by particular obstacle shapes [25].

In some particular applications (such as deburring, welding, screwing, or painting), the industrial robot can become functionally redundant. In such scenarios, the number of parameters needed to define the robot configuration is higher than the number of degrees of freedom of the task. Much research activity has been carried out on the topic. Sciavicco and Siciliano [27] studied the redundancy of a manipulator, considering collision avoidance and limited joint range. In particular, the inverse kinematic problem for constrained manipulators was investigated, using optimization techniques that considered all the aspects listed before. Panames-Garcia et al. [21] proposed a general formulation for the optimization of the path placement of redundant manipulators considering single or multiple objective optimizations. Redundancy and path planning can be merged also with heuristic searches, such as by using genetic algorithms [1,28,31] and neural networks [4]. However, while heuristics can be very valuable with a high number of degrees of freedom, they need to evaluate an objective function many times to obtain a suboptimal solution [31]. Doan and Lin [6] were able to optimize the task of a redundant robot while finding the best position of the robot base in relation to the task itself. This is a very interesting approach to task optimization: the design of the workcell is taken into account while improving the final robot behavior.

None of these works on redundant robots merge the collision detection algorithm with a suitable definition of the via points in the Cartesian space, so as to decouple the path planning problem from a kinematic point of view. With our work, we aim to assess if such an approach can provide benefits in terms of shorter movement times and/or of lower computational cost.

3. Functional Redundancy

A welding process can be performed with every possible orientation of the welding tool around the normal axis at the working point. The same feature can be found in other manufacturing processes, such as deburring, where the same task can be accomplished by rotating the spindle around its axis. In such examples, the welding axis and the spindle axis can be considered as an additional axis to the robot kinematic chain. This results in an $N+1$ degrees of freedom (DOF) manipulation, where N is the number of DOF of the robot. For example, a six axis robot like the one considered in the paper, equipped with a one DOF tool, equals a seven axis robot with a total of seven DOF. Therefore, this work applies also to seven axis robots, such as ABB YuMi and KUKA LBR iiwa, performing normal (non-redundant) tasks..

In this scenario, the configurations of the robot at the beginning and at the end of motion are not uniquely defined, so a proper choice of them can reduce the overall cycle time. In fact, since joint speed limits are upper bounded, a movement between very different configurations can be limited by the performances of the motors.

In this work, the redundant joint angle will be called θ_7. The redundant axis is normal to the surface of the workpiece and is off-axis with respect to the robot's sixth joint axis. The method, however, is general and can be applied to other scenarios.

4. Proposed Method

Let A and B be two given positions on a workpiece. Such positions are given in the Cartesian space and define the position and orientation of the end effector at the working point. To avoid collisions between the robot and the workpiece or environment, moving from A to B usually requires passing through several via points. Our idea is to create a grid of safe via points in the Cartesian space that the robot can use to move between the given positions, defining a suboptimal safe path between A and B. The final objective is to minimize the movement time between A and B. In our method, the via points describe the position of the wrist center; therefore, their location influences the first three joints of the robot only. In this way, the path planning problem is decoupled from a kinematic point of view, in the sense that the motion of the first three joints is calculated and optimized regardless of the wrist initial and final rotations. As a result:

- The definition of the via points around the workpiece is simplified; in fact, they are 3D points (three variables) instead of robot configurations (N variables);

- The rotations of Joints 4, 5, and 6 are minimized, since such joints are simply moved from the initial to the final values; in fact, they are not involved in the definition of the safe path surrounding the workpiece, so their rotations at each via point can be chosen in the most convenient way;
- The rotations of Joints 1, 2, and 3 may not be minimized, since the via points are at a non-negligible distance from the workpiece.

The main steps of the method (Figure 1), explained for a six DOF manipulator performing a redundant task around the tool axis (with redundant joint angle θ_7), are:

1. Place the devices in the workcell (the robot, the workpiece, and auxiliary systems) in the desired positions (or by finding the optimal positions [6]);
2. Define a safe volume around the workpiece (the Swept Sphere Volume (SSV) shown in green in Figure 2);
3. Create a cloud of via points for the wrist center around the safe volume (Section 4.1), whose distance from the safe volume is equal to the distance between the tool tip and wrist center; such points are safe in the sense that, if the wrist center lies in one of the points, the tool cannot collide with the workpiece for any Joint 4, 5, and 6 values;
4. Check if the via points are reachable by the robot from a kinematic point of view: if a collision occurs between the robot and the environment at some via points, they are removed from the cloud;
5. Connect the via points to the eight closest ones (in the Cartesian space) to form the branches that will be used in the search of the minimum time path;
6. Check the connections between the via points: if a collision occurs along the point to point motion between two connected via points, the connection is removed;
7. Choose the initial θ_7 values for starting and ending positions A and B ($\theta_{7,A}$ and $\theta_{7,B}$);
8. Find the suboptimal path that connects A and B without a collision using the Dijkstra algorithm to solve the graph whose nodes are the connected via points;
9. Change $\theta_{7,A}$ and $\theta_{7,B}$ within a fixed grid around the initial values and loop from Point 8 until all possible combinations are evaluated;
10. Find the combination $\theta_{7,A,opt}$ and $\theta_{7,B,opt}$ that yields the minimum time motion between A and B on the chosen grid.

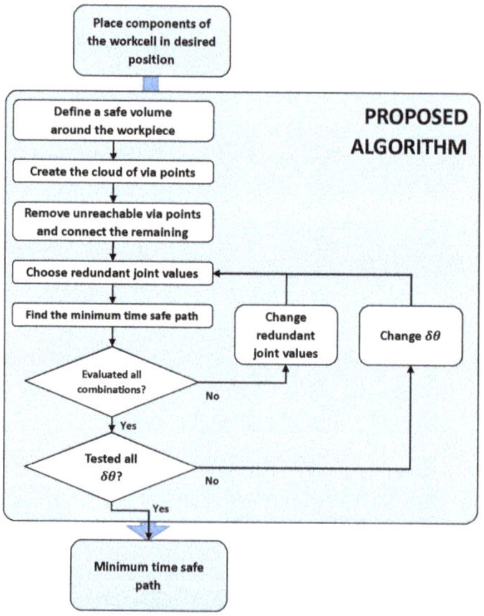

Figure 1. An overview of the proposed algorithm. At first, the operator has to place all the objects in the workspace. Then, the algorithm provides automatically the path between the working positions.

The points from 8 to 10 can be iterated by refining the grid around the final point, until a certain condition is met (e.g., a maximum number of iterations). At the end of the iterations, the final values of redundant joint angles ($\theta_{7,A,opt}$ and $\theta_{7,B,opt}$) are stored, and the corresponding graph is considered as the optimal path that connects A and B.

4.1. Via Points' Generation and Selection

A line Swept Sphere Volume (SSV) [16] is used to encapsulate the workpiece. The shape and orientation of the SSV are chosen in a way that minimizes its volume.

Since the configuration of the tool is unknown a priori, to be sure that the tool does not collide with the workpiece, we chose to place the wrist of the robot at a distance greater than R_{tool} from the SSV, where R_{tool} is the minimum radius of a sphere centered in the wrist center and containing the tool. By using this criterion, a net of wrist center points equally distributed is placed on a surface at a distance of R_{tool} from the SSV (Figure 2).

To create the point could, a section of the SSV, containing the SSV symmetry axis, is considered (Figure 2 on the left). On this plane, m_1 points are evenly placed at a distance R_{tool} from the border of the SSV. Then, the section is cloned m_2 times around the SSV symmetry axis. The first and last points on each section that lie on the symmetry axis are considered once, so the total number of points is:

$$N_p = m_2(m_1 - 2) + 2 \tag{1}$$

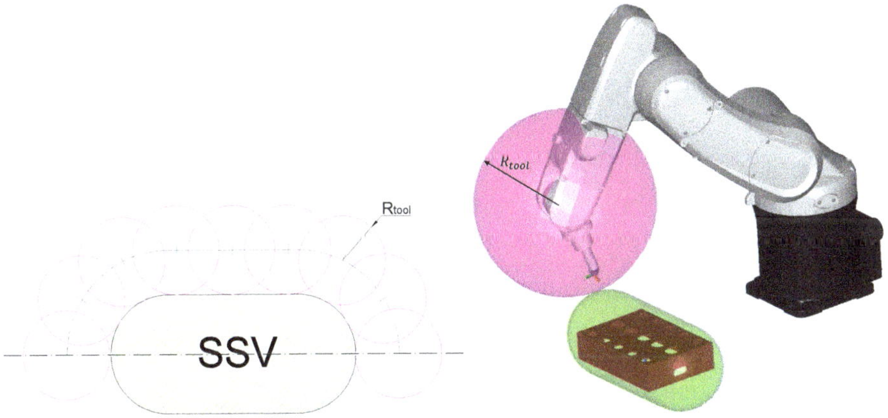

Figure 2. A simple scheme that shows the placement of the via points on a section of the Swept Sphere Volume (SSV) (to the left). Each section is then repeated many times around the symmetry axis of the SSV. All the via points are placed at a distance R_{tool} from the surface of the SSV, so regardless the orientation of the wrist, the end effector will not collide with the work piece (to the right).

The density of the net is a design choice (m_1 and m_2) and has a direct effect on the performance of the Dijkstra algorithm [10]. Each point of the net is connected to its eight neighbors as shown in Figure 3, similarly to the uniform space sampling method shown in [29].

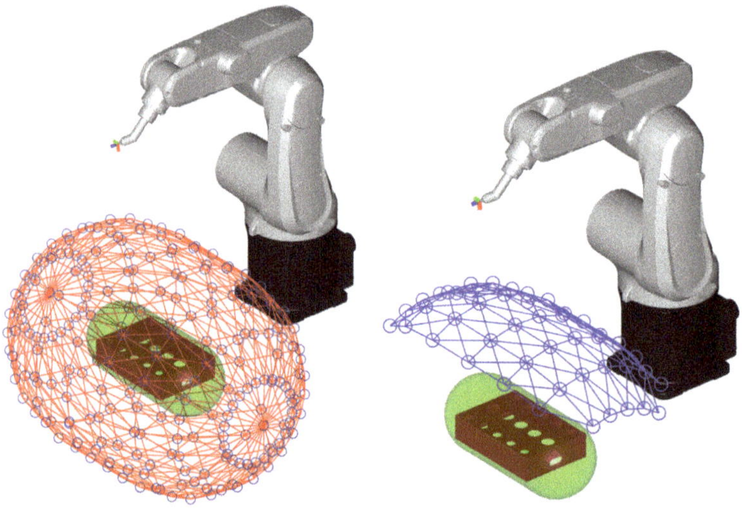

Figure 3. All the adjacent via points are connected to form the branches (to the left), and then, the unreachable via points and the branches that provide a collision are removed (to the right).

The first benefit provided by the proposed definition of via points is that, when the wrist center is in such points, the robot is in a safe position regardless of the orientation of the tool. In this way, if we include one of the via points in a motion between the starting and the ending points, such a via point will be safe for any configuration of the redundant axis in the given points. Thus, by creating a path that includes the given points and a subset of the calculated via points, such a path will be safe in all intermediate points for any choice of the initial and final configurations.

Secondly, since the via points are wrist center points, the inverse kinematic problem can be solved for the first three joints only (which is faster than solving the full problem), and an estimation of the motion time between connected points can be done by considering only such joints (Section 4.2, Equation (2)), thus drastically reducing the computational time. In fact, the estimated motion time on a given path depends only on the points chosen, regardless of the initial, pass-through, and final orientations of the tool. In this way, the time needed to move between different initial and final configurations of the robot can be evaluated without the need for re-calculating the inverse kinematics in the via points every time. To test the reduction of the computational time due to the decoupling of the inverse kinematic problem, a simple test was performed: the same position was used to calculate the entire configuration of the robot and only the first three joint values 100,000 times. The overall computational time for the whole inverse kinematic problem was 1.66 s, whilst the computational time required for the calculation of the first three joints was 1.06 s. As a result, the decoupling led to a reduction of the computational time of the inverse kinematic problem of around 36%.

Via points are checked for collision between the robot and the workpiece and the environment. If a collision is detected, the corresponding via point is deleted from the list. The collision test was performed as described in our previous work [3]: all the links were encapsulated inside SSVs, whilst all other objects were encapsulated inside Oriented Bounding Boxes (OBB [11]). In this way, it was possible to compute the interaction between the robot and the environment faster than with the usage of SSVs only [8].

4.2. Graph

The suboptimal path between two positions A and B is created using the Dijkstra algorithm [5]. To solve the algorithm, the branches and the nodes must be defined.

While the via points represent the nodes, the branches are represented by the connections between the via points. Considering the net of reachable via points, only adjacent via points (in Cartesian space) are connected to form branches (Figure 3 on the left), so that the total number of branches is reduced. To avoid unnecessary calculations, collision tests along the motions through all the branches are computed. If a collision is detected, the corresponding branch is removed (Figure 3 on the right).

Branches' weight equal the robot movement time between the nodes. Movement time between nodes h and k is estimated considering only the first three joints:

$$T_{hk} = \max_{j=1,2,3} \left\{ \frac{|\Delta q_{j,hk}|}{\dot{q}_{max,j}} c_v \right\} \tag{2}$$

where $\Delta q_{j,hk}$ is the total rotation of joint j between the nodes h and k, $\dot{q}_{max,j}$ is the maximum speed of joint j, and c_v is the velocity coefficient of the motion law [2].

Starting and ending nodes are defined by points A and B; A and B are connected to all the via points to form additional branches. In this way, it is possible to enter (from A) and exit (towards B) the cloud of via points from any one of the via points. The weights of the additional branches are calculated as the traveling times (as Equation (2)), and the connections are checked to find collisions: the branches that provide collision are removed. The Dijkstra algorithm calculates the graph using A as the first node and B as the last one.

The first $\theta_{7,A}$ and $\theta_{7,B}$ values are null to start the optimization algorithm around the position provided by the user. To find the optimal solution, the redundant joint angles θ_7 at A and B are changed by fixed steps ($\delta\theta$), from minimum values ($\theta_{7,i,min}$) to maximum ones ($\theta_{7,i,max}$). In this way, a fixed grid of possible combinations is created; the limits of the grid can be different for A and B (see Table 1 for an example).

To ensure that the calculated paths are feasible, a collision test along each suboptimal path is computed. To do so, the planning of robot motion is performed along the whole path, including wrist rotations. The angles of the first three joints are interpolated from A values to B values considering also the via points included in the path. The angles of the last three joints, on the other hand, are interpolated from A values to B values only, since their values are not assigned at the via points. If a collision is found, the graph is re-calculated. This may occur due to a possible colliding configuration of the robot provided by the values of Joints 4, 5, and 6 along one of the connections; in fact, wrist rotations have not been considered in the previous collision tests, and whilst the via points are safe regardless of wrist orientation, connecting paths (especially from A to the cloud and from the cloud to B) may not.

The final movement time of the path is given by the following expression:

$$T_{final}(\theta_{7,A}, \theta_{7,B}) = \max \left\{ T_{Dijkstra}(\theta_{7,A}, \theta_{7,B}), \max_{j=4,5,6} \left\{ \frac{|\Delta q_{j,AB}(\theta_{7,A}, \theta_{7,B})|}{\dot{q}_{max,j}} c_v \right\} \right\} \tag{3}$$

where $T_{Dijkstra}(\theta_{7,A}, \theta_{7,B})$ is the minimum time yielded by the Dijkstra algorithm and $\Delta q_{j,AB}(\theta_{7,A}, \theta_{7,B})$ is the total rotation of joint j between points A and B.

At the end of an iteration step, all the possible combinations of $\theta_{7,A}$ and $\theta_{7,B}$ are compared. The best combination of those two values, i.e., the one that minimizes T_{final}, provides $\theta_{7,A,opt}$ and $\theta_{7,B,opt}$ that are to be used in the next iteration step. To start the next iteration step, the value of $\delta\theta$ is reduced and the next boundaries of $\theta_{7,A}$ and $\theta_{7,B}$ are changed. The lower $\delta\theta$, the better the precision of the optimal position. In the first iteration, $\theta_{7,A,min} = \theta_{7,B,min} = -180°$ and $\theta_{7,A,max} = \theta_{7,B,max} = 180°$ to evaluate the entire workspace (Table 2).

At the end of the procedure, the optimal couple of redundant joint angles $\theta_{7,A,opt}$ and $\theta_{7,B,opt}$ is provided.

Table 1. An example of iterations changing $\delta\theta$ from 60° to 1°. Such values were used to run the algorithm in Test 1.

Step	$\delta\theta$ (°)	A Bounds (°) $(\theta_{7,A,min}, \theta_{7,A,max})$	B Bounds (°) $(\theta_{7,B,min}, \theta_{7,A,max})$	$\theta_{7,A,opt}$ (°)	$\theta_{7,B,opt}$ (°)
1	60	[−180, 180]	[−180, 180]	0	60
2	45	[−135, 135]	[−75, 195]	0	60
3	30	[−120, 120]	[−60, 180]	0	30
4	20	[−80, 80]	[−50, 110]	0	10
5	10	[−40, 40]	[−30, 50]	−10	20
6	5	[−30, 10]	[0, 40]	−15	20
7	2	[−23, −7]	[12, 28]	−15	22
8	1	[−19, −11]	[18, 26]	−14	23

Table 2. Values of $\theta_{7,A}$ and $\theta_{7,B}$ when the first iteration step has $\delta\theta = 60°$.

$\delta\theta$ (°)	60
$\theta_{7,A}$ values (°)	[−180, −120, −60, 0, 60, 120, 180]
$\theta_{7,B}$ values (°)	[−180, −120, −60, 0, 60, 120, 180]
Number of combinations	49

The optimal solutions are not to be intended as the globally optimal solutions to the problem. Since this is a nonlinear stochastic problem that has to face many constraints (collision avoidance, movement between the via points, discretization of the values of the redundant joint angle), the purpose of the algorithm is to find a local minima of the problem, which, however, can be good enough to satisfy the majority of the industrial scenarios.

5. Validation

A MATLAB® script was created through which an ADEPT VIPER s650 six axis robot was simulated (Figure 2) moving from one working position to another. The PC used for the simulation was powered by an Intel Core i7-2700K, 16 GB of RAM, and 64 bit Windows 10 Pro 1803 version.

The robot was moved from one side of the workpiece to another so that direct movement was impossible (the robot would collide with the workpiece). The collision check along the movement was performed at a frequency of 180 Hz.

The starting position was P_A (490, 167.5, 18.4) with the orientation defined by Cardan angles $\{x, y, z\} = \{0, 90, -90\}$. The ending position was P_B (410, −67.5, 48.4) with the orientation defined by Cardan angles $\{x, y, z\} = \{0, 90, 90\}$. The robot base frame was coincident with the global reference frame. The parameters of the planes defining the workpiece's OBB, used by the collision detection algorithm, are shown in Table 3. Moreover, another plane was added to simulate the table where the workpiece was placed. The parameters describing this plane are in the seventh column of Table 3.

Table 3. Parameters describing the 6 planes that define workpiece Oriented Bounding Box (OBB) (Columns 1–6) and the extra plane describing the table where the workpiece is placed (Column 7). The parameters refer to the point normal form of the equation of a plane: $ax + by + cz + \lambda = 0$.

Plane n°	1	2	3	4	5	6	7
a	1	−1	0	0	0	0	0
b	0	0	1	−1	0	0	0
c	0	0	0	0	1	−1	−1
λ (mm)	−368	532	67	167	1.6	55.4	1.6

The total net of via points was made of 227 points ($m_1 = 11$, $m_2 = 25$), while the number of feasible via points was 56, and the number of via points connected to the graph was 46 (Figure 3 to the right).

In this test, we set the range limits of θ_7 angles to ±3 times $\delta\theta$ (first two steps) and ±4 times $\delta\theta$ (for other steps, see Table 1). This ensured a good trade-off between algorithm performance and computational time, since for each iteration, it limited the number of computations of the graph.

At first, the lowest movement time at each iteration step was evaluated. In Figure 4 are shown the results: the lowest movement time decreased from the highest $\delta\theta$ (60°) to the lowest one (1°). In this case, the reduction of the movement time from the higher value of $\delta\theta$ to the lowest one was nearly 23% (Table 4). The computer took 24.88 s to find out $\theta_{7,A,opt}$ and $\theta_{7,B,opt}$ with eight steps. The best combinations at each step can be seen in Table 1.

Table 4. Results of the test changing $\delta\theta$ from 60° to 1°.

Step	$\delta\theta$	Lowest Movement Time (s)	Comparison to Step 1 (%)	Total Computational Time (s)
1	60	0.6035	100	1.77
2	45	0.6035	100	3.47
3	30	0.6016	99.7	6.54
4	20	0.5672	94.0	10.42
5	10	0.4893	81.1	14.35
6	5	0.4698	77.8	17.89
7	2	0.4689	77.7	21.37
8	1	0.4655	77.1	24.88

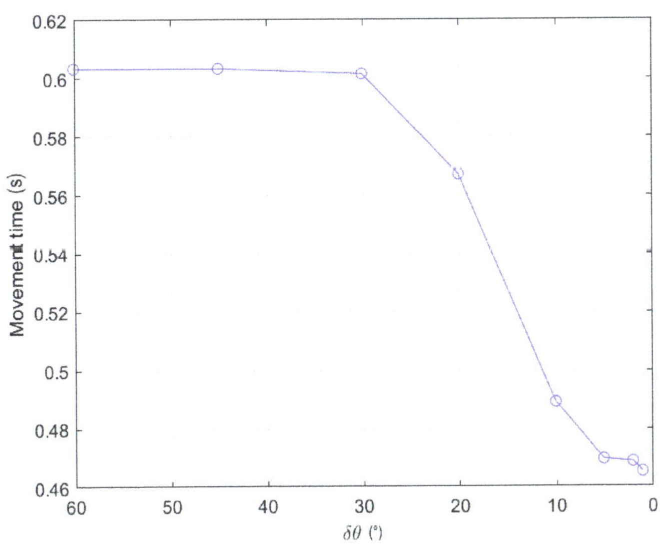

Figure 4. Minimum of moving total time for each step value. The x-axis direction has been reversed to show the iteration process from left to right.

Figure 5 shows the calculation times and the number of possible combinations of each $\delta\theta$ step. The calculation times nearly followed the trend of the number of combinations and was capped to the maximum of 81 starting from the third step. Then, at small $\delta\theta$ values, the calculation times slightly dropped due to a smaller amount of colliding paths: small variations of $\theta_{7,A}$ and $\theta_{7,B}$ around the optimal solution usually did not lead to many path collisions.

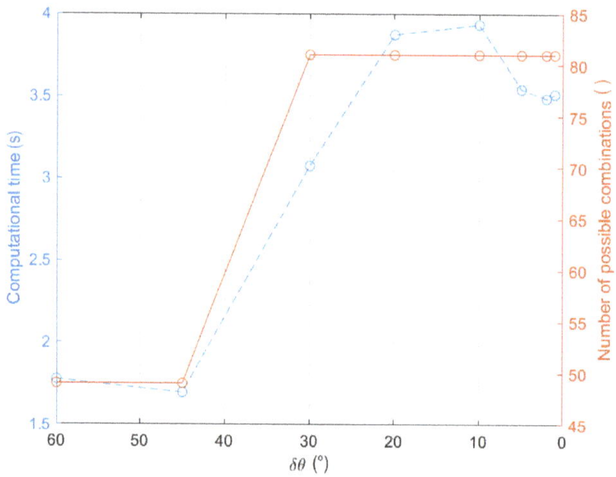

Figure 5. Computational time of the definition of the graph for each step value. The computational time nearly follows the number of possible combinations. The x-axis direction has been reversed to show the iteration process from left to right.

Even if a single calculation of 24.88 s reduced the movement time of nearly 23%, a similar result can be obtained using a reduced number of steps. We performed another test with the same starting and ending positions, but with only two iteration steps: $\delta\theta_1 = 30^\circ$ and $\delta\theta_2 = 5^\circ$. The results (Table 5) showed that with reduced computational times (only 11.41 s), the lowest movement time was nearly the same as in the previous test.

In Figure 6, the final movement provided by the algorithm with the parameters of Table 5 is shown. The shape of the path depended on the density of the via points defined in the workspace.

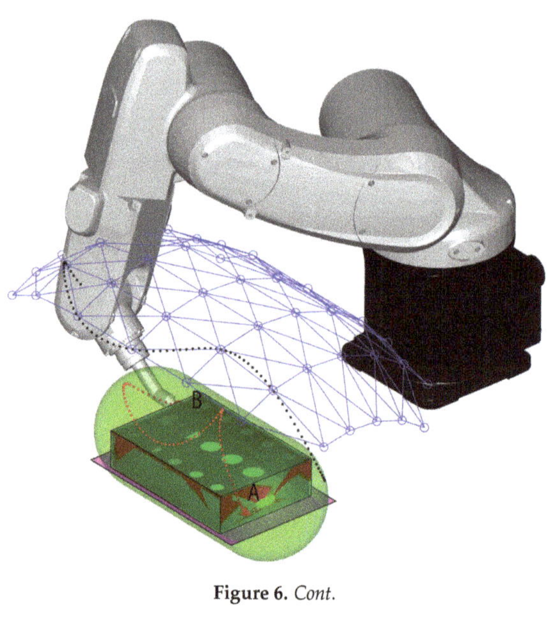

Figure 6. *Cont.*

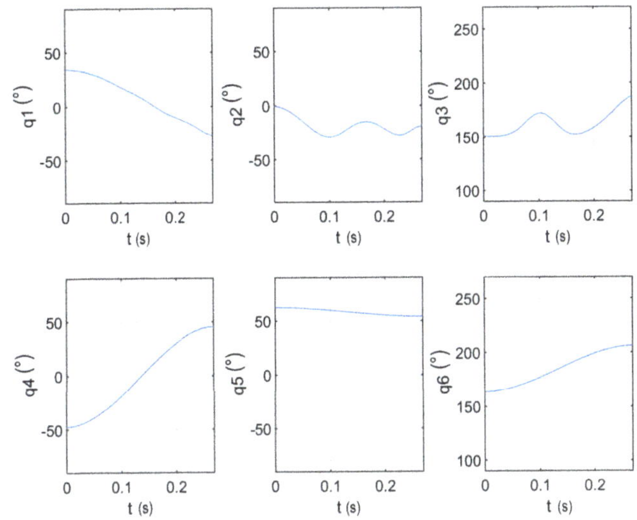

Figure 6. From the top: the final movement provided by the algorithm for Test 1, with the red dots that describe the end effector trajectory and the black dots that describe the wrist center trajectory. To the bottom: corresponding joint angles versus time.

Table 5. Results of the test changing $\delta\theta$ from 30° to 5°.

Step	$\delta\theta$	Lowest Movement Time (s)	Comparison to Step 1 (%)	Total Computational Time (s)
1	30	0.6016	100	5.05
2	5	0.4698	78.1	11.41

Comparison with PRM

A comparison of the proposed method with the Probabilistic Roadmap Method (PRM) was performed, to compare the solutions obtained and the computational times. The PRM is a sampling based method that uses the Dijkstra algorithm to connect two points defined in the configuration space by using randomly generated via points defined in the configuration space. The main differences with our method are that: (a) the via points were defined randomly (i.e., without considering workpiece actual encumbrance); (b) wrist rotations were assigned at the via points, and this would force the wrist joints to make wider rotations along the path with respect to our method.

In the comparison, the PRM was set as follows:

- The sampling of the configuration space of the robot was performed by randomly choosing 227 configurations (equal to the number of via points of our point cloud);
- The configuration space was searched within a limited range of Joint 1, 2, and 3 rotations to reduce the dispersion of the randomly generated points (which would result in worse performance of the PRM algorithm); Joint 4, 5, and 6 ranges were not limited, to avoid loosing dexterity;
- Each configuration was connected to the nearest eight configurations (in the configuration space).

With the aim of fitting the PRM to the method described in Section 4, the latter was modified as follows:

- Point 2 was eliminated;
- Point 3 was modified since the via points were provided by randomly generated samples in the six-dimensional configuration space;

- Point 5 was modified since the via points were connected to form branches by finding the eight closest points in the configuration space.

All the other points of the method were kept unchanged.

Since in the PRM the via points were randomly generated and may yield variable results, 400 instances of the PRM were performed (with two $\delta\theta$ steps as in Table 5). The simulations showed a mean computational time of 21.2 s (standard deviation of 3.38 s) and a mean robot movement time of 0.6915 s (standard deviation of 0.2177 s). An example of the PRM solution is shown in Figure 7, where among 227 random samples, only 133 via points were reachable and connected to the graph. In this particular case, both the computational time and mean movement times were greater than with our method.

This was due to the sparse location of the via points in the PRM, most of which were not useful in the definition of a path with a low movement time. Moreover, the search for the suboptimal path was performed within a six-dimensional space in the PRM (the configuration space), while the decoupling of the inverse kinematic problem in our approach allowed searching a three-dimensional space and then simply interpolating the start and end values of the last three joints, thus reducing the computational effort [22].

To better understand the difference between our method and the PRM, five different cases were considered, in which the starting/ending positions and/or the workcell layout were changed (Figure 8): Test 1 was the same as in Figure 6; Test 2 and Test 3 were very simple movements (Test 2 was a simple rotation around end effector position; Test 3 was a tool repositioning on the same surface on top of the work piece); Test 4 included a tool repositioning from one side to the top of the workpiece; Test 5 included the same movement of Test 1 with an additional obstacle to constrain the movement.

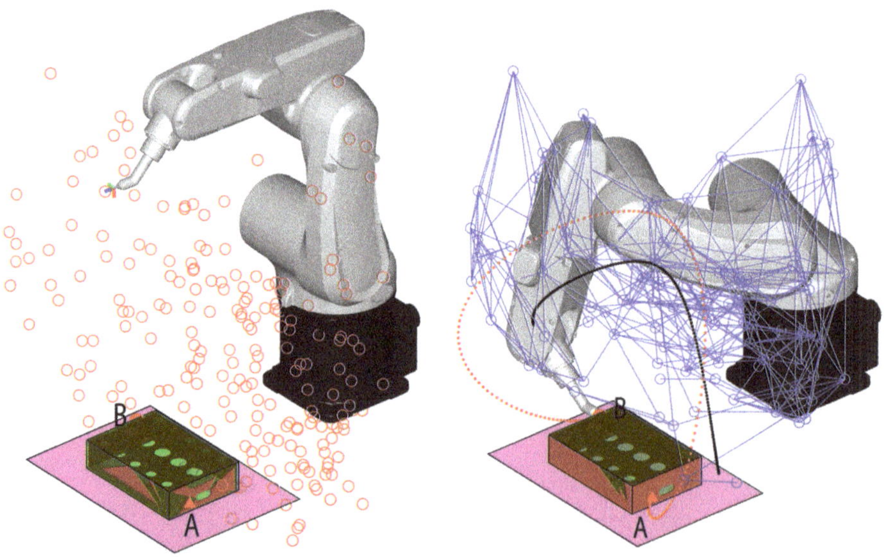

Figure 7. The wrist center points provided by Probabilistic Roadmaps (PRM) (to the left) and the corresponding final movement provided for Test 1, with the red dots that describe the end effector trajectory and the black dots that describe the wrist center trajectory (to the right). Blue circles are feasible and connected wrist center points.

Figure 9 shows optimal movement times provided by our method and by the PRM in the five tests (for the PRM, which was repeated 10 times for each test, minimum/maximum and mean values are depicted). Except from Test 2 (simple tool repositioning), where the results were comparable, our method outperformed the PRM in terms of the minimization of robot movement time. It may be that the PRM would perform better if we increased the number of the generated via points; however, this would dramatically increase computational time.

Figure 10 shows the total absolute joint displacements along the optimal path provided by our method and by the PRM in the five tests (for the PRM, which was repeated 10 times for each test, minimum/maximum and mean values are depicted). Cumulative displacement of the first three joints, which accounted for wrist center motion, and of the last three joints, which accounted for wrist rotation, are shown as well. This figure provides a possible explanation of the results shown in Figure 9: the PRM tended to produce wider displacements of the last three joints (i.e., wider rotations of the wrist), since it assigned their values in the via points; as a result, the motion time increased with respect to our method, where the displacement of the wrist joints was minimized (it equaled the difference between initial and final rotations in A and B). Even though our method may have sometimes provided wider rotations of the first three joints (i.e., larger motion of the wrist center), this seemed not to be the bottleneck in the tests performed.

Another interesting point is that, since the PRM is a stochastic method (in the sense that it produces a random set of via points), a single execution of it may yield very bad results in terms of robot movement time. On the other hand, our method provided a very good solution in one single shot.

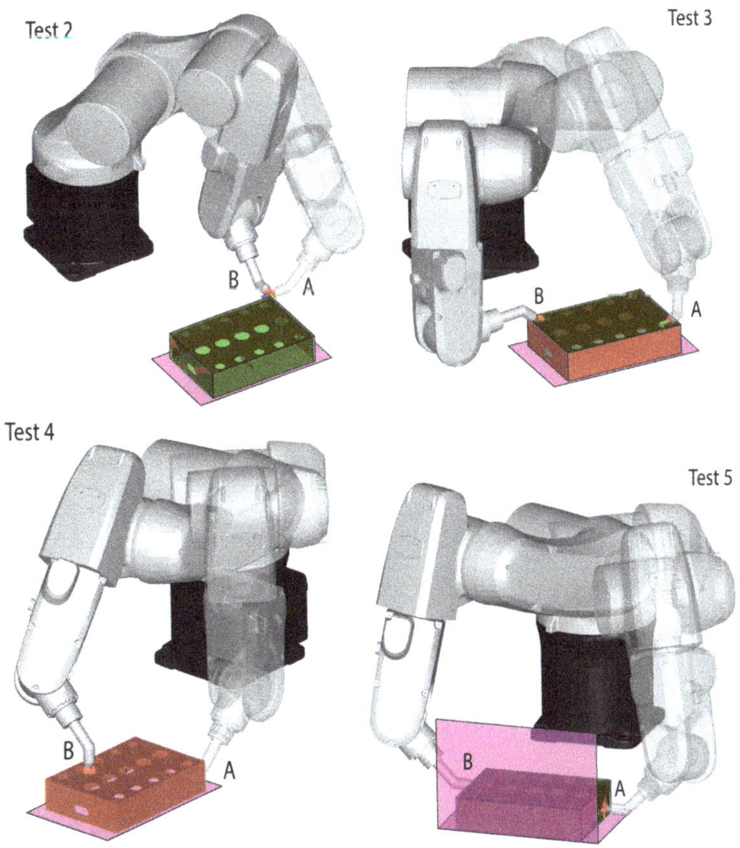

Figure 8. Starting (*A*) and ending (*B*) robot configurations of Tests 2 to 5; redundant angles are null in all the positions depicted ($\theta_{7,A} = 0$ and $\theta_{7,B} = 0$).

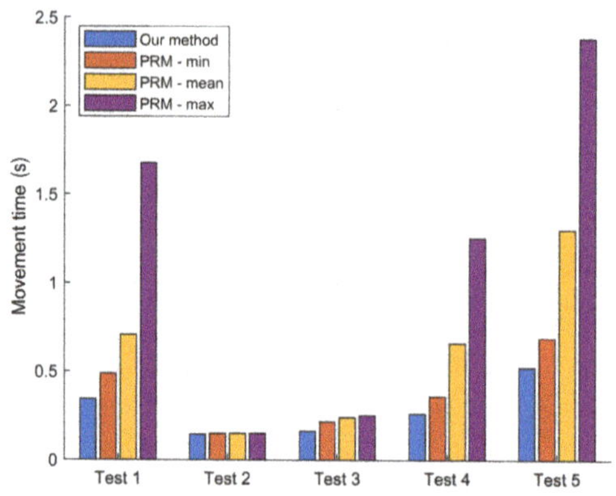

Figure 9. Optimized movement times retrieved from our method and from PRM. Since PRM planning was repeated 10 times for each test, the chart shows the minimum, mean, and maximum optimal times from PRM for each test.

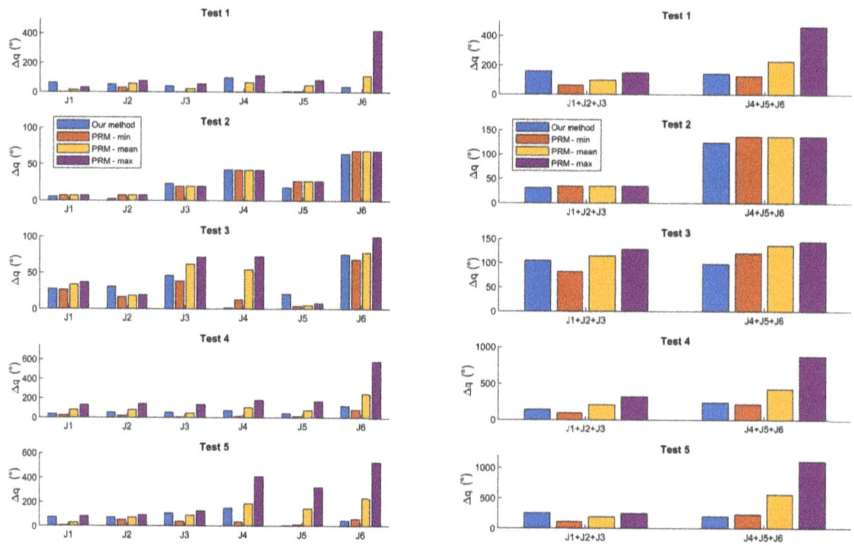

Figure 10. Joint displacement for each joint for each test (to the left) and the corresponding sum of joint triplets (to the right). Since PRM planning was repeated 10 times for each test, the chart shows the minimum, mean, and maximum displacements from PRM for each test.

6. Conclusions

In this paper, we dealt with the problem of defining a trajectory of an industrial robot from a given position to another, when the task is redundant around the tool axis. We proposed a new optimization technique, which finds the initial and final angles of the redundant axis and a proper sequence of via points that minimizes the movement time between the given positions while avoiding collisions with the obstacles at the same time.

The presented method was based on the automatic generation of a cloud of Cartesian via points around the workpiece for the wrist center, thus decoupling the inverse kinematic problem. The method used the Dijkstra algorithm to find the suboptimal path that connected the start and end positions passing through the via point cloud, and a final check for collision was performed to validate the solution. The algorithm can iteratively evaluate a high number of starting and ending configurations in low computational time, allowing one to perform a reasonably wide search of the suboptimal path within the infinite possible motions between the given points.

The particular choice of the via points in our algorithm allowed not only reducing the computational time, but also and mainly minimizing wrist rotations along the path, which in turn helped to obtain low robot motion times. A direct comparison of our method to the PRM in the case of an anthropomorphic industrial robot showed, in four out of five tested motion tasks, that our method could outperform the PRM in terms of robot optimal motion time, at a lower computational cost. Such results should be validated in a wider number of cases; however, their interpretation provided by the analysis on joint rotations suggested that they may be generalized.

Author Contributions: Both authors equally contributed to the paper.

Funding: This research was funded by University of Padua—Program BIRD 2018—Project. no. BIRD187930.

Conflicts of Interest: The authors declare no conflict of interest.

References

1. Ali, M.S.A.D.; Babu, N.; Varghese, K. *Offline Path Planning of Cooperative Manipulators Using Co-Evolutionary Genetic Algorithm*; NIST Special Publication SP: Gaithersburg, MD, USA, 2003; pp. 415–424.
2. Biagiotti, L.; Melchiorri, C. *Trajectory Planning for Automatic Machines and Robots*; Springer Science & Business Media: Berlin, Germany, 2008.
3. Bottin, M.; Boschetti, G.; Rosati, G. A novel collision avoidance method for serial robots. In *IFToMM Symposium on Mechanism Design for Robotics*; Springer: Berlin/Heidelberg, Germany, 2018; pp. 293–301.
4. Chen, D.; Li, S.; Lin, F.-J.; Wu, Q. New super-twisting zeroing neural-dynamics model for tracking control of parallel robots: A finite-time and robust solution. *IEEE Trans. Cybern.* **2019**. [CrossRef] [PubMed]
5. Dijkstra, E.W. A note on two problems in connexion with graphs. *Numer. Math.* **1959**, *1*, 269–271. [CrossRef]
6. Doan, N.C.N.; Lin, W. Optimal robot placement with consideration of redundancy problem for wrist-partitioned 6r articulated robots. *Robot. Comput. Integr. Manuf.* **2017**, *48*, 233–242. [CrossRef]
7. Elbanhawi, M.; Simic, M. Sampling-based robot motion planning: A review. *IEEE Access* **2014**, *2*, 56–77. [CrossRef]
8. Ericson, C. *Real-Time Collision Detection*; CRC Press: Boca Raton, FL, USA, 2004.
9. Ferrara, A.; Incremona, G.P. Design of an integral suboptimal second-order sliding mode controller for the robust motion control of robot manipulators. *IEEE Trans. Control. Syst. Technol.* **2015**, *23*, 2316–2325. [CrossRef]
10. Fredman, M.L.; Tarjan, R.E. Fibonacci heaps and their uses in improved network optimization algorithms. *J. ACM* **1987**, *34*, 596–615. [CrossRef]
11. Gottschalk, S.; Lin, M.C.; Manocha, D. Obbtree: A hierarchical structure for rapid interference detection. In Proceedings of the 23rd Annual Conference on Computer Graphics and Interactive Techniques, New Orleans, LA, USA, 4–9 August 1996; pp. 171–180.
12. Han, D.; Nie, H.; Chen, J.; Chen, M. Dynamic obstacle avoidance for manipulators using distance calculation and discrete detection. *Robot. Comput. Integr. Manuf.* **2018**, *49*, 98–104. [CrossRef]
13. Kavraki, L.E.; Svestka, P.; Latombe, J.-C.; Overmars, M.H. Probabilistic roadmaps for path planning in high-dimensional configuration spaces. *IEEE Trans. Robot. Autom.* **1996**, *12*, 566–580. [CrossRef]
14. Khatib, O. Real-time obstacle avoidance for manipulators and mobile robots. In *Autonomous Robot Vehicles*; Springer: New York, NY, USA, 1986; pp. 396–404.
15. Klosowski, J.T.; Held, M.; Mitchell, J.S.B.; Sowizral, H.; Zikan, K. Efficient collision detection using bounding volume hierarchies of k-dops. *IEEE Trans. Vis. Comput. Graph.* **1998**, *4*, 21–36. [CrossRef]

16. Larsen, E.; Gottschalk, S.; Lin, M.C.; Manocha, D. *Fast Proximity Queries with Swept Sphere Volumes*; Technical Report, Technical Report TR99-018; Department of Computer Science, University of North Carolina: Chapel Hill, NC, USA, 1999.
17. LaValle, S.M. *Planning Algorithms*; Cambridge University Press: Cambridge, UK, 2006.
18. LaValle, S.M.; Kuffner, J.J., Jr. Randomized kinodynamic planning. *Int. J. Robot. Res.* **2001**, *20*, 378–400. [CrossRef]
19. Lin, H.C.; Liu, C.; Fan, Y.; Tomizuka, M. Real-time collision avoidance algorithm on industrial manipulators. In Proceedings of the 2017 IEEE Conference on Control Technology and Applications (CCTA), Mauna Lani, HI, USA, 27–30 August 2017; pp. 1294–1299.
20. Mac, T.T.; Copot, C.; Tran, D.T.; de Keyser, R. Heuristic approaches in robot path planning: A survey. *Robot. Auton. Syst.* **2016**, *86*, 13–28. [CrossRef]
21. Pamanes-García, J.A.; Cuan-Durón, E.; Zeghloul, S. Single and multi-objective optimization of path placement for redundant robotic manipulators. *INGENIERÍA Investigación y Tecnología* **2008**, *9*, 231–257. [CrossRef]
22. Plaku, E.; Kavraki, L.E. Quantitative analysis of nearest-neighbors search in high-dimensional sampling-based motion planning. In *Algorithmic Foundation of Robotics VII*; Springer: Berlin/Heidelberg, Germany, 2008; pp. 3–18.
23. Ratiu, M.; Prichici, M.A. Industrial robot trajectory optimization—A review. *MATEC Web Conf.* **2017**, *126*, 02005. [CrossRef]
24. Redon, S.; Kheddar, A.; Coquillart, S. An algebraic solution to the problem of collision detection for rigid polyhedral objects. In Proceedings of the IEEE International Conference on Robotics and Automation, ICRA'00, San Francisco, CA, USA, 24–28 April 2000; Volume 4, pp. 3733–3738.
25. Ribeiro, M.I. *Obstacle Avoidance*; Instituto de Sistemas e Robótica, Instituto Superio Técnico: Lisbon, Portugal, 2005; p. 1.
26. Rodriguez-Garavito, C.H.; Pati no-Forero, A.A.; Camacho-Munoz, G.A. Collision detector for industrial robot manipulators. In Proceedings of the 13th International Conference on Soft Computing Models in Industrial and Environmental Applications, San Sebastian, Spain, 6–8 June 2018; Springer: Cham, Switzerland, 2018; pp. 187–196.
27. Sciavicco, L.; Siciliano, B. A solution algorithm to the inverse kinematic problem for redundant manipulators. *IEEE J. Robot. Autom.* **1988**, *4*, 403–410. [CrossRef]
28. Tian, L.; Collins, C. Motion planning for redundant manipulators using a floating point genetic algorithm. *J. Intell. Robot. Syst.* **2003**, *38*, 97–312. [CrossRef]
29. Tsardoulias, E.G.; Iliakopoulou, A.; Kargakos, A.; Petrou, L. A review of global path planning methods for occupancy grid maps regardless of obstacle density. *J. Intell. Robot. Syst.* **2016**, *84*, 829–858. [CrossRef]
30. Weller, R.; Debowski, N.; Zachmann, G. kdet: Parallel constant time collision detection for polygonal objects. In *Computer Graphics Forum*; Wiley Online Library: Hoboken, NJ, USA, 2017; Volume 36, pp. 131–141.
31. Xidias, E.K. Time-optimal trajectory planning for hyper-redundant manipulators in 3d workspaces. *Robot. Comput. Integr. Manuf.* **2018**, *50*, 286–298. [CrossRef]
32. Yang, L.; Qi, J.; Song, D.; Xiao, J.; Han, J.; Xia, Y. Survey of robot 3d path planning algorithms. *J. Control. Sci. Eng.* **2016**, *2016*, 5. [CrossRef]

robotics

MDPI

Article

Design of Multiple Wearable Robotic Extra Fingers for Human Hand Augmentation

Monica Malvezzi [1,*], Zubair Iqbal [1], Maria Cristina Valigi [2], Maria Pozzi [1,3], Domenico Prattichizzo [1,3] and Gionata Salvietti [1,3]

1 Department of Information Engineering and Mathematics, University of Siena, 53100 Siena, Italy; mxuberiqbal@gmail.com (Z.I.); pozzi@diism.unisi.it (M.P.); prattichizzo@ing.unisi.it (D.P.); salviettigio@dii.unisi.it (G.S.)

2 Department of Engineering, University of Perugia, Via G. Duranti, 1–06125 Perugia, Italy; mariacristina.valigi@unipg.it

3 Department of Advanced Robotics, Istituto Italiano di Tecnologia, 16163 Genoa, Italy

* Correspondence: malvezzi@dii.unisi.it

Received: 4 November 2019; Accepted: 8 December 2019; Published: 11 December 2019

Abstract: Augmenting the human hand with robotic extra fingers is a cutting-edge research topic and has many potential applications, in particular as a compensatory and rehabilitation tool for patients with upper limb impairments. Devices composed of two extra fingers are preferred with respect to single finger devices when reliable grasps, resistance to external disturbances, and higher payloads are required. Underactuation and compliance are design choices that can reduce the device complexity and weight, maintaining the adaptability to different grasped objects. When only one motor is adopted to actuate multiple fingers, a differential mechanism is necessary to decouple finger movements and distribute forces. In this paper, the main features of a wearable device composed of two robotic extra fingers are described and analyzed in terms of kinematics, statics, and mechanical resistance. Each finger is composed of modular phalanges and is actuated with a single tendon. Interphalangeal joints include a passive elastic element that allows restoring the initial reference configuration when the tendon is released. The stiffness of each passive element can be customized in the manufacturing process and can be chosen according to a desired closure movement of the fingers. Another key aspect of the device is the differential system connecting the actuator to the fingers.

Keywords: wearable robots; underactuated robots; robotic manipulation

1. Introduction

1.1. Robotic Hands

Robotic hands have represented a challenge for designers and engineers for at least three decades [1] due to the complexity of grasping and manipulation tasks and device limits [2]. Although several interesting solutions have been presented in the literature, there are still open challenges. There are hands that have an anthropomorphic structure [3,4], and others that exploit underactuation and parallel kinematic structures to gain adaptability to the grasped objects [5]. Applications range from humanoids [3,6], to prostheses [7], to space applications [5]. Different types of actuations have been implemented, not limited to electro-mechanic transmissions. In [8], for instance, pneumatic systems are adopted, while in [9], Shape Memory Alloy were used. Underactuation [10–12] and modularity [13] are aspects that are investigated to reduce the complexity of the hand by maintaining a suitable level of performance. Also, transmission systems have a great importance in robotic hands, as described in [14], where different transmission systems are compared for different robotic hands. Tendon-driven mechanisms have been widely used in articulated-finger robotic hands. In [15], several interesting issues on tendon-driven mechanisms are discussed, with a focus on tendon redundancy and joint stiffness adjustability for a robotic mechanism driven with redundant tendons.

1.2. Wearable Robotic Extra Fingers

More recently, a new branch in robotics research, involving both robotic hands and human-robot interaction problems, was born: it is referred to as human augmentation and consists in developing wearable robotic systems that are supplementary with respect to human body structure [16,17]. A particularly interesting application of wearable robotic extra limbs is introduced in [18]: the proposed device is a modular robotic extra finger that can be worn by a user as a bracelet and augments the human hand in terms of workspace and manipulation capabilities. An underactuated version of this device has been introduced in [19] to assist patients with limited hand and arm functions for instance after a stroke event. Another implementation of wearable robotic extra fingers is presented in [20], in which a pair of soft robotic fingers driven by tendons and servomotors are worn on a human hand and lay on the same plane of the palm, working as two additional thumbs. In applications in which wearable extra fingers are used as compensatory devices, although the availability of a wearable robotic extra-finger opposed to the paretic limb allows the patient a stable hold for a large number of objects, a solution with two or multiple fingers could further help the user when performing some activities requiring higher payloads and grasp stability [21].

1.3. Underactuation in Robotic Extra Fingers

Underactuation is an important aspect to be considered in designing safe and robust robotic hands and fingers. Reducing the number of actuators is particularly important in wearable devices, in which the weight and the complexity should be reduced as much as possible. In general, an underactuated mechanism is defined as a mechanism that has fewer actuators than degrees of freedom. In robotic hands, it is important to notice that underactuation can provide interesting properties to the device, like for instance self adaptability [22]. However, reducing the number of actuated degrees of freedom (DoFs) may decrease the overall manipulability properties and the capability to adapt to different shapes and dimensions of grasped objects. For single-finger robotic devices, these properties can be partially recovered by means of compliance, that in the wearable device introduced in [19] has been implemented by passive elastic elements in interphalangeal joints.

1.4. Differential Mechanisms

In devices composed of more than one finger, however, joint compliance is not enough to guarantee the adaptability to the different shapes of grasped objects. If one motor is employed directly to drive simultaneously the opening and closing motion of the fingers, when a finger is blocked, due to the contact with an object or a surface, the other one will stop, too. A differential mechanism is therefore necessary in devices composed of multiple fingers actuated with a single motor. For a double robotic extra finger, that is a device with two outputs and one input, a simple differential mechanism can be used to decouple finger motions when one of them is constrained. In general, a differential is a mechanism with two-degrees of freedom able to transform a single input into two outputs. The role of the differential mechanism is to distribute an input force/torque F_a between two outputs, F_1^a and F_2^a, so that $F_a = F_1^a + F_2^a$. In the literature, there are many applications of differential mechanisms for robotic fingers and hands [23]. In [24], a differential system based on gears is used for a novel architecture of robotic hand and the properties of differential mechanisms arranged in cascade via parallel or serial connections is studied. In [25], a planetary gear solution and a fluid T-pipe scheme are described. In [26], a moving pulley differential mechanism was used, while in [3] a differential with a T-shape fluid mechanism and the connected seesaw circuit is presented. In [27], an underactuated anthropomorphic gripper for prosthetic applications was presented, in which a mechanical lever inside the palm allowed to extend the grasping capabilities and improve the force transmission ratio of the gripper. This mechanism was further developed in [28], in which the differential mechanism included a set of locking buttons allowing the user to stop the motion of each finger.

1.5. Paper Contribution

In this paper, the main criteria that were followed to design a double robotic extra finger for compensating and augmenting human hand abilities (Figure 1) are described. Differently from previous works [19,21], all the mechanical components of the fingers were designed in a parametric way, to further exploit the modularity of the device and adapt it to different users and applications. A double wearable robotic extra finger was preliminarily introduced in [29], however, in that device both fingers were actuated with the same motor, using a single Y-shaped tendon that constrained fingers' motions. This solution was an improvement with respect to a single finger, as it allowed to increase the grasping quality in terms of number of contacts, stability, and grasp stiffness. However, the wearable robot introduced in [29] was not able to adapt to grasp objects with irregular shapes or multiple objects, due to the coupling between fingers generated by the single tendon. The problem can be solved by inserting a differential system between the actuator and the fingers, allowing to decouple their motions. In this paper, we analyze the requirements and main design aspects of the differential mechanism that is necessary to guarantee the proper and decoupled closure movement of each of the two fingers.

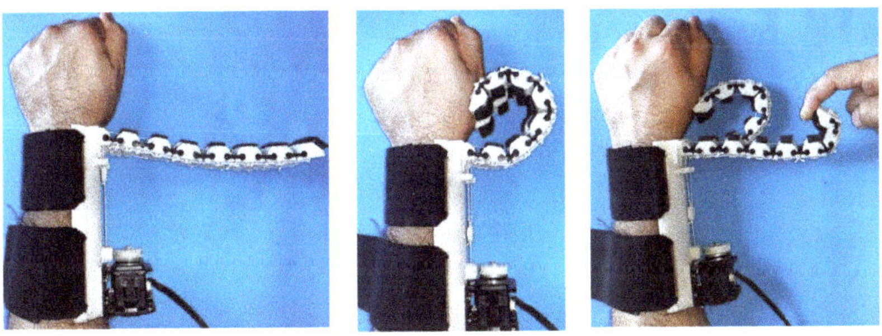

(a) (b) (c)

Figure 1. The prototype of the device, consisting in two wearable extra fingers, worn by a user. (**a**) Reference open configuration. (**b**) Closed configuration. (**c**) Example of how the differential mechanism decouples the motion of the fingers, so that the device can adapt to different surfaces and object shapes.

The device that is described in the following sections consists of two main parts: the support base and the mobile part, composed of two modular robotic fingers actuated by means of two tendons and a single motor. The end edge of each tendon is fixed on the distal module of the finger, while the opposite side is fastened to an element sliding inside the differential box. Fingers are composed by rigid phalanges connected by compliant interphalangeal joints, that can be approximated as simple 1-DoF revolute joints. When the motor actuates the differential by pulling the sliding element, both tendons flex the fingers, that reach a configuration that is suitable for grasping objects (closed configuration, Figure 1b), while when the torque of the motor is released, the passive elastic elements in the joints restore the fingers to their initial extended configuration (reference open configuration, Figure 1a).

The paper is organized as follows. After an overview of the finger structure (Section 2), a detailed mechanical analysis of the device is given in Section 3 and a first prototype is described in Section 4. Sections 5 and 6 draw the conclusions of the paper, summarizing envisaged applications of the proposed wearable robotic device.

2. Device Overview

Figure 1 shows the prototype of the double robotic extra finger presented in this work, constituted by two fingers. Each finger has 7 modules, each of which has a part that can be approximately considered rigid (white phalanges in Figure 1) and a part that is passively deformable (black interphalangeal joints in Figure 1). Finger orientation with respect to the fixed support can be manually adapted by the user according to the specific task (see the video S1 in Supplementary Materials).

Rigid elements of the finger can be created with standard Fused Deposition Modelling (FDM) techniques and materials, as for instance Acrylonitrile Butadiene Styrene (ABS) or Polylactic Acid (PLA). For the deformable passive interphalangeal elements, a material that can undergo high deformations when subject to external forces is needed: for these elements, we employed Thermoplastic Polyurethane (TPU) [30], a plastic material that has this desirable mechanical property and can be also manufactured with standard FDM techniques. Furthermore, by changing some manufacturing parameters, as for instance the infill density percentage or pattern, it is possible to modify mechanical properties of the component. The main elements composing modular fingers, as detailed in Figure 2, are rigid elements (Figure 2a–c) and deformable elements (Figure 2d). Three different rigid elements have been designed for the proximal (Figure 2a), intermediate (Figure 2b), and distal (Figure 2c) phalanges. The structure is modular and solutions with different numbers of phalanges can be easily realized. In this paper, we considered fingers composed of seven modules, since we verified that this value is an acceptable trade off to guarantee wearability and suitable grasp capability of several objects related to activities of daily living (ADL).

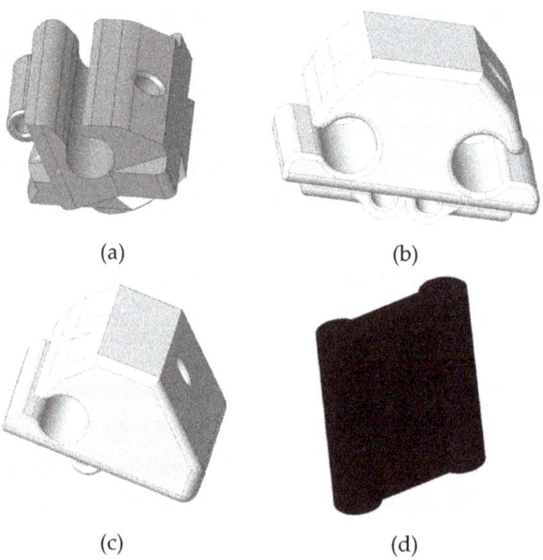

Figure 2. Main components of the modular underactuated finger. (**a**) Proximal element connected to finger base; (**b**) intermediate phalanx; (**c**) distal phalanx; (**d**) deformable interphalangeal element.

Each finger is actuated with a single tendon, connected on one side to the actuation and transmission part, and on the other side to the distal phalanx. When the motor actuation pulls the tendon, fingers close following a movement that depends on passive deformable element stiffness. The next section presents the method followed for choosing the mechanical properties (stiffness) of passive elements in interphalangeal joints to obtain a desired closure motion, and how it was applied to different configurations of single and double extra fingers.

The motor and differential mechanism of the double extra finger are installed in a support fixed on user's forearm (Figure 1a), the overall encumbrance of the motor, differential, and tendon routing system is 130 × 50 × 40 mm. The size of this support can be customized and modelled according to user's specific requirements. In particular, in future developments of this work we will use 3D scanner techniques [31,32] to reconstruct user's forearm geometry and to customize the design of the fixed support.

The motor, fixed at the base of the box, rotates the drive pulley. The first tendon connects the motor pulley to a sliding element. Two secondary tendons connect the pulleys of the sliding element to the fingers and are fixed to their distal phalanges.

The main components of the device are shown in the CAD model represented in Figure 3 and are: A) the actuator, B) the drive pulley, C) the fixed pin necessary to provide a suitable direction to the tendon, D) the sliding element, E) two elements necessary to provide the right directions to tendons connected to the fingers, F) the tendons, the finger modules, constituted by G) a rigid element, H) connected to a compliant element, J) through cylindrical guides.

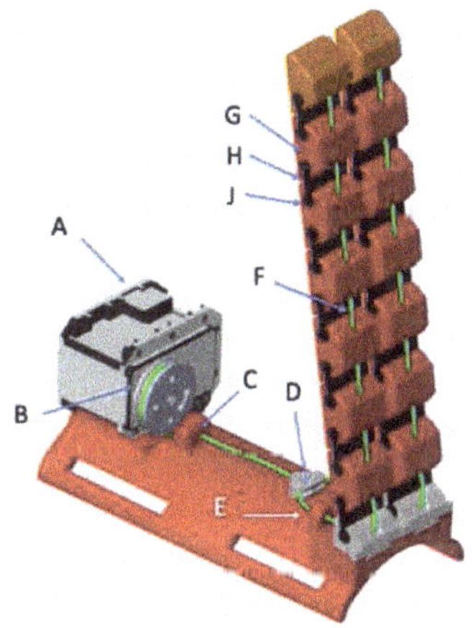

Figure 3. Main components of the device. Tendons are represented in green. A—actuator. B—drive pulley. C—fixed pin necessary to provide a suitable direction to the tendon. D—Sliding element. E—elements necessary to provide the right directions to finger tendons. F—Tendon. G—Rigid interphalangeal element. H—Flexible joint element. J—cylindrical connection between rigid and flexible elements.

Assembling and de-assembling fingers to adapt them to specific needs or to substitute some of the components is a task that does not need specific knowledge or skills, even if bi-manual operations are required. If the device is used as a rehabilitation or compensation tool by a patient with upper limb impairments limiting its functionality, this operation can be managed for instance by the physiotherapist, a family member, a collaborator, etc. Referring to the elements sketched in Figure 3, the needed operations are: *i*) disconnect the tendon (F) of the finger that needs to be adapted; *ii*) de-assemble and re-assemble rigid/flexible elements (G,H) of the finger according to the new configuration (flexible joint passive elements can be connected/disconnected to the rigid links by sliding them along the cylindrical guides in the rigid links (J)); *iii*) if the number of phalanges has been modified, also tendon length needs to be adapted; *iv*) re-connect the tendon to the distal phalanx and to the differential sliding element, passing through the rigid links.

3. Analysis

3.1. Design of Finger Passive Elements

As introduced in the previous section, each finger has a single tendon and then, when actuated, can follow only one predefined closure motion. The characteristics of closure motion depend on both geometrical properties of finger components and on the stiffness of the passive element in the interphalangeal joints. When a pulling force is applied to the tendon, stiffer joints will bend less than softer ones. By properly regulating the stiffness of each of the passive elements, it is therefore possible to shape the closure motion. As previously introduced, the modular structure of the finger allows to easily add, remove, or substitute the passive interphalangeal elements, to adapt the finger to different applications.

The first step consists of choosing the closure motion that has to be followed by the extra finger. In this paper, we focus on a double extra finger device, however the procedure that we present is general and can be adapted to any number of fingers. Let us indicate with $\mathbf{r_{1des}}(z) = [x_{1des}(z), y_{1des}(z), z_{1des}(z)]^{\mathrm{T}}$ and $\mathbf{r_{2des}}(z) = [x_{2des}(z), y_{2des}(z), z_{2des}(z)]^{\mathrm{T}}$ two vectors containing the desired trajectories of fingertips, with respect to a reference frame $\{S_0\}$ fixed on fingers' base. Fingertip coordinates are defined as a function of the variable z, describing the "closure" ($z = 0$ for completely open fingers, $z = 1$ for completely closed fingers). For the sake of compactness, let us also collect both finger trajectories in a unique vector $[\mathbf{r_{des}}(z)] = [[\mathbf{r_{1des}}(z)]^{\mathrm{T}}, [\mathbf{r_{2des}}(z)]^{\mathrm{T}}]^{\mathrm{T}}$.

To let the extra fingers be able to grasp a wide range of objects, cooperating with the human hand, it is important to choose the most suitable finger closure movement. Different methods can be used to define the desired trajectories $\mathbf{r_{des}}(z)$. In the simulations presented in the following we adopt the procedure based on bio-artificial synergies proposed in [33]. The concept of bio-artificial synergies was introduced in [34] to control wearable extra fingers. Synergies were derived, similarly to [35], by performing the Principal Component Analysis (PCA) on a set of grasps performed by the human hand augmented by two robotic extra fingers. In this paper, we exploit the concept of synergies but we could not extract directly the augmented hand synergy as in [34], since the methodology therein proposed was based on the analysis of a dataset of grasps, that is not available in the design phase of the device. Here we extend the idea of mapping human hand synergies introduced for the human hand in [35], based on the definition of a virtual object, that we presented in some previous works. The method to map human hand first postural synergy to robotic extra fingers has been previously introduced in [33] and in [36]. The output of the mapping procedure are the fingertip trajectories that we want to obtain when actuating the robotic extra fingers.

We therefore choose a set of discrete samples for the variable z describing fingers' closure motion, $\mathbf{z} = [z_1, \ldots, z_{Ns}]$, we indicate with z_i, with $i = 1, \ldots, N_s$ the generic sample, so that $z_1 = 0$ and $z_{Ns} = 1$. We indicate with $\mathbf{q}$ a vector containing all the rotations of interphalangeal joints and with $\mathbf{q}_0$ its value corresponding to the initial (completely open) configuration. The value of $\mathbf{q}$ at the generic configuration z_i can be evaluated by means of a standard inverse kinematic procedure, in particular, in this work we adopted an approximated algorithm based on the Jacobian pseudo-inverse, so that:

$$\mathbf{q}_i = \mathbf{q}_{i-1} + [\mathbf{J}(\mathbf{q}_{i-1})]^{\#}(\mathbf{r}_{des}(z_i) - \mathbf{r}_{des}(z_{i-1})) \tag{1}$$

where $\mathbf{J}(\mathbf{q})$ represent the Jacobian matrix of the fingers, relating joint rotation time derivatives to fingertip velocities, i.e.,

$$\dot{\mathbf{r}} = \mathbf{J}(\mathbf{q})\dot{\mathbf{q}} \tag{2}$$

and # indicates its pseudo-inverse, i.e., $[\mathbf{J}]^{\#} = [\mathbf{J}]^{T}\left([\mathbf{J}][\mathbf{J}]^{T}\right)^{-1}$.

Indicating with $\Delta\mathbf{l}$ is a vector collecting the variation of tendon lengths with respect to the initial reference configuration, choosing the closure motion means specifying the relationship:

$$\mathbf{q} = \mathbf{q}(\Delta\mathbf{l}), \tag{3}$$

assuming that no external forces, except the tendon force, are applied to the finger elements. In other terms, in this phase we do not consider any contact between finger elements and an object or the environment. For the prototype that has been introduced in the previous section, $\mathbf{q}$ is a vector with 14 elements (7 joints per finger), and $\Delta\mathbf{l}$ is a vector with 2 elements (lengths of the tendons connected to each finger). In general, in an underactuated system, such a relationship cannot be defined, however, in our case, it can be established since in the joints there are passive elastic elements. The evaluation of the relationship in Equation (3) is possible if the stiffness values of the passive elements are known, as detailed in [33].

As we introduced above, in an underactuated mechanism it is not possible to define the relationship introduced in Equation (3), but it is easy to define, by means of simple geometrical considerations, its dual, i.e.,

$$\Delta\mathbf{l} = \mathbf{M}\mathbf{q}. \tag{4}$$

It is interesting to notice that this relation can be approximated with a linear map, and the elements of matrix $\mathbf{M}$ (whose dimension is 2×14 for the device introduced in Section 2) depend on finger geometrical properties [15]. By applying the Principle of Virtual Works to the fingers, it is easy to show that:

$$\boldsymbol{\tau} = \mathbf{M}^T\mathbf{f}, \tag{5}$$

where $\boldsymbol{\tau}$ is a vector collecting joint stiffness values, while $\mathbf{f}$ is the tendon force vector. Since in each joint a passive element is present, from the static analysis of the finger, it results that

$$\boldsymbol{\tau} - \mathbf{K}_q\left(\mathbf{q} - \mathbf{q}_0\right) = 0, \tag{6}$$

where $\mathbf{q}_0$ indicates the reference joint configurations, i.e., the configuration corresponding to $\boldsymbol{\tau} = 0$, and $\mathbf{K}_q$ is the stiffness matrix. $\mathbf{K}_q$ elements represent the stiffness of passive elements of the interphalangeal joints. Since these elements are realized with a material with a high flexibility, in general to describe its overall compliance, a 6×6 matrix should be necessary. However, it is reasonable to neglect the linear terms (normal and shear deformations), while regarding the rotation elements, the bending stiffness with respect to an axis perpendicular to the finger closure plane is much lower than the value evaluated in the lateral direction, and it is also higher than the torsional stiffness. It is therefore reasonable to model the passive element as a single DoF revolute joint, with an equivalent stiffness value k_{qi}. Matrix $\mathbf{K}_q$ is a diagonal matrix collecting all the k_{qi} values.

Once the desired closure motion $\mathbf{r}_{des}(z)$ is defined, through the inverse kinematics algorithm in Equation (1) is possible to define the corresponding reference values for interphalangeal joint rotations, $\mathbf{q}_r = \mathbf{q}_r(\Delta\mathbf{l})$, and therefore, from Equations (3)–(6) it is possible to find the stiffness matrix

$$\mathbf{K}_q = \mathbf{K}_q\left(\mathbf{q}_r, \mathbf{q}_0, \mathbf{f}\right). \tag{7}$$

The second step of the procedure consists in the definition of the stiffness value that each passive element of finger joints has to assume so to realize the desired closure motion. For a more detailed description of the procedure, the reader can refer to [33].

In this paper, we report the results obtained by applying this procedure to three different configurations of human hand and extra fingers. For the evaluation, we realized the mathematical models of the hand augmented with the robotic extra fingers using SynGrasp [37], a Matlab toolbox for the simulation of robotic hand grasping and manipulation, that can manage synergy-actuated and more in general underactuated structures. The human hand biologically has 27 DoF, in this study we adopted a model with 20 DoF, similar to the one described in [38], in which we neglected the carpometacarpal (CMC) articulation of index, middle, ring, and little fingers. We choose this simplified model because we could apply to it the quantitative results, in terms of bio-artificial synergies, evaluated in [35] and available in the Hand Corpus repository (https://www.handcorpus.org/) [39].

Figure 4a shows a link-joint representation of the kinematic structure of the hand augmented with two extra fingers, each extra finger is composed of seven modules. Figure 4b shows the augmented hand configuration when the human fingers are actuated according to the first postural synergy evaluated in [35] and the motion of the robotic ones are evaluated with the mapping procedure reported in [36].

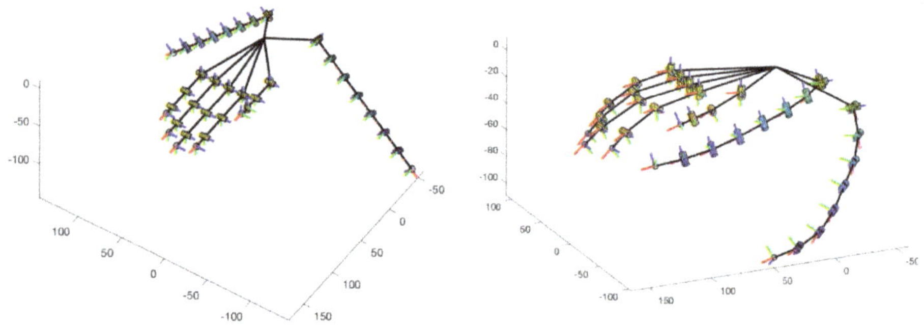

(a) (b)

Figure 4. SynGrasp model of the human hand augmented with two robotic extra fingers. Black lines represent rigid links, cylinders represent revolute joints. The human hand is modelled as a 20 degrees of freedom (DoF) system, augmented fingers have a modular structure with seven elements. (**a**) Reference initial configuration. (**b**) Configuration of the hand when the first synergy is activated, in an intermediate configuration between the initial reference and the completely closed ones.

According to the procedure that we previously described, stiffness values of the passive elastic interphalangeal elements of robotic fingers depend on the trajectory that is mapped, but also on the structure of the robotic fingers, and on the force applied on the tendons by the motor. Table 1 summarizes stiffness values obtained with extra fingers composed of 7 modules, in the three configurations, shown in Figure 5: single extra finger opposite to human hand palm (Figure 5a), double configuration with two parallel fingers opposite to human hand palm (Figure 5b), double configuration, with two opposite fingers, one close to human thumb and the other one close to the little finger (Figure 5c). This last configuration has been considered in the analysis phase to verify the effect of finger configuration on closure motion and of the corresponding joint stiffness values. As previously introduced, the orientation of the fingers with respect to the hand can be manually regulated by the user to adapt to different objects and tasks. A configuration similar to the one described in Figure 5c is the one described in [20]. Values in Table 1 have been normalized with respect to the stiffness of the proximal joint, in this way it is possible to compare single and double finger configurations. Since the analyzed configurations are symmetric with respect to the hand, for the sake of brevity in double finger configurations the results have been reported only for one of the fingers. As it can be seen, single and double parallel configurations present similar stiffness values, while the double opposite configuration presents some differences.

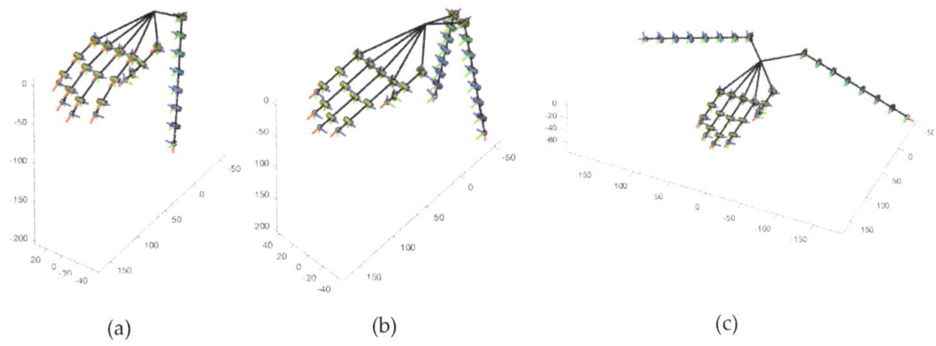

Figure 5. Configurations that we compared for the evaluation of passive joint stiffness with the mapping procedure applied to the first human hand postural synergy. (a) Single finger, opposite to human hand. (b) Double configuration, fingers opposite to human hand palm. (c) Double configurations, fingers are aligned with hand palm.

Table 1. Passive elements of interphalangeal joints, stiffness values normalized with respect to the first (proximal) element, evaluated for the three configurations shown in Figure 6.

Configuration	k_1/k_1	k_2/k_1	k_3/k_1	k_4/k_1	k_5/k_1	k_6/k_1	k_7/k_1
Single (Figure 5a)	1.00	1.17	1.39	1.74	2.32	3.50	6.98
Double, opposite to the palm (Figure 5b)	1.00	1.10	1.33	1.65	2.20	3.31	6.61
Double, aligned with the palm (Figure 5c)	1.00	1.43	1.71	2.14	2.85	4.28	8.57

Figure 6a shows, for one of the fingers of the configuration in Figure 5b, the stiffness values of the passive interphalangeal elements $\mathbf{K}_q$, that have been evaluated on the basis of the motion that we imposed on the fingers, represented as a function of the variable z, describing finger closure, that for this type of mapping is also related to synergy actuation [36] ($z = 0$ in the reference open configuration, $z = 1$ when the hand is closed).

It is interesting to notice that:

- $\mathbf{K}_q$ values vary for the different joints: their values determine the closure motion of the fingers and are defined on the basis of the mapping procedure proposed in [36] and in [33].
- $\mathbf{K}_q$ values vary as a function of hand configuration, however, for the motion that we selected, such a variation is not very high.

To highlight the effect of joint stiffness values in the closure motion of the finger, in Figure 6b we simulated, for one of the robotic extra fingers, the trajectory of the fingertips obtained with the variable stiffness values defined by the previously described procedure, exactly resembling the "ideal" desired trajectory, (blue curve), the one that we obtain if we consider, for each joint, the mean stiffness value (red curve), and the one that we would obtain if all the joints of the finger had the same stiffness value (magenta curve). As it can be seen, the difference between the blue curve and the red one is quite limited, compatible with the precision required for this type of device (few millimeters), so considering a stiffness value different for all the joints, but constant for each joint, does not introduce significant difference between the closure motion that was planned and the one that can be obtained with the underactuated finger. On the other hand, the difference between ideal fingertip trajectory and the one obtained with the same joint stiffness for all the modules (in the simulation we assumed the mean stiffness value among all the joints) is very high.

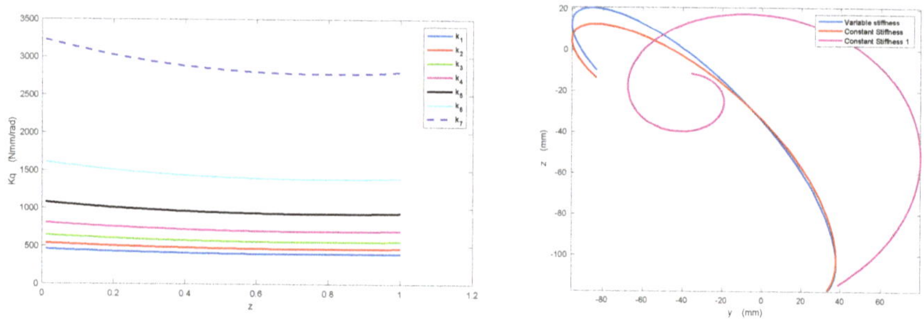

(a) (b)

Figure 6. (**a**) Joint stiffness values for one of the extra fingers of the configuration in Figure 5b, evaluated as a function of synergy activation. (**b**) For the same configuration, trajectories of one of the fingertips, during a finger closure motion, projected on a plane. Different trajectories are obtained by considering variable stiffness profile evaluated by means of the mapping procedure (blue curve), the mean stiffness value (constant) for each joint (red curve), the mean stiffness value for all the joints (magenta curve).

The third step consists in creating the passive elements so that their stiffness is the one calculated in the previous phase. We analyzed the possibility of tuning finger joint stiffness values by exploiting the potentialities of 3D printing fabrication methods, that are nowadays rapidly improving and offering interesting opportunities. In particular, choosing a material as the Thermoplastic Polyurethane (TPU) for realizing the flexible parts, we can get different stiffness values, while maintaining the same geometric shape, by regulating the percentage of infill density. This parameter affects primarily material density, but also its mechanical properties [30]. As an example, Table 2 summarizes the variation of Young's modulus E of TPU as a function of the infill percentage density.

Table 2. Properties of Thermoplastic Polyurethane (TPU) as a function of infill density percentage.

Infill Density %	E (MPa)
10	1.07
30	1.38
50	2.07
70	6.53
90	9.45
100	10.5

In a more general framework, the overall passive stiffness of a generic joint, indicated with k, depends mainly on joint geometry and material structural properties, and, considering a linear elastic behavior, on Young's modulus E, i.e.,:

$$k = k(\mathbf{d}, E), \tag{8}$$

where $\mathbf{d}$ is a vector containing all the parameters defining joint geometry (e.g., for a parallelepiped joint, its length l, width w and thickness t). Young's modulus E depends on material parameters and fabrication methods, i.e.,:

$$E = E(p_1, p_2, \ldots, p_n), \tag{9}$$

where each value p_i indicates one specific material property. In this work, we exploited the dependency of k with respect to infill density percentage to define robotic finger properties. Future developments of this study will consider other manufacturing parameters, as for instance the infill pattern profile, and also composite materials.

3.2. Mechanical Transmission and Differential Mechanism Analysis

Concerning the differential system, the solution proposed in this paper to actuate a double extra finger device for human hand augmentation is inspired by Dollar's work on underactuated robotic hands [40], using a moving pulley system; in Figure 7, a simplified and general scheme of this kind of differential is presented [23].

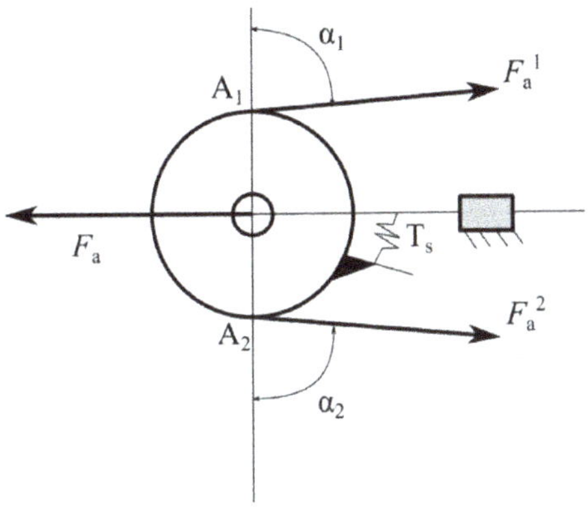

Figure 7. Simplified scheme of a typical differential mechanism with moving pulley.

The motivation for this choice is the compactness and excellent adaptability to the human's forearm dimensions and, since the robotic fingers that are connected through the mechanism use tendons, the mobile pulley mechanism is the most effective system for force transmission. A spring is often used to keep the two outputs in the same reference configuration when the system is not actuated, and the fingers are not externally constrained.

The pulley has 2 DoF: it can translate along a sliding guide (realizing a prismatic joint) and it can rotate around an axis perpendicular to the plane of the figure.

In this scheme, the input force is F_a and the output ones are F_1^a and F_2^a. The spring applies a torque T^s. In static equilibrium condition the following relationship holds:

$$\mathbf{F} = \left[\mathbf{T}^f\right]\mathbf{t}^* \tag{10}$$

with

$$\mathbf{F} = \begin{bmatrix} F_1^a \\ F_2^a \end{bmatrix} \quad \mathbf{t}^* = \begin{bmatrix} Fa \\ T^s \end{bmatrix} \tag{11}$$

where the transformation matrix $\left[\mathbf{T}^f\right]$ is

$$\left[\mathbf{T}^f\right] = \frac{1}{c}\begin{bmatrix} r & \sin\alpha_2 \\ r & -\sin\alpha_1 \end{bmatrix} \tag{12}$$

where c is the sum of the distances of A_1 and A_2 points from the sliding guide (or prismatic joint)

$$c = r\,(\sin\alpha_1 + \sin\alpha_2) \tag{13}$$

and α_1 and α_2 are the angles shown in Figure 6. If the stiffness of the spring is negligible, Equation (10) can be simplified as:

$$F_1^a = F_2^a = \frac{Fa}{\sin\alpha_1 + \sin\alpha_2} \tag{14}$$

An important property of this differential mechanism is that it is isotropic, and thus the two output forces are equal for any value of α_1 and α_2. The ratio F_i^a/F_a for i = 1,2 is influenced by α_1 and α_2, and c must be positive for a proper device working.

The distance between the pin and the sliding has to be at least equal to the maximum distance that the pulley has to travel while both the fingers are completely closed from an initial extended configuration. Let us consider one of the fingers and let us indicate with Θ the generic rotation between the x_n axis, defined on the distal phalanx of one finger and the reference direction x_0 on the device base. Reference frames are defined so that $\Theta = 0$ in the initial extended configuration, while $\Theta = \Theta_c$ in the configuration corresponding to complete finger closure. Let us indicate with ϑ_i the rotation of the i-th joint, so that

$$\Theta_c = \sum_{i=1}^{n} \vartheta_i \tag{15}$$

To obtain the rotation ϑ_i the tendon is pulled by the motor and its length variation is indicated with Δl_i (Figure 8). When the joint is rotated, the elastic element bends; indicating with l its length and assuming that after the deformation its profile is circular, its curvature radius can be evaluated as $r_i = \frac{l}{\vartheta_i}$. The corresponding arc cord can be evaluated as:

$$a_i = 2r_i \sin\left(\frac{\vartheta_i}{2}\right) \tag{16}$$

and tendon length variation is therefore (Figure 8)

$$\Delta l_i = l - \frac{a_i}{l}\,(r_i - h) = l - 2r_i\,\frac{(r_i - h)}{r_i}\sin\left(\frac{\vartheta_i}{2}\right) \tag{17}$$

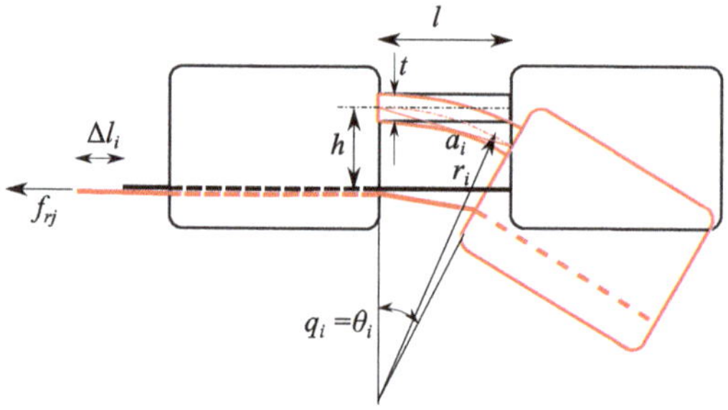

Figure 8. Deformations of interphalangeal joints, scheme, and main geometrical parameters.

To close the fingers, it is necessary to vary the length of the tendon by Δl_{tot}, that can be evaluated as

$$\Delta l_{tot} = \sum_{1}^{n} \Delta l_i \tag{18}$$

Assuming $l_i = 17.3\ mm$ for all the elements, $h = 5\ mm$ and $\Theta_c = 2\pi$ we obtain $\Delta l_{tot} = 28\ mm$: this is the maximum length variation of the tendon and corresponds to the maximum displacement of the sliding element. This result is necessary to define the minimum size of the box containing the differential mechanism.

Δl_{tot} represents also the maximum difference between fingers' tendon lengths when the differential is working, for example when one of the fingers is blocked in its straight initial configuration and the other one is free and can complete the closure.

Two possible solutions can be realized (Figure 9): in the first one (Figure 9a) the sliding element of the differential is constrained to move in one fixed direction by means of a prismatic joint, the two tendons are connected in a unique tendon that can slide over the element. In this case at least one pulley is necessary to reduce the friction. In the second solution (Figure 9b), the sliding element is able both to slide and to tilt inside the box, and the output tendons are connected directly to its edges, the differentiation between output motions is possible thanks to the tilting motion of the sliding element.

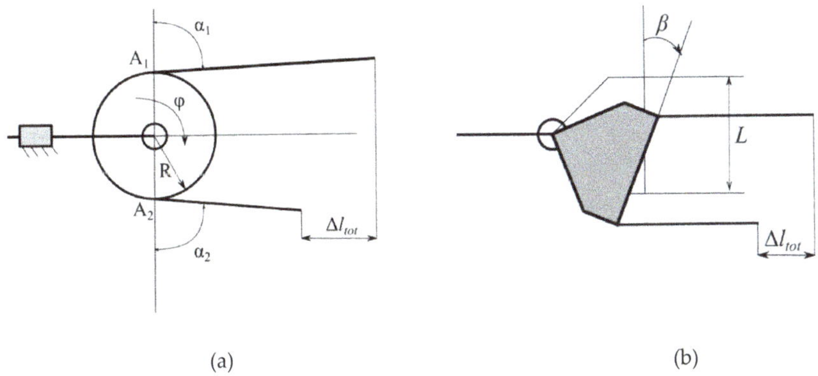

Figure 9. Two different schemes for the differential mechanism. (**a**) differential based on a pulley mechanism translating along a fixed direction; (**b**) differential based on the tilting motion of an element.

The first solution allows a more compact implementation; however, it requires that the sliding element is not a rigid body (at least one pulley, connected to the main body of the sliding element through a revolute joint, is needed).

Considering a single pulley, as shown in Figure 9a, indicating with R its radius, when a differential motion Δl_{tot} is required, pulley rotation is simply given by

$$\varphi = \frac{\Delta l_{tot}}{R} \tag{19}$$

The second solution needs larger components, but simpler from the mechanical and manufacturing point of view. In a preliminary design phase, we analyzed both the solutions, in the following prototyping phase we realized the second solution because it resulted mechanically simpler and more robust. Indicating with L the width of the sliding element, i.e., the distance between the points in which fingers tendons are connected to it and with β its rotation generated by a difference Δl_{tot} between the deformation of fingers' tendons, it is easy to verify that (Figure 9b):

$$\frac{\Delta l_{tot}}{L} = \sin\beta \tag{20}$$

assuming $\boldsymbol{L = 32\ mm}$, we get $\boldsymbol{\beta = 64^\circ}$.

3.3. Structural Analysis of Rigid Elements

For the realization of the prototypes we exploited standard additive manufacturing techniques, that allow to use light materials with quite low price, and to easily adapt the design to specific user's need. The differential mechanism was designed for a standard 3D printing considering generic Acrylonitrile Butadiene Styrene (ABS) material, that has a limited density, good structural properties, good resistance to heat and impacts [41]. The main mechanical properties of ABS material are summarized in Table 3.

Table 3. Acrylonitrile Butadiene Styrene (ABS) main mechanical properties.

Property	Value	Unit
Density	1.03	kg/m^3
Elastic modulus	2000	MPa
Poisson's coefficient	0.394	
Yield stress	45–60 MPa	MPa
Specific heat	1386 J/(kg K)	J/(kg K)
Thermal conductivity	0.2256 W/(m k)	W/(m k)

Considering the limited dimensions of the mechanical components and the loads that they have to resist; a static analysis of the most solicited elements was performed testing their resistance by means Finite Element Method (FEM) analysis.

3.3.1. Force Analysis

When the motor applies the maximum torque (we considered a Dynamixel MX-28T— Robotis, South Korea—able to apply a torque of 3.1 Nm @ 12 V), assuming a motor pulley radius of 11 mm, we get an equivalent force on the tendon equal to F = 282 N. In quasi-static conditions the elements of the differential mechanism are subject to the forces sketched in Figure 10.

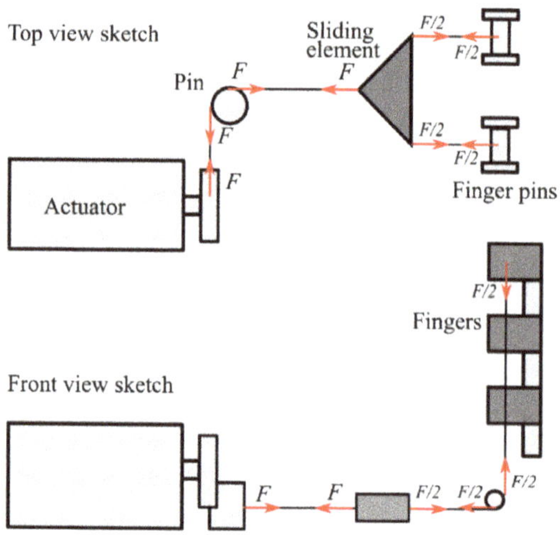

Figure 10. Sketch of the forces acting on the main elements of the finger actuation and transmission system in quasistatic conditions. When the motor applies its maximum torque, 3.1 Nm at 12 V, the force applied to the tendon driving the differential is F = 282 N. The forces applied to the tendons driven by the differential and connected to the fingers is therefore F/2 = 141 N.

3.3.2. Stress Analysis Results

Given the previously introduced force distribution, we analyzed stress distribution on the main elements of the finger transmission. The most stressed element results to be the sliding element. The dimensions of the sliding element have been defined based on geometrical considerations summarized in the preceding sections. On the motor side, the central pivot is tied to the tendon connected to the driving pulley, that allows sliding. On the finger side, two tendons are connected to the finger pins. Figure 11a shows Von Mises equivalent stress distribution. It can be noticed that the central pivot is the most stressed part of the sliding element. There are peaks of stress concentration in the edges that are not compatible with a standard ABS material. This drawback can be easily mitigated in the design and prototyping phase by increasing fillet radii. Since this is the most stressed element of the transmission system and it is not physically connected to other components, we can also consider different types of material to realize it, as for instance ABS reinforced with carbon or metal fiber [30], or metal alloys.

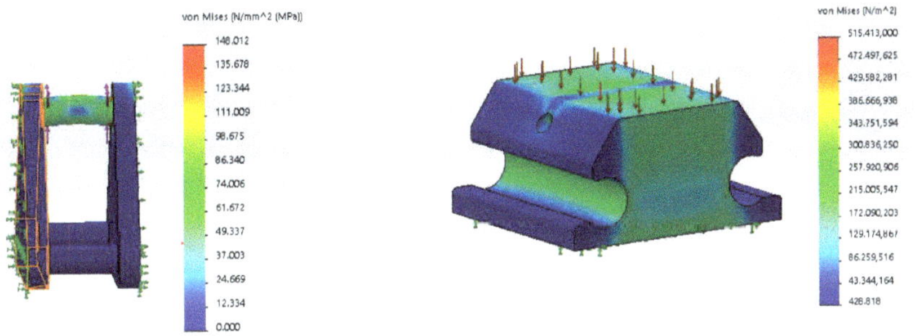

(a) (b)

Figure 11. Main results of the static FEM analysis on some elements of the transmission and differential. Equivalent Von Mises stress distributions are reported. (**a**) Sliding element, (**b**) rigid element of finger phalanges.

Rigid elements of finger phalanges have a quite robust shape; in Figure 11b we report as an example the results of a FEM static simulation in which we applied a compressive loading of 40 N. As it can be seen, the overall stress level is low.

4. Prototype Presentation

A simplified demonstrative prototype of the proposed double finger device is shown in Figure 1. As previously explained, the device has been designed to be wearable, robust and compliant to adapt to different object shapes. The device consists of two main parts, a support base and two flexible fingers. The flexible fingers have a modular structure, each module is composed of a 3D printed ABS polymeric part that acts as a rigid link and a TPU part that realizes the flexible joint. Soft rubber pads are glued to the rigid links to increase the friction at possible contact areas. The modules are assembled by sliding the flexible part in the rigid one (Figure 2). The support base of the finger is realized in ABS. It contains the actuator, the differential and two elastic bands that allows the user to wear it on the forearm.

The actuator moves the fingers through a tendon connected to the sliding element that constitute the differential. A hole in the rigid links allows the passage of a cable (Dyneema polyethylene fiber, Japan) which is used to realize the tendon driven actuation. The tendon wires run through the fingers and are attached on one side to the fingertips and on the other to the sliding element. Another tendon connects the sliding element to a pulley connected to the actuator. The actuator used, as previously introduced, is a Dynamixel servo AX12-A (Robotis, Korea) [42]. Principal details on the motor and device features are reported in Table 4. We use ArbotiX-M Robocontroller to drive the Dynamixel motor [43]. This control solution for Dynamixel motors incorporates an AVR microcontroller, Xbee wireless radio and the motor driver.

Table 4. Main technical properties of the prototype.

Description	Value	Unit
Rigid module, dimension	20 × 31 × 12	mm
Flexible module, dimension	20 × 18 × 2	mm
Support base, dimension (approx.)	170 × 50 × 40	mm
Actuator, dimension	71 × 71 × 45	mm
Actuator, weight	146	g
Max torque (@ 12V)	3.1	Nm
Pulley radius	11	mm
Max current (@ 12 V)	1.4	A
Operating angles	300	°
Max unloaded velocity	684	°/s
Total weight	210	g

Interfaces for controlling wearable extra fingers have been developed within this research activity, for the sake of brevity we report in this paper a brief summary on the main solutions. The key aspects that were considered in the design of these interfaces were reliability, easiness and intuitiveness of use also by untrained people, accessibility from people with partial upper limb impairments, and limited costs. Different wearable interfaces have been developed to control wearable extra fingers, and also to let the user feel a haptic stimulus communicating a specific state of the robotic extra finger. An interesting example control device is the eCap, an Electromyography (EMG) interface embedded in a cap: wearing the eCap the user can control the flexion/extension of the robotic finger by contracting the frontalis muscle by moving his or her eyebrows upwards. The motion of the robotic extra finger is controlled using a trigger signal based finite state machine (FSM), as detailed in [44].

Another possible interface is the so-called hRing, a ring that can be worn by the user on the healthy hand, provided by buttons for opening/closing the wearable extra fingers and with a vibrating motor providing a haptic cutaneous feedback stimulus to the user's healthy hand [44].

For control design, a simple and robust solution has been implemented. Device actuator is position controlled, a linearly increasing/decreasing reference position value is sent to the actuator according to inputs provided by the users through the interface or state variations, as detailed in the following. The prototype proposed in this paper has not position/force sensors on the fingers, control system behavior is developed on the basis of motor internal sensors only. In particular, it is possible to use the internal controller available in the servomotor to control the exerted torque. This controller uses a torque estimation based on the measured load current. The torque control may be used to limit the grasping force on the objects or to check whether an object is firmly grasped. This information can be haptically communicated to the user as a vibration signal if the hRing interface is employed. Examples on the use of torque control can be found in [45].

Through the interfaces, the users control the motion/stop of the finger with a specific command on the control interface (e.g., a single muscle contraction with the eCap, the activation of one of the two buttons on the hRing). Once the finger is stopped, another input (e.g., two contractions of the muscle, the activation of the second button in the hRing) switches the direction of motion from flexion to extension and vice versa. A software defined trigger stops the actuator's motion once the object is considered as grasped, to avoid a torque overloading situation. Object grasping is detected by continuous monitoring of the actuator's shaft position and the exerted torque.

In particular, when flexion command is selected, a desired position for the servomotor ϑ_{des}, corresponding to a configuration in which, both the extra fingers are completely flexed. For instance, for the device previously introduced, with 7 modules for both the fingers, from Equation (16), it results that the overall displacement of the main tendon, needed to fully close the fingers, is $\Delta l_{tot} = 28\ mm$. Considering that the pulley connected to the servomotor is $r_m = 11$ mm, the corresponding rotation for the motor is $\vartheta_{des} = 150°$. The desired value of the angular velocity ($\omega_{des} = 33.3$ °/s) is also set in the position control system. When extension/release command is selected by the user, the desired position is set to 0°, which results in the finger completely extended, with the same desired velocity.

The lack of precision in sensing and control is compensated by the adaptability properties of the device, inherently provided by joint passive compliance and by the differential mechanism that decouples fingers' closure motions [7]. More complex control strategies were tested in some preliminary works [18], but they were suitable for fully actuated devices and needed sophisticated and expensive sensors for human hand tracking, for example a Cyberglove [46]. Although these works provided us interesting results, we recognized that they were not suited for the application outside research laboratories, with patients dealing with everyday activities.

Figure 12 shows some examples of grasps realized with the wearable device. As it can be seen, even if the device is actuated by only one motor, the underactuated structure of the fingers and the differential system allow the device to adapt to different object shapes, realizing stable grasps of large objects, having a weight up to 2 kg, and also to grasp multiple objects. Device performance, in terms of payload (e.g., maximum weight of the grasped object, maximum fingertip force, maximum and minimum diameter of the grasped object), are summarized in Table 5. It is worth to notice that some of the grasps reported in Figure 12 would be weak or impossible with a single finger device (e.g., Figure 12c–e) or with a double extra finger without differential mechanism as the one presented in [29] (Figure 12c,e). Other grasp examples have been presented in the video S1 in Supplementary Materials.

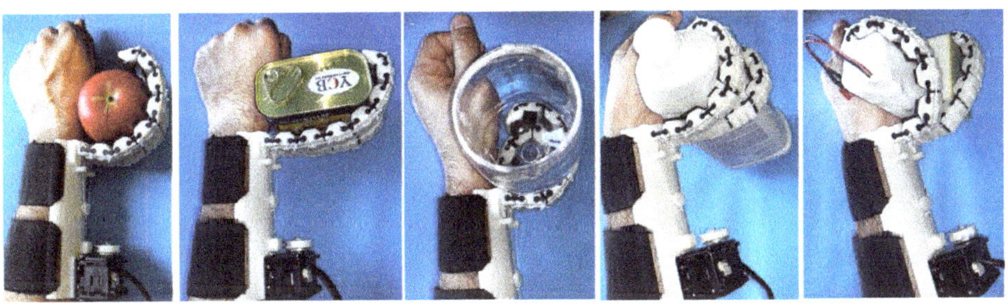

(a) (b) (c) (d) (e)

Figure 12. Examples of grasps realized with the wearable device: (**a**) apple, (**b**) rectangular box, (**c**) glass, (**d**) bottle, (**e**) two different objects (battery packs).

Table 5. Main performance parameters of the prototype.

Description	Value	Unit
Max payload	2	kg
Max force at the fingertip (closed)	12	N
Max force at the fingertip (open)	6	N
Diameter of the smallest graspable object	17	mm
Diameter of the largest graspable object	180	mm

5. Discussion on Potential Applications, Links

Wearable robots present interesting design challenges, since they are expected to closely work and collaborate with users, in an unstructured and dynamical environment. Wearable devices require additional design requirements, concerning weight, adaptability to user's body specificity, safety, robustness, and easiness of use. A rather new research branch in wearable robotics deals with human body augmentation. In this work, we analyzed a device for human hand augmentation, focusing on its mechanical design, rather than on its control and interfacing with the human, that was the focus of other works [44]. Human hand augmentation by means of robotic extra fingers represents an interesting research topic, involving engineering as well as ethical and philosophical challenges.

With respect to other solutions available in the literature [16,17,20], the device that we propose presents the following features:

- Actuation: while other devices available in the literature have a rigid and fully actuated structures, in the device presented in this work only one motor is necessary to actuate both the single and the double finger configurations. This feature limits the weight, complexity of the device, and improve its wearability and user comfort.
- Modularity: as discussed in Section 2, the structure of the fingers is modular at the phalanx level, the same modules can be used both for the single and double finger configurations. The device can be easily adapted to user's specific needs and features. For instance, a smaller hand would need a device with smaller dimensions that can be easily obtained realizing fingers with less modules.
- Robustness, safety for the user: in this type of devices the fingers are the elements more sensitive to unexpected contacts with the environment and shocks. Passive elements present in the interphalangeal joints of the fingers are realized by TPU, a material that presents high resistance to impacts and elongation at break. As a result, the fingers are quite robust and can resist to uncertainties and unpredictable impacts that may occur during activities of daily living (ADL). At the same time, their compliance limits the risks for the user and for other people.
- Adaptability: the compliant structure provided by deformable elements in interphalangeal joints and the differential mechanism allow the device to automatically adapt to different objects with different shapes and dimensions, without the need of specific sensors on the fingers and complex control strategies.
- Costs: the simple but versatile mechanical structure, the choice of widely diffused and affordable manufacturing technologies and materials, and the modularity, lead to a cost of the device that is quite limited.

We believe that the most impactful application of the device presented in this paper is rehabilitation and assistance of patients with upper limb impairments, for example after a stroke [47]. Stroke is a brain attack, affecting 17 million people worldwide each year, it is the second most common cause of death and a leading cause of adult physical disability (http://strokeeurope.eu/). Impairment of the hand, and in particular of its grasping and dexterous manipulation capabilities is one of the common deficits after a stroke. Approximately 60% of stroke survivors suffer from some form of sensorimotor impairment associated with their hand. Rehabilitation processes could take advantage from robotic devices, that can be used to practice intense movement training at home. Supernumerary fingers enable patients to execute grasp and release exercises and practice intensively using repetitive movements.

Different configurations for the robotic extra fingers, with one or two fingers, have been realized following the principles that we summarized in the previous sections and are now available for testing. Double configurations allow more stable grasps and the possibility of carrying heavier loads. In addition to rehabilitation purposes, supernumerary robotic fingers can increase patients' grasping capabilities, especially in objects manipulation, thereby improving their independence in ADL, and simultaneously decreasing the need of compensatory unnatural motor strategies for solving everyday tasks. Supernumerary limbs will provide novel opportunities to recover missing abilities, resulting in improvements of patients' quality of life.

The devices were tested with chronic stroke patients through qualitative experiments based on ADL and preliminary results are widely discussed in [19,45]. The goal of the tests was to evaluate how quickly and easily the patients could learn to use the device in ADL. Different applications were considered (e.g., kitchen scenario, consisting in preparing breakfast and lunch, some tools activities, using the extra finger as an active hook, etc.). The initial tests gave encouraging results; users could complete the tasks reducing the time and with good success rates.

6. Conclusions

In this paper, we summarized the work regarding the mechanical design of supernumerary robotic fingers that can be used for augmenting and compensating the human hand manipulation abilities. Wearability requirements set constraints on the hardware structure, that needs to be light, robust and safe. We focused in particular on the analysis of the transmission system that is needed when more than one finger is employed and all the fingers are actuated with the same motor. We are still optimizing the design, by further studying the possibility of exploiting additive manufacturing technology to change the stiffness of the finger's joints. We are working on integrating the extra fingers with other devices, as for instance assistive tools for wrist motion and arm support, investigating the possibility of using our devices in patients affected by other neurological diseases affecting hand grasping.

Supplementary Materials: The following are available online at http://www.mdpi.com/2218-6581/8/4/102/s1, Video S1: Design of multiple wearable extra fingers for human hand augmentation.

Author Contributions: Conceptualization, G.S.; Investigation, M.M., Z.I. and M.C.V.; Methodology, M.M., M.C.V., M.P., D.P. and G.S.; Supervision, M.P. and D.P.

Funding: This research was funded by the European Union EU H2020 projects under grant n. 780073 of the project "INBOTS Inclusive Robotics for a Better Society.

Conflicts of Interest: The authors declare no conflict of interest.

References

1. Mason, M.T.; Salisbury, J.K. *Robots Hands and the Mechanics of Manipulation*; MIT Press: Cambridge, MA, USA, 1985; ISBN 9780262132053.
2. Carbone, G. *Grasping in Robotics*; Springer: Dordertch, The Netherlands, 2013. [CrossRef]
3. Fukaya, N.; Toyama, S.; Asfour, T.; Dillmann, R. Design of the TUAT/Karlsruhe humanoid hand. In Proceedings of the IEEE/RSJ International Conference on Intelligent Robots and Systems, Takamatsu, Japan, 31 October–5 November 2000; pp. 1754–1759. [CrossRef]
4. Butterfass, J.; Grebenstein, M.; Liu, H.; Hirzinger, G. DLR-hand II: Next generation of a dexterous robot hand. In Proceedings of the IEEE International Conference on Robotics and Automation, Seoul, Korea, 21–26 May 2001; pp. 109–114. [CrossRef]
5. Martin, E.; Desbiens, A.L.; Laliberté, O.T.; Gosselin, C. SARAH hand used for space operation on STVF robot In Proceedings of the Intelligent Manipulation and Grasping, Genoa, Italy, 1–2 July 2004; pp. 279–284.
6. Roccella, S.; Carrozza, M.C.; Cappiello, G.; Dario, P.; Cabibihan, J.J.; Zecca, M.; Hiwa, H.; Itoh, K.; Matsumoto, M.; Takanishi, A. Design, fabrication and preliminary results of a novel anthropomorphic hand for humanoid robotics: RCH-1. In Proceedings of the IEEE/RSJInternational Conference on Intelligent Robots and Systems, Sendai, Japan, 28 September–2 October 2004; pp. 266–271, ISBN 0780384636.
7. Dechev, N.; Cleghorn, W.L.; Nauman, S. Multiple finger, passive adaptive grasp prosthetic hand. *Mech. Mach. Theory* **1999**, *36*, 1157–1173. [CrossRef]
8. Raparelli, T.; Mattiazzo, G.; Mauro, S.; Velardocchia, M. Design and development of a Pneumatic anthropomorphic hand. *J. Robot. Syst.* **2000**, *17*, 1–15. [CrossRef]
9. Maffiodo, D.; Raparelli, T. Three-Fingered Gripper with Flexure Hinges Actuated by Shape Memory Alloy Wires. *Int. J. Autom. Technol.* **2017**, *11*, 355–360. [CrossRef]
10. Niola, V.; Penta, F.; Rossi, C.; Savino, S. An underactuated mechanical hand: Theoretical studies and prototyping. *Int. J. Mech. Control* **2015**, *16*, 11–19.
11. Cosenza, C.; Niola, V.; Savino, S. Analytical study for the capability implementation of an underactuated three-finger hand. *Mech. Mach. Sci.* **2019**, *65*, 161–168. [CrossRef]
12. Niola, V.; Rossi, C.; Savino, S.; Timpone, F. Study of an underactuated mechanical finger driven by tendons. *Int. J. Autom. Technol.* **2017**, *11*, 344–354. [CrossRef]
13. Maffiodo, D.; Raparelli, T. Comparison among different modular SMA actuated flexible fingers. *Mech. Mach. Sci.* **2019**, *68*, 324–331. [CrossRef]
14. Carbone, G.; Rossi, C.; Savino, S. Performance comparison between FEDERICA hand and LARM hand. *Int. J. Adv. Robot. Syst.* **2015**, *12*, 90. [CrossRef]

15. Kobayashi, H.; Hyodo, K.; Ogane, D. On tendon-driven robotic mechanisms with redundant tendons. *Int. J. Robot. Res.* **1998**, *17*, 561–571. [CrossRef]
16. Parietti, F.; Asada, H. Supernumerary robotic limbs for human body support. *IEEE Trans. Robot.* **2016**, *32*, 301–311. [CrossRef]
17. Wu, F.Y.; Asada, H.H. "Hold-and-manipulate" with a single hand being assisted by wearable extra fingers. In Proceedings of the 2015 IEEE International Conference on Robotics and Automation (ICRA), Seattle, WA, USA, 26–30 May 2015.
18. Prattichizzo, D.; Malvezzi, M.; Hussain, I.; Salvietti, G. The sixth-finger: A modular extra-finger to enhance human hand capabilities. In Proceedings of the 23rd IEEE International Symposium on Robot and Human Interactive Communication, Edinburgh, UK, 25–29 August 2014; pp. 993–998. [CrossRef]
19. Hussain, I.; Salvietti, G.; Spagnoletti, G.; Prattichizzo, D. The soft-sixthfinger: A wearable emg controlled robotic extra-finger for grasp compensation in chronic stroke patients. *IEEE Robot. Autom. Lett.* **2016**, *1*, 1000–1006. [CrossRef]
20. Hu, Y.; Leigh, S.W.; Maes, P. Hand Development Kit: Soft Robotic Fingers as Prosthetic Augmentation of the Hand. In Proceedings of the Adjunct Publication of the 30th Annual ACM Symposium on User Interface Software and Technology, Québec City, QC, Canada, 22–25 October 2017.
21. Hussain, I.; Salvietti, G.; Spagnoletti, G.; Malvezzi, M.; Cioncoloni, D.; Rossi, S.; Prattichizzo, D. A soft supernumerary robotic finger and mobile arm support for grasping compensation and hemiparetic upper limb rehabilitation. *Robot. Auton. Syst.* **2017**, *93*, 1–12. [CrossRef]
22. Laliberté, T.; Gosselin, C.M. Underactuation in space robotic hands. In Proceedings of the International Symposium on Artificial Intelligence, Robotics and Automation in Space, Montréal, QC, Canada, 18–22 June 2001.
23. Birglen, L.; Gosselin, C.M. Force Analysis of Connected Differential Mechanisms: Application to Grasping. *Int. J. Robot. Res.* **2006**, *25*, 1033–1046. [CrossRef]
24. Zappatore, G.A.; Reina, G.; Messina, A. Analysis of a highly underactuated robotic hand. *Int. J. Mech. Control* **2017**, *18*, 17–24.
25. Birglen, L.; Gosselin, C.M. On the force capability of underactuated fingers. In Proceedings of the IEEE International Conference on Robotics and Automation, Taipei, Taiwan, 14–19 September 2003; pp. 1139–1145. [CrossRef]
26. Massa, B.; Roccella, S.; Carrozza, M.C.; Dario, P. Design and development of an underactuated prosthetic hand. In Proceedings of the 2002 IEEE International Conference on Robotics and Automation, Washington, DC, USA, 11–15 May 2002; pp. 3374–3379. [CrossRef]
27. Baril, M.; Laliberté, T.; Gosselin, C.; Routhier, F. On the design of a mechanically programmable underactuated anthropomorphic prosthetic gripper. *J. Mech. Des.* **2013**, *135*, 121008. [CrossRef]
28. Kontoudis, G.P.; Liarokapis, M.V.; Zisimatos, A.G.; Mavrogiannis, C.I.; Kyriakopoulos, K.J. Open-source, anthropomorphic, underactuated robot hands with a selectively lockable differential mechanism: Towards affordable prostheses. In Proceedings of the 2015 IEEE/RSJ International Conference on Intelligent Robots and Systems (IROS), Hamburg, Germany, 28 September–2 October 2015; pp. 5857–5862.
29. Hussain, I.; Spagnoletti, G.; Salvietti, G.; Prattichizzo, D. Toward wearable supernumerary robotic fingers to compensate missing grasping abilities in hemiparetic upper limb. *Int. J. Robot. Res.* **2017**, *36*, 1414–1436. [CrossRef]
30. Ning, F.; Cong, W.; Qiu, J.; Wei, J.; Wang, S. Additive manufacturing of carbon fiber reinforced thermoplastic composites using fused deposition modeling. *Compos. Part B Eng.* **2015**, *80*, 369–378. [CrossRef]
31. Valigi, M.C.; Logozzo, S.; Canella, G. A new automated 2 DOFs 3D desktop optical scanner. In *Mechanisms and Machine Science*; Springer: Berlin, Germany, 2017; Volume 47, pp. 231–238. [CrossRef]
32. Valigi, M.C.; Logozzo, S.; Canella, G. A robotic 3D vision system for automatic cranial prostheses inspection. *Mech. Mach. Sci.* **2018**, *49*, 328–335. [CrossRef]
33. Salvietti, G.; Hussain, I.; Malvezzi, M.; Prattichizzo, D. Design of the passive joints of underactuated modular soft hands for fingertip trajectory tracking. *IEEE Robot. Autom. Lett.* **2017**, *2*, 2008–2015. [CrossRef]
34. Wu, F.Y.; Asada, H. *Bio-Artificial Synergies for Grasp Posture Control of Supernumerary Robotic Fingers*; MIT Press: Cambridge, MA, USA, 2014.
35. Santello, M.; Flanders, M.; Soechting, J.F. Postural Hand Synergies for Tool Use. *J. Neurosci.* **1998**, *18*, 10105–10115. [CrossRef]

36. Malvezzi, M.; Valigi, M.C.; Salvietti, G.; Iqbal, Z.; Hussain, I.; Prattichizzo, D. Design criteria for wearable robotic extra–fingers with underactuated modular structure. *Mech. Mach. Sci.* **2019**, *68*, 509–517. [CrossRef]
37. Malvezzi, M.; Gioioso, G.; Salvietti, G.; Prattichizzo, D. SynGrasp: A Matlab Toolbox for Underactuated and Compliant Hands. *IEEE Robot. Autom. Mag.* **2015**, *22*, 52–68. [CrossRef]
38. Cobos, S.; Ferre, M.; Uran, M.S.; Ortego, J.; Pena, C. Efficient human hand kinematics for manipulation tasks. In Proceedings of the 2008 IEEE/RSJ International Conference on Intelligent Robots and Systems, Nice, France, 22–26 September 2008; pp. 2246–2251.
39. Bianchi, M.; Liarokapis, M.V. HandCorpus, a new open-access repository for sharing experimental data and results on human and artificial hands. In Proceedings of the IEEE World Haptics Conference (WHC), Daejeon, Korea, 14–17 April 2013.
40. Dollar, A.M.; Howe, R.D. The highly adaptive sdm hand: Design and performance evaluation. *Int. J. Robot. Res.* **2010**, *29*, 585–597. [CrossRef]
41. Ahmed, M.; Islam, M.; Vanhoose, J.; Rahman, M. Comparisons of Elasticity Moduli of Different Specimens Made Through Three Dimensional Printing. *3D Print. Addit. Manuf.* **2017**, *4*, 105–109. [CrossRef]
42. Robotis. Dynamixel Mx-28t Robot Actuator. 2012. Available online: http://www.trossenrobotics.com/dynamixel-mx-28-robotactuator.aspx (accessed on 10 December 2019).
43. ArbotiX. Arbotix-m Robocontroller, Open Source. 2012. Available online: http://www.trossenrobotics.com/p/arbotix-robotcontroller.aspx (accessed on 10 December 2019).
44. Spagnoletti, G.; Hussain, I.; Pacchierotti, C.; Salvietti, G.; Prattichizzo, D. The hRing: A haptic interface for an extra robotic finger. In Proceedings of the IEEE Haptic symposium, Philadelphia, PA, USA, 8–11 April 2016.
45. Salvietti, G.; Hussain, I.; Cioncoloni, D.; Taddei, S.; Rossi, S.; Prattichizzo, D. Compensating Hand Function in Chronic Stroke Patients Through the Robotic Sixth Finger. *Trans. Neural Syst. Rehabil. Eng.* **2016**. [CrossRef]
46. Cyberglove. Cyberglove Systems. 2019. Available online: http://www.cyberglovesystems.com/ (accessed on 10 December 2019).
47. Mozaffarian, D.; Benjamin, E.J.; Go, A.S.; Arnett, D.K.; Blaha, M.J.; Cushman, M.; Howard, V.J. Heart disease and stroke statistics—2016 update: A report from the American Heart Association. *Circulation* **2015**. [CrossRef]

MDPI
St. Alban-Anlage 66
4052 Basel
Switzerland
Tel. +41 61 683 77 34
Fax +41 61 302 89 18
www.mdpi.com

Robotics Editorial Office
E-mail: robotics@mdpi.com
www.mdpi.com/journal/robotics

www.ingramcontent.com/pod-product-compliance
Lightning Source LLC
Chambersburg PA
CBHW061816210926
51850CB00006B/718
9783039289318